CLEAN SHIPS
CLEAN PORTS
CLEAN OCEANS

Controlling Garbage and
Plastic Wastes at Sea

Committee on Shipborne Wastes

Marine Board
Commission on Engineering and Technical Systems
National Research Council

NATIONAL ACADEMY PRESS
Washington, D.C. 1995

NATIONAL ACADEMY PRESS • 2101 Constitution Avenue, N.W. • Washington, D.C. 20418

NOTICE: The project that is the subject of this report was approved by the Governing Board of the National Research Council, whose members are drawn from the councils of the National Academy of Sciences, the National Academy of Engineering, and the Institute of Medicine. The members of the panel responsible for the report were chosen for their special competencies and with regard for appropriate balance.

This report has been reviewed by a group other than the authors according to procedures approved by a Report Review Committee consisting of members of the National Academy of Sciences, the National Academy of Engineering, and the Institute of Medicine.

The program described in this report is supported by cooperative agreement No. 14-35-0001-30475 between the Minerals Management Service of the U.S. Department of the Interior and the National Academy of Sciences and by interagency cooperative agreement No. DTMA91-94-G-00003 between the Maritime Administration of the Department of Transportation and the National Academy of Sciences.

v

The National Academy of Sciences is a private, nonprofit, self-perpetuating society of distinguished scholars engaged in scientific and engineering research, dedicated to the furtherance of science and technology and to their use for the general welfare. Upon the authority of the charter granted to it by the Congress in 1863, the Academy has a mandate that requires it to advise the federal government on scientific and technical matters. Dr. Bruce M. Alberts is president of the National Academy of Sciences.

The National Academy of Engineering was established in 1964, under the charter of the National Academy of Sciences, as a parallel organization of outstanding engineers. It is autonomous in its administration and in the selection of its members, sharing with the National Academy of Sciences the responsibility for advising the federal government. The National Academy of Engineering also sponsors engineering programs aimed at meeting national needs, encourages education and research, and recognizes the superior achievements of engineers. Dr. Harold Liebowitz is president of the National Academy of Engineering.

The Institute of Medicine was established in 1970 by the National Academy of Sciences to secure the services of eminent members of appropriate professions in the examination of policy matters pertaining to the health of the public. The Institute acts under the responsibility given to the National Academy of Sciences by its congressional charter to be an adviser to the federal government and, upon its own initiative, to identify issues of medical care, research, and education. Dr. Kenneth I. Shine is president of the Institute of Medicine.

The National Research Council was organized by the National Academy of Sciences in 1916 to associate the broad community of science and technology with the Academy's purposes of furthering knowledge and advising the federal government. Functioning in accordance with general policies determined by the Academy, the Council has become the principal operating agency of both the National Academy of Sciences and the National Academy of Engineering in providing services to the government, the public, and the scientific and engineering communities. The Council is administered jointly by both Academies and the Institute of Medicine. Dr. Bruce M. Alberts and Dr. Harold Liebowitz are chairman and vice-chairman, respectively, of the National Research Council.

Preface

In 1987, the U.S. government ratified Annex V (Garbage) of the International Convention for the Prevention of Pollution from Ships (1973) and its 1978 Protocol, known jointly as MARPOL 73/78. That same year, the U.S. Congress enacted the Marine Plastic Pollution Research and Control Act (MPPRCA) (P.L. 100-220) to implement the agreement domestically. Both the treaty and the law address the need to curtail the debris littering oceans and beaches, particularly by restricting the age-old practice of tossing garbage overboard from vessels. The regulated garbage includes solid wastes (other than sewage) generated during normal operations at sea.

While the congressional action denotes official U.S. acceptance of MARPOL Annex V, additional work is required to realize the related goals and objectives. A national implementation plan is needed to convert Annex V and the domestic legislation into a tangible regime through which the United States can encourage, monitor, report, and enforce compliance with the new standards. In this way, the work of diplomats and legislators can be translated into the duties of agencies, government personnel, business persons, educators, advocates, and private citizens. The U.S. implementation strategy must put into action the words of Annex V within the context of the international law of the sea, which places some constraints on unilateral action but also offers many opportunities for use and study of the oceans, control of pollution, and settling of disputes.

ORIGIN OF THE STUDY

No single federal agency is responsible for the comprehensive implementation of Annex V in the United States. Instead, the duties are distributed among the

Coast Guard, the Environmental Protection Agency, the National Marine Fisheries Service of the National Oceanic and Atmospheric Administration, and, indirectly, the Maritime Administration and others.[1] In addition, the Congress instructed the Navy to comply with the MPPRCA. In examining the issues affecting the implementation of Annex V, these agencies identified the lack of strategic planning and organization as a major obstacle. These agencies therefore requested that the National Research Council (NRC) undertake an assessment of U.S. activities and evaluate how well Annex V implementation was progressing. Accordingly, the NRC Commission on Engineering and Technical Systems assembled a committee under the auspices of the Marine Board to conduct a comprehensive assessment of U.S. implementation of Annex V.

Committee members were selected for their expertise and to achieve balanced experiences and viewpoints. (Biographical information is presented in Appendix A.) The principle guiding the constitution of the committee and its work, consistent with NRC policy, was not to exclude any bias that might accompany expertise vital to the study, but to seek balance and fair treatment. The resulting committee membership balanced the technical, scientific, and legal professional disciplines and encompassed the diverse commercial and recreational communities that must comply with Annex V.

The committee sought the assistance of the federal agencies that have duties and undertake activities in conjunction with the national Annex V implementation effort. As a result, in addition to the aforementioned project sponsors, contact was maintained with the Marine Mammal Commission, the Department of State's Bureau of Oceans and International Environmental and Scientific Affairs, the Department of the Interior's National Park Service and Minerals Management Service, and the Department of Agriculture's Animal and Plant Health Inspection Service.

SCOPE OF THE STUDY

The task of the Committee on Shipborne Wastes was to focus on the preparations the federal government must make after accepting an international standard for environmental protection. The objective was to devise a strategy to help promote and compel compliance with Annex V by surface vessels in all U.S. maritime sectors[2] and promote the elimination of ocean pollution from garbage.

---

[1]While the Maritime Administration is not assigned specific duties by law with respect to Annex V, the agency administers federal laws and programs designed to promote and maintain the U.S. merchant marine and carries out promotional, research, and training programs that can assist in Annex V implementation.

[2]The study encompassed all U.S.-flag surface vessels, fixed and floating manned platforms in U.S. waters, and foreign-flag vessels that transit U.S. waters out to 200 nautical miles from shore (the Exclusive Economic Zone). While excluded from this study, U.S. Navy submarines are required by the MPPRCA to comply with certain provisions of Annex V by 2008.

Although the emphasis was on vessel garbage,[3] in some respects this problem could not be separated from the problem of marine debris in general, as noted in several sections of this report (such as those relating to ecological effects). Thus, elements of the committee's analysis and recommendations are applicable to the broader problem of marine debris as well as the specific objective of the study.

The committee made no recommendations going beyond Annex V or the International Maritime Organization (IMO) guidelines for implementation, even when there were compelling arguments for doing so. For example, several committee members argued that vessel operators and crews should halt all littering of the oceans, even that which is permitted by Annex V. Despite the appeal of a total ban,[4] the committee adhered to the limits of Annex V, which imposes a total discharge prohibition only in certain sea areas, and the IMO implementation guidelines, which recommend discharging garbage in port reception facilities "whenever practicable."

The committee's study encompassed all aspects of the U.S. implementation of Annex V. The committee addressed all vessel operations—all fleets, all ports and terminals, and all pertinent public and private institutions. It was charged with

• examining the roles and responsibilities of the agencies, organizations, fleets, and ports in a national implementation of the convention;
• identifying institutional, administrative, or policy changes that could contribute to the implementation of MARPOL Annex V, including proposals needing further research or application;
• reviewing the state of practice for marine debris controls, shipboard waste handling, and shoreside waste reception facilities;
• suggesting strategies for integrating waste management practices;
• identifying technology or science areas that could contribute to the implementation of MARPOL Annex V, including methods needing further research or development; and
• developing elements of a strategy to improve the authorities' abilities to compel compliance with MARPOL Annex V.

---

[3]The committee focused on the disposal of vessel garbage regulated under Annex V and the MPPRCA. The study did not address the transportation of material for the specific purpose of dumping it into the ocean, regulated under the Marine Protection, Research and Sanctuaries Act of 1972 (P.L. 92-532).

[4]Such a ban may be justifiable scientifically on the basis of evidence that garbage discharged legally far from shore can drift into areas (even all the way to the shoreline) where discharge is prohibited. On the other hand, there are practical and scientific reasons for not pursuing a total ban. First, not all vessels are technically capable of holding all garbage on board for disposal ashore. Second, there has been no comprehensive, multimedia study comparing the environmental effects of discharging garbage overboard to those of other disposal options, such as incineration or off-loading at an island port that lacks proper landfills.

No similar strategic analysis has been conducted by any nation signing Annex V or earlier MARPOL annexes. Therefore, the committee's effort may establish a precedent for examining how to incorporate a global environmental treaty into national governmental responsibilities. The present focus on vessel garbage notwithstanding, the committee's overall approach may be applicable to the broader roster of MARPOL annexes, which address prevention of pollution by oil, hazardous substances, and sewage from ships, as well as a future annex that will address air pollution.

STUDY METHODS AND REPORT ORGANIZATION

Over a two-year period the committee met six times, including four meetings in working ports on the Atlantic, Pacific, and Gulf of Mexico coasts. The committee received briefings from representatives of all major domestic fleets, as well as port operators, waste haulers, environmental advocates and scientists studying marine debris, technologists developing garbage disposal methods and equipment, and a variety of state and local government officials working to incorporate Annex V into the duties and responsibilities of their organizations.

The meetings were supplemented by individual interviews and site visits at waterfront facilities and waste hauling firms. A brief questionnaire was sent to a variety of port officials, port users, and waste haulers. The committee also conducted international correspondence to keep abreast of other national implementation regimes, especially with regard to port reception facilities and emerging developments in regional Annex V enforcement arrangements. A broad literature search assisted the committee in gathering information from a variety of private and government sources, from the well known to the obscure.

As part of the study, the committee reviewed earlier estimates of garbage generated by vessels (National Research Council, 1975; Eastern Research Group, 1988; Cantin et al., 1990) and examined other data of potential use in developing new estimates. While all available data sets are flawed, the committee drew on a variety of sources to develop its own rough estimates of the garbage generated by each U.S. maritime sector. The committee also sought to characterize, to the degree possible, current disposal practices and options for improving garbage management. As part of this effort, the committee commissioned a background paper on the U.S. Navy's garbage disposal practices and proposals (Swanson et al., 1994).[5]

The report is organized into three general sections: background, analysis, and synthesis. Chapter 1 provides background by summarizing the history and mandates of Annex V and progress in U.S. implementation efforts to date. The

---

[5]Copies of this unpublished background paper may be obtained from the Marine Board, National Research Council, 2101 Constitution Avenue, N.W., Washington, D.C. 20418.

analysis begins in Chapter 2, which defines the scope of the problem by outlining what is known about the sources, fates, and effects of vessel garbage. In addition to compiling the findings of others with respect to these topics, the committee conducted original analyses of garbage sources.

Chapters 3–8 build the foundation for the design of an Annex V implementation program. Chapter 3 outlines the hazard evolution model employed by the committee. This model is applied to the various maritime sectors in Chapter 4, which identifies opportunities for intervening in the evolution of the hazard (marine debris). The committee found it essential to examine each fleet separately, because their characteristics varied so widely. Chapter 5 examines the interface between vessels and ports, viewing vessel garbage management as a system. Chapter 6 addresses Annex V education and training. Chapter 7 examines several overarching issues, including the need for leadership and problems related to Annex V enforcement. Chapter 8 reviews opportunities for measuring progress in implementation of Annex V.

The last two chapters synthesize the findings from the analysis to outline a strategy that, in the committee's judgment, can lead to more complete U.S. compliance with and implementation of the mandates of Annex V. Chapter 9 contains fleet-specific advice, recommending objectives and tactics to be used within each maritime sector. Chapter 10 presents conclusions and recommendations for action by the federal government to improve overall implementation of Annex V in multiple maritime sectors.

The volume also contains, in addition to the biographies of the committee members, five other appendixes, which supplement the committee's report. Appendix B contains copies of Annex V and the IMO standards for on-board incinerators. Appendix C is a paper written by a committee member on the international law of the sea. The remaining three appendixes, which were written or commissioned by the committee, summarize background information compiled from multiple sources that may be difficult for readers to gather themselves. Appendix D lists key milestones in U.S. implementation of Annex V. Appendix E, an excerpt from the background paper commissioned by the committee, outlines the characteristics of the eight special areas designated under Annex V. Appendix F provides details on the harm caused by marine debris to supplement the summary of ecological effects at the end of Chapter 2.

The report is organized so that readers interested in specific maritime sectors or federal agencies can find relevant sections easily. Each sector is examined individually in chapters 2, 4, and 9. These sections also address related federal activities. Federal officials also will be interested in chapters 5–8 and 10. Recommendations for federal action are organized by agency in the Executive Summary.

The recommendations in chapters 9 and 10 represent the committee's consensus concerning the best use of the disparate skills and authorities of government, industry, and community-based individuals and organizations to improve management of an activity that, while seemingly mundane, can have far-reaching effects—disposal of vessel garbage.

ACKNOWLEDGEMENTS

The committee wishes to thank the dozens of individuals who contributed their time and effort to this project, whether in the form of presentations at meetings, correspondence, or telephone calls. Invaluable assistance was provided to both the committee and the Marine Board staff by representatives of federal agencies, private companies in various maritime sectors, citizen and environmental groups, and waste management industries.

In particular, the committee wishes to acknowledge its liaisons with the project sponsors: Commander Jeff Beach and Lieutenant Commander J.M. Farley, Marine Environmental Protection Division, U.S. Coast Guard; James Coe and John Clary of the Marine Entanglement Research Program, National Marine Fisheries Service, National Oceanic and Atmospheric Administration; Lawrence J. Koss, Office of the Chief of Naval Operations, U.S. Navy; Daniel W. Leubecker, Office of Technology Assessment, U.S. Maritime Administration; Steve Levy, Municipal Solid Waste Program, U.S. Environmental Protection Agency; and David Redford, Oceans and Coastal Protection Division, U.S. Environmental Protection Agency.

Special thanks also are due to Robert Blumberg, Office of Ocean Affairs, Department of State; Ronald B. Caffey, Plant Protection and Quarantine, Animal and Plant Health Inspection Service; William Eichbaum, World Wildlife Fund and Marine Board liaison to the committee; and John Twiss, executive director, and David Laist, policy and program analyst, of the Marine Mammal Commission.

Finally, the chairman wishes to recognize members of the committee, not only for their hard work during meetings and in reviewing drafts of this report but also for their many individual efforts in gathering information and writing sections of the report.

REFERENCES

Cantin, J., J. Eyraud, and C. Fenton. 1990. Quantitative Estimates of Garbage Generation and Disposal in the U.S. Maritime Sectors Before and After MARPOL Annex V. Pp. 119-181 in Proceedings of the Second International Conference on Marine Debris, 2–7 April 1989, Honolulu, Hawaii (Vol. I), R.S. Shomura and M.L. Godfrey, eds. NOAA-TM-NMFS-SWFSC-154. Available from the Marine Entanglement Research Program of the National Marine Fisheries Service (National Oceanic and Atmospheric Administration), Seattle, Wash. December.

Eastern Research Group (ERG). 1988. Development of Estimates of Garbage Disposal in the Maritime Sectors. Final report prepared for the Transportation Systems Center, Research and Special Programs Administration, U.S. Department of Transportation. Arlington, Mass.: ERG. December. (ERG is now in Lexington, Mass.)

National Research Council (NRC). 1975. Assessing Potential Ocean Pollutants. Ocean Affairs Board, NRC. Washington, D.C.: National Academy Press.

Swanson, R.L., R.R. Young, and S.S. Ross. 1994. An Analysis of Proposed Shipborne Waste Handling Practices Aboard United States Navy Vessels. Paper prepared for the Committee on Shipborne Wastes, Marine Board, National Research Council, Washington, D.C.

Terminology and Acronyms

TERMINOLOGY

garbage:* food, domestic, and operational waste (excluding fresh fish and parts thereof, sewage, and drainage water) generated during normal operations and liable to be disposed of continuously. Garbage thus includes solid wastes often identified as "trash".

ocean(s): all waters where Annex V is in force, including seas, estuaries, coastal waters, and, in the United States (under domestic law), inland waterways.

marine environment: same as ocean.

port: any landing area (port, marina, pier, dock, or ramp) for vessels.

port reception facility:* any receptacle, from trash cans to dumpsters to barges, maintained by or at a port to receive garbage generated on vessels.

ship: a large vessel, such as a cargo or passenger cruise ship.

special area:* a sea area subject to special Annex V restrictions on garbage discharges.

vessel: any water craft or structure, from small boats to ships to oil drilling platforms, that carries humans.

zero discharge: no garbage is discharged overboard except, under certain conditions, food waste.

---

*Denotes terms for which the meaning is essentially the same as in Annex V.

ACRONYMS

APHIS	Animal and Plant Health Inspection Service
CDC	Centers for Disease Control and Prevention
CMC	Center for Marine Conservation
COA	Certificate of Adequacy
DOS	Department of State
EPA	Environmental Protection Agency
FDA	Food and Drug Administration
GOMP	Gulf of Mexico Program
IMO	International Maritime Organization
IOC	Intergovernmental Oceanographic Commission
ISWMS	Integrated Solid Waste Management System
MARAD	Maritime Administration
MDIO	Marine Debris Information Office
MERP	Marine Entanglement Research Program
MMC	Marine Mammal Commission
MMS	Minerals Management Service
MPPRCA	Marine Plastics Pollution Research and Control Act
NMFS	National Marine Fisheries Service
NOAA	National Oceanic and Atmospheric Administration
SPA	Shore Protection Act
USDA	U.S. Department of Agriculture

Contents

APPENDIXES

INDEX   343

List of Tables and Figures

FIGURES

CLEAN SHIPS
CLEAN PORTS
CLEAN OCEANS

Executive Summary

The international maritime community has taken steps to restrict garbage discharged overboard from vessels to curb environmental harm. The fundamental restrictions were laid out by the International Maritime Organization (IMO) in Annex V of the International Convention for the Prevention of Pollution from Ships (1973) and its 1978 Protocol, together known as MARPOL 73/78. MARPOL Annex V bans all overboard disposal of plastics and limits other discharges based on the form of the material and the vessel's location and distance from shore. The regulated garbage includes solid wastes (other than sewage) generated during normal operations at sea.

The U.S. Congress ratified Annex V in 1987 and enacted the Marine Plastics Pollution Research and Control Act (MPPRCA) (P.L. 100-220). The Coast Guard is responsible for enforcing Annex V and the MPPRCA, but many federal agencies are involved in implementing the convention and the domestic law. These agencies, while making some progress in implementation, identified the lack of strategic planning and organization as a major obstacle. As a step toward improving national implementation of Annex V, the agencies asked the National Research Council (NRC) to conduct a comprehensive assessment and recommend a national strategy. Accordingly, the Committee on Shipborne Wastes was convened under the auspices of the NRC's Marine Board. The committee focused on vessel garbage, but in some respects this problem could not be separated from the problem of marine debris in general. Thus, elements of the committee's analysis and recommendations are applicable to the problem of marine debris in general as well as the specific objective of the study.

1

SOURCES OF VESSEL GARBAGE

The committee examined individually each fleet operating in U.S. waters, because Annex V implementation constraints and opportunities vary so widely within the overall maritime community. The nine fleets examined were recreational boats; commercial fishing vessels; cargo ships; passenger day boats and ferries; small public vessels (Coast Guard and naval auxiliaries); offshore oil platforms, rigs, and supply vessels; U.S. Navy surface combatant vessels; passenger cruise ships; and research vessels.

Considerable amounts of garbage are generated by most if not all sectors, but available data concerning garbage generated and disposal practices are imprecise and incomplete. Detailed, comprehensive data on garbage generation have been collected only for the Navy. Neither U.S. nor international Annex V compliance and enforcement programs support the gathering of such data for other sectors.

FATES AND EFFECTS OF MARINE DEBRIS

Knowledge concerning the movement of marine debris is derived primarily from beach surveys; little data is available on debris that ends up in the sea or on the seabed. It is difficult to obtain such data without a systematic, worldwide effort involving the cooperation of multiple maritime nations, so an international data collection effort would be useful. The harmful effects of marine debris, particularly plastics, are all too evident, albeit not documented in a comprehensive and systematic manner. Plastics are causing considerable harm, including mortality among individual marine mammals, turtles, birds, and fish, as a result of either entanglement or ingestion. However, the overall ecological effects of marine debris cannot be established on the basis of surveys and other information-gathering efforts conducted to date, due primarily to the lack of a common framework for data collection, centralized data analysis, and information exchange. Scientists suspect that entire populations of animals may be affected adversely by debris in the water or washed up on shore, and that debris accumulations in the benthos may interfere with dissolved gas exchange between the pore waters of the sediment and the overlying waters, leading to hypoxic or anoxic[1] environments that can kill some organisms.

The committee concludes that (1) statistically valid long-term programs are needed to monitor the flux of plastics in the oceans, assess the accumulation of debris in the benthos, and monitor interactions of marine species with debris in the oceans and the impact of debris on pristine areas; and (2) the National Oceanic and Atmospheric Administration (NOAA) is best equipped of all federal agencies to lead a monitoring effort, because its Marine Entanglement Research Program (MERP) has collected much of the existing knowledge on marine de-

---

[1]An hypoxic environment is oxygen deficient; anoxia results when oxygen is absent entirely.

bris, and because its Status and Trends Program could be expanded readily to monitor plastic debris.

THE COMMITTEE'S ANALYSIS

The committee adapted a hazard evolution model from the literature and used it to identify opportunities for enhancing implementation of Annex V in each maritime sector. The model establishes a framework for examining each stage of hazard evolution, from the satisfaction of human needs (such as the need for food, which may be wrapped in packaging that ends up as garbage) through the mitigation of consequences (such as through physical removal of debris during beach cleanups). The model also provides parameters to aid in the selection of interventions to halt or slow the evolution of the hazard (marine debris). Chief among these parameters are *intelligence* and *control*; the extent of available information and the means of influence determine in large part whether an intervention can be successful. The committee applied the model, with minor modifications, to each of the nine maritime sectors, seeking to identify means of intelligence and control as well as opportunities for intervention to enhance Annex V implementation. These sector-specific analyses resulted in the establishment of Annex V implementation *objectives* for each fleet. These objectives are summarized in Table ES-1.

THE VESSEL GARBAGE MANAGEMENT SYSTEM

The committee examined vessel garbage management as a system. One part of the system encompasses on-board garbage handling techniques and treatment technologies. The other, often-neglected component of the system is port reception facilities, which need to be linked to the local scheme for managing land-generated waste. The committee found that the link between the vessel and port components of the system is generally clumsy and sometimes non-functional.

Source control (i.e., reducing amounts of packaging and other waste materials brought on board) is an important aspect of garbage management. For garbage that is generated, a range of on-board treatment technologies—including compactors, pulpers, shredders, and incinerators—is available or under development. However, these units generally are designed only for certain types of ships (e.g., the Navy's or passenger cruise ships) and, due to their size and operating features, are not appropriate to every type of vessel. Some fleets, such as fisheries, may need financial assistance in order to purchase and install appropriate equipment. In addition, several obstacles may be impeding safe and efficient on-board garbage management: the lack of federal guidelines on shipboard sanitation[2] for

---

[2]In this context, sanitation refers specifically to the promotion of hygiene and prevention of disease through proper handling and storage of garbage (not sewage).

TABLE ES-1   National Strategy for Annex V Implementation: Objectives for Each Maritime Sector[a]

Sectors	Objectives
Recreational boats and their marinas	• Achieve zero-discharge capability • Assure adequacy of port reception facilities • Assure that boaters are provided with appropriate Annex V information and education
Commercial fisheries and their fleet ports	• Achieve zero-discharge capability for fishing vessels that operate as day boats • Provide adequate port reception facilities • Assure access to appropriate on-board garbage handling and treatment technologies • Provide comprehensive vessel garbage management system • Assure that seagoing and management personnel are provided with appropriate Annex V information, education, and training • Improve Annex V enforcement • Extend U.S. cooperation to encourage compliance by foreign-flag vessels
Cargo ships and their itinerary ports	• Improve access to on-board garbage handling and treatment technologies • Provide comprehensive vessel garbage management system, including adequate port reception facilities • Assure that seagoing and management personnel are provided with appropriate Annex V information, education, and training • Fully exercise U.S. authority to improve compliance by foreign flag vessels and by all vessels in foreign waters
Passenger day boats, ferries, and their terminals	• Achieve zero-discharge capability, integrating the handling of vessel garbage into local solid waste management systems
Small public vessels and their home ports	• Improve on-board garbage handling and treatment technology • Assure adequacy of port reception facilities • Assure that seagoing and management personnel are provided with appropriate Annex V information, education, and training • Develop model Annex V compliance program
Offshore platforms, rigs, supply vessels, and their shore bases	• Achieve zero discharge at sea • Assure comprehensive garbage management system, including adequate port reception facilities

TABLE ES-1 Continued

Sectors	Objectives
	• Assure that seagoing and management personnel are provided with appropriate Annex V information, education, and training
Navy surface combatant vessels and their home ports	• Develop plans for full Annex V compliance, including capability to achieve zero discharge in special areas, making the best use of existing technologies and strategies • Develop model Annex V implementation program
Passenger cruise ships and their itinerary ports	• Increase use of on-board garbage handling and treatment technologies • Assure comprehensive vessel garbage management system, including adequate port reception facilities • Assure that seagoing and management personnel are provided with appropriate Annex V information, education, and training • Exploit U.S. authority to improve compliance by foreign-flag vessels and by all vessels in foreign waters
Research vessels and their ports of call	• Provide model Annex V compliance program • Improve on-board garbage handling and treatment technology • Assure that seagoing and management personnel are provided with appropriate Annex V information, education, and training

aIn developing these objectives, the committee screened possible alternatives informally using six criteria: effectiveness, cost effectiveness, efficiency, timeliness of results, equity, and sustainability. The committee further emphasized actions "upstream" in the hazard evolution model (and therefore most effective from an environmental standpoint), and actions that would promote achievement of zero-discharge capability where feasible or required. The committee wishes to emphasize that an objective is something to be pursued, as opposed to an absolute requirement (as would be established by law), and that existing obstacles to Annex V compliance, however onerous, should not serve as justification for abandoning an objective.

any sector other than cruise ships; the lack of quarantine standards based on compacted waste; and the lack of federal standards on shipboard incinerators.

The committee concludes that (1) vessel garbage management must be viewed as a system that includes port reception facilities, and this system needs to be combined with the integrated solid waste management system for land-generated waste; (2) there is a need for new and improved on-board garbage treatment technologies, a problem that may be resolved in part by adapting commercial equipment used in homes, retail establishments, and industry; (3) demonstration projects, research on operations and maintenance issues, and information exchange are needed; (4) the Maritime Administration (MARAD) is the logical agency to coordinate development and deployment of on-board garbage handling technologies, due to its ongoing, broad-based marine technology assessment and development efforts; and (5) steps must be taken to resolve issues that may be impeding safe garbage storage and expanded use of compactors and incinerators.

On the port side of the system, there is little evidence of strategic planning to support the provision of "adequate" garbage reception facilities as required by Annex V. The Coast Guard issues Certificates of Adequacy (COA) to large commercial and fishing ports and requires that reception facilities be provided at many other ports, but there are no technical standards for judging adequacy. Other shortcomings of this part of the system include the poorly developed infrastructure for recycling; the need to address the authorities of the Coast Guard, the Environmental Protection Agency (EPA), and the states concerning the integration of vessel garbage into the regional solid waste management system; the lack of full integration of the Annex V regime and the U.S. Department of Agriculture's Animal and Plant Health Inspection Service (APHIS) program, which oversees quarantine of garbage from foreign sources that may harbor diseases; and the need to address economic issues, including who should pay for vessel garbage services, and how—questions that may require some federal attention to resolve.

The committee concludes that (1) there is a need to assure accountability of both vessel operators and port operators; (2) recycling of vessel garbage needs to be promoted; (3) the EPA is the logical agency to establish the overall framework for improving the vessel/shore interface, due to its expertise in and authority for national management of solid waste; (4) the handling of APHIS waste needs to be integrated as fully as possible with the Annex V regime and the system for managing land-generated waste; and (5) there is a need to address economic issues, including the cost of technologies to vessel operators, trade-offs with garbage disposal services, and who should pay for garbage services and how.

EDUCATION AND TRAINING

Education has a strategic role to play in Annex V implementation because the oceans are too vast to monitor comprehensively. Seafarers therefore must be convinced to comply voluntarily and given the knowledge, training, and motivation to do so. A number of education and training programs have been carried out in support of Annex V implementation, most notably through MERP. While these efforts have been instrumental in the progress of Annex V implementation to date, they have been neither comprehensive nor long term. These features will be needed to raise Annex V implementation to a higher level. A successful national Annex V education and training program would need to include research, execution, evaluation, and innovation.

The committee concludes that (1) a sustained national program of Annex V education and training is needed that reaches all levels of all maritime sectors (including visitors and other members of the public, employees, and management) as well as non-traditional target groups such as the packaging industry and government officials; provides for information exchange, both domestically and internationally; and stimulates innovation; and (2) a publicly chartered, independent foundation offers the most promise for coordinating and enhancing a successful, long-term program of education, training, and information exchange.

OVERARCHING ISSUES

Development of a successful Annex V implementation strategy demands attention to three overarching issues that affect all fleets, require effective national coordination, and involve international aspects.

The first issue is the need for overall national leadership in Annex V implementation. Many strategies for improving Annex V implementation require the cooperation of multiple agencies and organizations and diverse maritime sectors.

The committee concludes that (1) U.S. government and government-supported fleets, to set an example, need to work systematically to comply with Annex V, upgrade crew training and provisioning practices, and encourage transfer of successful experiences to other fleets; (2) centralized oversight, direction, and coordination of Annex V implementation is needed; (3) the United States needs to continue to take a leadership role in the international community with respect to Annex V implementation; (4) a permanent national commission offers the most promise as a means of providing consistent, independent, expert oversight and coordination of Annex V and MPPRCA implementation, as well as international leadership; and (5) memoranda of understanding (MOUs) need to be negotiated between relevant agencies and observed.

The second issue is enforcement[3] of Annex V. The Coast Guard is taking steps to expand its use of internationally recognized authorities over foreign-flag vessels. The committee identified a number of additional opportunities for improving enforcement. U.S. authorities could work through IMO to resolve ambiguities concerning the extent of port state[4] authorities with respect to Annex V enforcement; extend the requirement for garbage logs to foreign-flag vessels;

---

[3]Enforcement, for purposes of this report, includes all actions taken to obtain some remedy for violations of Annex V. Such actions may include pursuit of a civil or criminal case against an alleged violator, referral of a case involving a foreign-flag vessel to the appropriate flag state, and record keeping undertaken as a means of keeping track of repeat violators.

[4]A port state is a nation in which foreign-flag vessels make port calls.

streamline enforcement by issuing "tickets" in civil cases, particularly in the fisheries and recreational boating sectors; require that ports provide receipts for garbage off-loaded into their reception facilities, and then compare the receipts to vessel garbage logs; require that cargo and cruise ships off-load garbage at U.S. port calls; enlist the assistance of additional government agencies in reporting Annex V violations; encourage vessel operators to report inadequate reception facilities; and conduct public awareness campaigns urging citizens to report illegal garbage disposal.

The committee concludes that (1) enforcement action must be taken and followed up in every case where the United States can assert jurisdiction, even when the violator is a foreign-flag vessel; (2) the Coast Guard is the appropriate agency to lead expanded enforcement efforts; and (3) the Coast Guard needs to take additional steps to enhance enforcement where most needed.

Accurate record keeping and analysis of garbage records could be useful in determining where special enforcement efforts are needed as well as in measuring progress in Annex V implementation. The most easily implemented record-keeping system may be a combined Coast Guard/APHIS database on vessel garbage handling, making use of existing APHIS records of vessel boardings and garbage off-loading, and information from garbage logs and Coast Guard enforcement reports.

The committee concludes that, to make the best use of existing information and enforcement assets, systematic government record keeping and analysis is needed.

The third issue is special areas, which must be taken into account in devising a U.S. strategy for Annex V implementation. These are areas designated under Annex V where, because of heavy vessel traffic and/or highly sensitive ecosystems, IMO prohibits overboard discharges of all garbage except food waste.[5] These restrictions mean that vessels operating in special areas need to achieve zero-discharge capability. In addition, the United States needs to find ways to help assure that sufficient numbers of adequate port reception facilities exist in the nearby Wider Caribbean special area.

RECOMMENDATIONS FOR GOVERNMENT ACTION

While it is the responsibility of individual mariners to conform with international standards on garbage management and disposal, the federal government can take important steps to facilitate, promote, and compel compliance. Recom-

---

[5]Eight special areas have been designated under Annex V. The requirements are in force in the Antarctic Ocean, Baltic Sea, and the North Sea. Once IMO determines that sufficient numbers of adequate port reception facilities have been provided, the mandates will take effect in the Black Sea, the Mediterranean Sea, the Persian Gulf, the Red Sea, and—of chief concern to the United States— the Wider Caribbean, which includes the Gulf of Mexico.

mendations for government action were derived from the analyses of each maritime sector, as well as examination of the vessel garbage management system and issues related to education and training, national leadership, Annex V enforcement, special areas, and measuring progress in Annex V implementation.

Legislative Actions

Improve Management of Vessel Garbage

To improve management of vessel garbage and meet U.S. national and international commitments to implement Annex V, the Congress should direct EPA to use its current resources to establish an overall framework that (1) incorporates the vessel garbage management system into the system for managing land-generated waste; (2) requires states to include in their solid waste management plans the disposal of garbage from vessels docked at their ports; (3) establishes technical standards for reception facilities appropriate to each type of port; (4) provides for accountability by requiring commercial ports to issue receipts for garbage discharged at their facilities, and by assuring that states follow up reports of inadequate port reception facilities; and (5) promotes recycling of vessel garbage. The EPA should obtain assistance from the Coast Guard, the states, port and terminal operators, the private sector, and the maritime communities and should make use of the forthcoming IMO manual on reception facilities.

National Leadership

The Congress should establish a permanent national commission with a clear legislative mandate establishing its authority to oversee the national Annex V and MPPRCA implementation effort. The panel should be modeled on other national commissions, such as the Marine Mammal Commission, established to address major issues of concern. The legislation should outline the commission's responsibilities and authorize funding sufficient for execution of its duties.

The commission should (1) review information on the sources, amounts, effects, and control of vessel garbage; (2) work with federal agencies to assure they carry out their roles and responsibilities and share relevant information; (3) assure that MOUs for Annex V implementation are negotiated and observed; (4) make recommendations to federal agencies on actions or policies related to identification and control of sources of vessel garbage; (5) provide support for research, regulatory, and policy analyses; (6) provide the Congress with periodic reports on the state of the problem, progress in research and management measures, and factors limiting the effectiveness of implementation; (7) oversee an Annex V educational foundation; and (8) oversee international aspects of Annex V implementation.

Sustained Education and Training

The Congress should charter and endow a foundation to coordinate a sustained, long-term, national program that would assure development and execution of focused Annex V education and training programs for all maritime sectors as well as non-traditional target groups and provide for domestic and international exchange of information on Annex V compliance strategies. The program should include research, execution, and evaluation components and should promote innovation. To develop and carry out projects, the foundation should award grants to private industry and associations, academic institutions, public agencies, and non-profit organizations.

Model Programs

The Congress should require that federal and federally supported fleets, to set an example, work systematically toward full Annex V compliance, upgrade crew training and provisioning practices, and encourage transfer of successful experiences to commercial fleets.

Federal Agency Actions

Coast Guard

The Coast Guard should require cargo and cruise ships lacking comprehensive on-board garbage management systems to off-load garbage at each U.S. port call. Vessel garbage logs and on-board garbage handling and treatment technologies should be examined during routine inspections. The Coast Guard also should require vessel operators to report inadequate port reception facilities using the IMO forms and should follow up these reports to ensure that the necessary changes are made. If ports are required to issue receipts for garbage discharged into their reception facilities, then the Coast Guard should examine these receipts when reviewing vessel garbage logs. In addition, the Coast Guard should require ports to have the necessary state permits as a condition of granting a COA. And, unless and until the COA program is merged with EPA's vessel garbage management effort, the Coast Guard should incorporate into the program requirements that port reception facilities meet EPA technical standards and have any requisite state and EPA approvals.

The Coast Guard, together with the Department of State (DOS) and Department of Justice, should continue, consistent with the nation's international obligations, to enforce Annex V aggressively against foreign-flag violators and should pursue efforts at the international level to resolve any outstanding ambiguities concerning the rights and obligations of port states with respect to control of pollution from vessels. Requirements for garbage logs should be extended to foreign-flag vessels. The Coast Guard also should adopt a policy of issuing tickets in civil cases if pilot projects already under way show this streamlined enforcement approach to be successful. In addition, the Coast Guard should request the assistance of the National Marine Fisheries Service (NMFS), Minerals Management Service, and state marine police in reporting Annex V violations. Annex V information should be distributed through the Coast Guard's voluntary fishing vessel examination program, and the agency should pursue aggressively its campaign to encourage reports of violations by the public.

The Coast Guard and APHIS should collaborate to develop, maintain, and use for enforcement purposes an Annex V record-keeping system incorporating records from vessel boardings, vessel garbage logs, enforcement reports, and, if a receipt system is instituted, port receipts for off-loaded garbage.

The Coast Guard should issue a periodic report listing Annex V enforcement actions and the assistance provided by other federal agencies and marine police units in the states. Analyses of data from the Coast Guard/APHIS record-keeping system should be included. Such reports would allow the Congress to evaluate the adequacy of appropriations for Annex V implementation projects and enforcement.

Department of State (DOS)

The DOS should try to resolve, through IMO or other avenues, the procedural obstacles that block garbage off-loading at some foreign ports. The DOS also should draw attention to the need for an international data collection effort through IMO and the Intergovernmental Oceanographic Commission.

Environmental Protection Agency (EPA)

The EPA should comply with the congressional mandate (recommended earlier) to oversee the port side of the vessel garbage management system. The EPA also should adopt IMO standards for shipboard incinerators.

National Oceanic and Atmospheric Administration (NOAA)

With the assistance of EPA, NOAA should establish statistically valid, long-term monitoring programs to gather data on the flux of marine debris, the physical transport and fate of marine debris, accumulation of plastic on beaches and in the benthos, wildlife interactions with debris, and the impact of debris on pristine areas. NOAA also should assure that the results of its monitoring programs are communicated to agencies responsible for Annex V implementation and enforcement.

The NMFS should offer financial assistance to fisheries fleets for research on and investments in on-board garbage handling and treatment technology. The

NMFS should waive policy conditions, such as minimum cost requirements, that limit access to these programs. The NMFS also should discourage abandonment of fishing gear, particularly in intensively fished areas. And, where appropriate and feasible, fisheries observers should be enlisted to monitor garbage disposal practices.

U.S. Department of Agriculture (USDA)

The APHIS regime should be integrated as fully as possible with the Annex V implementation program and the system for managing land-generated waste. Cargo and cruise ships should be required to off-load APHIS waste at U.S. port calls. In addition, APHIS should consider developing standards based on compacted waste.

Maritime Administration (MARAD)

The Maritime Administration should develop and execute an R&D program that addresses needs for on-board garbage treatment equipment; alteration of commercial equipment; technology demonstration and information exchange; and operational, maintenance, and cost issues. MARAD should obtain technical support from the Navy and maintain contact with the various fleets through NOAA's Sea Grant Marine Advisory Service and the NMFS. The technology development program should be responsive to the needs of the Coast Guard, NOAA, and other government fleets, as well as the private sector.

1

Dimensions of the Challenge and U.S. Progress

Human use of the oceans is extensive and varied, and one of the by-products is shipborne garbage. For centuries, as most land-generated waste was discarded in open dumps, vessel-generated garbage was discharged overboard. To do otherwise was to transport unnecessary weight and to invite the ever-present vermin to prosper. When population density was low and waste consisted primarily of food items and inert inorganic materials (i.e., metal, glass, or china), the land and sea environments were used freely as convenient dumps without apparent damage. Indeed, until recent years it was assumed that discharging garbage into the marine environment was not harmful, because the oceans were so vast that their capacity to absorb waste was infinite. Discarding waste in the ocean was seen as complementary to disposal on land (Goldberg, 1976), and many coastal communities legally barged garbage to sea for disposal.

Most food wastes and garbage thrown overboard disappeared without a trace, but mariners long have observed that such debris sometimes floats on the surface of the sea or washes up on beaches. Some of these fragments are deposited on the shoreline and near-coastal zone by wind and wave action. The long-held assumption that such debris was benign began to change in the 1970s, as scientists documented the accumulation of garbage in the sea and the resulting harm to the marine environment. Part of the problem was the changing composition of garbage, which increasingly contained durable, synthetic materials such as plastic packaging, cargo nets, packing straps, and synthetic-fiber fishing lines and nets (Recht, 1988; Alig et al., 1990). Even in the most remote locations, observers tallied accumulations of debris that could have come only from maritime sources (Amos, 1993; Ryan and Moloney, 1993). Such evidence, along with the resulting

Scenes like this drew public attention to the marine debris problem and stimulated efforts to control disposal of vessel garbage. Credit: John Miller, National Park Service.

harm to wildlife (Marine Mammal Commission, 1986, 1987, 1988, 1989, 1990, 1991, 1992, 1993) as well as beach closings, eventually provided the basis for international and U.S. action to restrict overboard disposal of garbage.

Apart from the environmental harm attributed to garbage discarded from vessels, numerous accounts have reported direct damage to human activities (O'Hara and Debenham, 1989; O'Hara and Younger, 1990; Debenham and Younger, 1991; Younger and Hodge, 1992; Hodge and Glen, 1993) and described the loss of the aesthetic and recreational value of beaches accumulating substantial amounts of debris (Roehl and Ditton, 1993). As a result, oceanfront communities and federal and state agencies throughout the United States now spend public monies or rely on volunteers to clean debris from beaches on a regular basis. The direct and indirect costs of marine debris—including the costs of beach cleanups, lost tourism, maintenance and repairs to damaged vessels, lost fishing time, and "ghost fishing" by lost nets and traps—cannot be appraised without an assessment of the quantities and types of marine debris, but the total could be in the billions of dollars.[1] Thus, from many perspectives, improperly discarded vessel garbage and other types of marine debris are a burden on society.

INTERNATIONAL AND U.S. MANDATES

A linchpin of early international efforts to control disposal of vessel garbage was the International Convention for the Prevention of Pollution from Ships (1973) and its 1978 Protocol, known collectively as MARPOL 73/78. The convention was developed under the auspices of the International Maritime Organization (IMO), a specialized, multilateral United Nations agency that serves as the principal global forum for negotiating treaties and convening diplomatic conferences related to maritime safety and pollution control. MARPOL is administered primarily by IMO's Marine Environment Protection Committee (MEPC), to which the United States regularly sends participants. As of mid-1994, MARPOL had been signed by 83 nations, including the United States; the first part of the convention, Annex I, entered into force in 1983.[2]

MARPOL currently includes five annexes, each addressing the control of a different type of pollutant: Annex I (oil), Annex II (noxious liquid substances), Annex III (packaged goods), Annex IV (sewage), and Annex V (garbage). Still under development is Annex VI (air pollution). All parties to MARPOL must adhere to Annex I and Annex II but have the option of ratifying the other annexes; once a nation ratifies an additional annex, compliance with it becomes mandatory. This report focuses solely on Annex V, which first entered into force on December 31, 1988 and by the end of 1993 had been ratified by 65 nations. Even though ratification of Annex V is optional, MARPOL signatories have

---

[1] The costs of routine beach cleanup alone may justify the effort to reduce marine debris (although not necessarily the effort to manage vessel garbage, which is only one source of beach debris). An informal survey conducted in 1993 for the Center for Marine Conservation revealed annual costs for beach cleanup ranging from $24,240 per mile in Virginia Beach to $119,530 per mile in Atlantic City, New Jersey. The costs to coastal communities can escalate further when debris problems capture public attention. According to one study, medical waste appearing on beaches during the summers of 1987 and 1988 caused an estimated $1 billion in tourism losses in New Jersey (R.L. Associates, 1988).

[2] MARPOL took effect once signed by 15 nations representing more than 50 percent of the world fleet.

moved forward in an effort to protect their shores and coastal waters from the harmful effects of vessel garbage and other types of marine debris.

Annex V addresses solid waste generated during normal vessel operations at sea, on fixed and floating platforms, and in port, as well as the solid waste generated by economic activities, such as fishing and oil and gas production, carried out on these vessels and structures. (The full annex and the IMO implementation guidelines are reproduced in Appendix B.) The key components of solid waste are domestic garbage, including galley waste and food packaging; operational wastes, such as old fishing gear, fish processing materials, and items generated through vessel maintenance; and cargo-related garbage, such as packaging materials and dunnage.[3]

The Annex V control strategy emphasizes performance rather than specific techniques; discharges are restricted by location and material but the regulations do not specify how compliance should be accomplished. Figure 1-1 summarizes the at-sea garbage discharge restrictions. The performance standards vary depending on how harmful particular materials are believed to be and how long they persist in the marine environment. The most notable standard is for plastics: No plastic may be discarded overboard, except in rare cases such as emergencies. This means all plastic must be stored on board for disposal in port reception facilities; incineration is also an option, with disposal of the resulting ash in an appropriate shore facility. (On vessels entering U.S. ports from foreign shores, domestic regulations require that "food-contaminated" plastics be stored separately, because the organic residues could harbor disease and pests.)

In practice, the plastics prohibition is key to the implementation of Annex V worldwide; until all mariners can comply with this standard, implementation is incomplete. In addition, Annex V provides for the designation of special areas in the seas where no garbage may be discharged except, under certain conditions, food waste.[4] Thus, vessels that transit special areas must have zero-discharge capability. Proper garbage handling practices need to be devised and followed because plastics are highly functional materials and will continue to be available. The basic approaches employed by fleets are waste reduction, which includes reducing amounts of plastics and packaging brought on board; installation of on-

---

[3]Dunnage is timber, pallets, and other packing material used to protect cargo from damage during transport.

[4]The additional protection given to special areas is as follows: No discharges are allowed of plastics, dunnage, lining and packing materials, or other garbage, including paper, rags, glass, metal, bottles, and crockery. Only food wastes may be discharged, as far as practicable from shore but in no case less than 12 nautical miles from the nearest land (except in the Wider Caribbean special area, where comminuted [i.e., ground] food waste may be discharged outside 3 nautical miles from shore). Mixtures of garbage and/or other discharges must be treated in accordance with the most stringent requirements applicable to any part of the mixture.

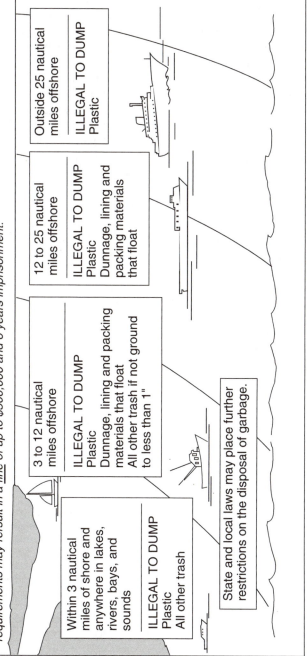

Under the MARPOL agreement and U.S. federal law, it is illegal for any vessel to discharge plastics or garbage containing plastics into any waters. Additional restrictions on dumping non-plastic waste are outlined below. All discharge of garbage is prohibited in the Great Lakes or their connecting or tributary waters. Each knowing violation of these requirements may result in a fine of up to $500,000 and 6 years imprisonment.

Outside 25 nautical miles offshore

ILLEGAL TO DUMP
Plastic

12 to 25 nautical miles offshore

ILLEGAL TO DUMP
Plastic
Dunnage, lining and packing materials that float

3 to 12 nautical miles offshore

ILLEGAL TO DUMP
Plastic
Dunnage, lining and packing materials that float
All other trash if not ground to less than 1"

Within 3 nautical miles of shore and anywhere in lakes, rivers, bays, and sounds

ILLEGAL TO DUMP
Plastic
All other trash

State and local laws may place further restrictions on the disposal of garbage.

FIGURE 1-1 Summary of the at-sea garbage discharge restrictions. Source: Center for Marine Conservation.

board garbage treatment technologies, such as compactors, pulpers, and incinerators; and return of materials to shore for disposal or recycling.

Implementation of MARPOL among signatories has been monitored poorly (U.S. General Accounting Office, 1992) and the monitoring methods now available seem ineffective.[5] Still, certain problems are evident. Most notably, enforcement has been hampered by ambiguities concerning the rights of port states to pursue violations by foreign-flag vessels (a concept known as port state enforcement).[6] Port states have extensive powers to either impose their own rules or enforce international conventions. The United States recently changed its MARPOL enforcement policy to expand its exercise of port state enforcement authorities with respect to violations by foreign-flag vessels within the U.S. Exclusive Economic Zone (EEZ), a 200-nautical-mile-wide band around the coastline. But the United States may not be exercising fully its rights to control pollution from vessels. Port state authority to enforce international rules and standards *outside* the EEZ is established by the Third United Nations Convention on the Law of the Sea (UNCLOS III), which was adopted in 1982 and entered into force in late 1994. (The implications of UNCLOS III, including effects on port state authorities, are addressed in more detail in Appendix C. Enforcement of Annex V in general is discussed in Chapter 7.)

The United States became the 21st signatory to Annex V in 1987, and the regulations took effect a year later. As is routine with international conventions, each signatory nation is responsible for enacting domestic laws to implement the convention and effectively pledges to comply with the Annex V-related laws of other nations. As a world leader, the United States is expected not only to comply with Annex V, but also to lead efforts to develop and implement standards worldwide. Accordingly, numerous steps have been taken to implement Annex V (see

---

[5]The General Accounting Office (GAO) found, for example, that only 13 of (at that time) 57 parties to MARPOL had satisfied treaty obligations to provide IMO with information on MARPOL Annex I violations and penalties imposed. (Additional results of the GAO study are summarized in Chapter 7.)

[6]A port state is a nation in which foreign-flag vessels make port calls. Under the Third United Nations Convention on the Law of the Sea (UNCLOS III), port state enforcement refers to the right of a state, when a foreign vessel is voluntarily in its ports or at an offshore terminal, to undertake investigations and, if warranted, institute proceedings with respect to violations of applicable international rules and standards. Enforcement, for purposes of this report, includes any actions taken to obtain some remedy for violations of Annex V. Such actions may include pursuit of a civil or criminal case against an alleged violator, referral of a case involving a foreign-flag vessel to the appropriate flag state, and record keeping as a means of keeping track of repeat violators. The flag state is the nation where a vessel is registered; flag states have primary responsibility for ensuring that penalties for MARPOL violations are assessed. The United States may act as either a port state or a flag state, depending on the facts of a situation, including whether the vessel in question is registered in the United States. See Appendix C for a more complete explanation of the rights and responsibilities of port states and flag states under UNCLOS III.

TABLE 1-1   Fleets Examined

Recreational boats
Commercial fisheries
Cargo ships
Passenger day boats and ferries
Small public vessels
Offshore platforms/rigs/supply vessels
U.S. Navy surface combatant vessels
Passenger cruise ships
Research vessels

Appendix D). But U.S. implementation of Annex V has been complicated and frustrated by four major factors in addition to those challenging other signatories.

First, unlike a domestic environmental law, Annex V was not crafted to fit neatly into the federal governance structure. When domestic legislation is drafted, its substance typically reflects knowledge of which agencies can bring resources and authority to the problem at hand. In sharp contrast, an international agreement must be accepted in its generic, all-purpose form, leaving the signatory nation to devise a manageable implementation program. The challenge of transforming a sweeping international mandate into a national regime was particularly formidable in the case of Annex V, because the requirements affect a community so broad as to exceed the boundaries of the conventional U.S. regulatory regime. IMO rules typically affect only commercial mariners, who are regulated by the Coast Guard; Annex V rules extend to most seafarers, meaning that, in the United States, numerous federal agencies have some role in implementing the convention across a number of fleets. Nine fleets are addressed in this report (see Table 1-1).[7]

The second complicating factor has been the expansion of the international mandate by the U.S. implementing law, the Marine Plastics Pollution Research and Control Act (MPPRCA) of 1987 (P.L. 100-220). Warships are exempt from MARPOL requirements. But the MPPRCA applies to all vessels on virtually all U.S. waters[8] and to all U.S.-flag vessels anywhere in the world, specifically imposing Annex V standards on the Navy fleet, which was recognized as a major producer of garbage. The MPPRCA did incorporate a grace period for Navy compliance, to allow for an orderly shift in practices and equipment. The Navy

---

[7]Seven of the nine fleets are obvious choices. In addition to those seven, the committee considered offshore oil and gas drilling platforms, rigs, and supply vessels to be a fleet. The other choice that requires some explanation is "small public vessels," which includes the Coast Guard, naval auxiliaries, and other small government vessels. These were grouped together because they have comparable mission and operating constraints. Additional details about all the fleets may be found in Chapters 2 and 4.

[8]The exception is waters under the exclusive jurisdiction of a state.

has pursued both managerial and technical initiatives and can comply with basic requirements for discharge of nonplastic garbage. The challenge is so great that the grace period has been extended to 1998 for the plastics ban and the year 2000 for special area requirements.

The third factor is the U.S. requirement for the quarantine inspection and disposal of food-contaminated garbage from any vessel or aircraft arriving from a foreign port.[9] Quarantine serves an important public health purpose, and these requirements, enforced by the Animal and Plant Health Inspection Service (APHIS), have been in place since the mid-1950s. Unfortunately, the APHIS requirements, while independent of Annex V, may have hindered its implementation by creating confusion and added burdens for vessel and port operators. These problems exist in part because the APHIS regime has not been integrated fully with either the Annex V implementation program or the land-based waste management system.

Finally, implementation of Annex V also has been delayed by ambiguous requirements for port reception facilities, which are critical to proper management of vessel garbage. Annex V requires only that such facilities be "adequate," and U.S. port operators are on their own in determining precisely what that means.[10] The United States does not have a national port authority as most other nations do and, furthermore, it has not integrated management of vessel garbage with the disposal system for land-generated waste.

Port reception facilities are regulated, but with limited effectiveness. A Certificate of Adequacy (COA) verifying compliance with MARPOL must be obtained by ports or terminals serving ocean-going vessels of 400 gross tons or more carrying oil or noxious liquid substances, or fishing vessels landing more than 500,000 pounds of commercial catch per calendar year. The Coast Guard has legal authority to close a port that fails to comply. Reception facilities (but not COAs) also are required at other U.S. ports and terminals, including commercial fishing piers, shore bases for the offshore oil and gas industry, and marinas capable of providing wharfage or other services for 10 or more recreational boats. However, neither the COA program nor the non-COA requirements have resulted in any significant improvement in portside garbage management facilities or operations because there are no technical standards for judging what is adequate. Furthermore, the many small, unattended piers and launch ramps throughout the United States are not required to have reception facilities.

---

[9]In theory, APHIS requirements apply to all vessels that have visited a foreign port before arriving in the United States; in practice, the standards are enforced only for cargo ships and passenger cruise ships.

[10]General guidance is provided (see *Code of Federal Regulations*, Title 33, Section 158) but there are no technical standards. Proposed MPPRCA amendments would require Coast Guard inspections of port reception facilities, but, even if these requirements were adopted, the absence of technical standards would allow for wide variations in "adequacy." (Henceforth, references to the *Code* will be abbreviated using the format 33 C.F.R. §158).

Missions of U.S. Government Agencies

Implementation of Annex V requires the combined resources and skills of an eclectic mix of federal agencies, going well beyond the roles assigned in the MPPRCA.

The MPPRCA gives the Secretary of Transportation, through the Coast Guard, sole authority to enforce Annex V. The Coast Guard is to consult with the Environmental Protection Agency (EPA) in establishing standards for shipboard equipment. The Coast Guard and EPA are to consult with the Department of Commerce (specifically the National Oceanic and Atmospheric Administration [NOAA]) in reporting on the effects of marine debris. In addition, to enforce the law against foreign-flag vessels, the Coast Guard is to cooperate with the Department of State's Office of Ocean Affairs.

Since the MPPRCA was enacted, a number of other agencies also have been recognized as playing important roles, including the Minerals Management Service, which regulates the offshore oil and gas industry; the U.S. Department of Agriculture, which through APHIS is responsible for ensuring quarantine of certain types of garbage; the Maritime Administration, which promotes the U.S. maritime industry and has a technology assessment program that could help meet compliance needs; the National Park Service, which conducts beach debris monitoring programs at national seashores; and the Marine Mammal Commission, which reports annually to the Congress on marine mammal protection issues.

Even with multiple federal agencies striving to accomplish Annex V objectives within their domains of expertise, they have not been able to reach and influence all segments of the highly diverse and dispersed maritime community. Not only are the fleets affected by Annex V highly varied in terms of their practices and accessibility, but also, even within a given fleet, operators may have no common topics to discuss and may not meet regularly on a national or regional level. As a result, a number of years after the ratification of Annex V, it is clear that a comprehensive strategy for integrating Annex V into the national environmental protection regime remains elusive.

PROGRESS IN U.S. IMPLEMENTATION OF ANNEX V

More than seven years have passed since the United States ratified Annex V and enacted the MPPRCA, yet some plastics continue to be discharged overboard. The Navy has obtained permission to do so temporarily, but federal officials suspect that other fleets routinely violate the law. While a minority of vessels apparently off-load garbage at U.S. port reception facilities, Coast Guard boarding officers often "find no trace of garbage, separated plastics, or incinerated ash on ships that doubtlessly generate large quantities of garbage" (Federal

Register, Vol. 59, p. 18,700 [1994][11]). Thus, it is clear that the United States has yet to implement Annex V fully. Full compliance will be difficult to attain, and measuring progress will be a major challenge.

The Coast Guard has reported many difficulties with Annex V compliance and MPPRCA enforcement. External constraints include the vastness of the oceans, which makes comprehensive federal surveillance impossible; the difficulty of obtaining first-hand accounts from witnesses; the lack of follow-up prosecution of foreign-flag vessels by flag states; and economic disincentives, in that large penalties for violations are offset by the perceived low risk of detection (Eastern Research Group, 1992). Internal limitations include the low priority assigned to the problem of shipborne wastes; the complexity of administrative procedures for proceeding against violators; and shortcomings of Coast Guard training with regard to international shipping (Eastern Research Group, 1992).

The EPA and NOAA have had to contend with similar internal constraints. No additional personnel or funding was allocated initially for either Coast Guard or EPA implementation efforts, although NOAA has received limited funding for its Marine Entanglement Research Program (MERP). The Coast Guard has suggested that Annex V compliance depends on factors other than government efforts, specifically the levels of environmental consciousness in the maritime industry and among the general public (Eastern Research Group, 1992).

In spite of these challenges, some steps have been taken to implement Annex V, and there is reason for optimism about their effectiveness, due principally to the exemplary efforts of a network of dedicated advocates. Some of these individuals are federal employees or contractors, while others are private citizens acting on behalf of companies, advocacy groups, or trade associations. Whether motivated by a desire to halt the environmental damage caused by marine debris or by pressure for compliance with the law, many of these isolated initiatives have demonstrated, on a limited scale, that Annex V can succeed.

Perhaps as a result of the combined efforts, compliance with Annex V may be increasing, as can be inferred from national statistics for APHIS garbage offloading for the fiscal years 1988 through 1991. The annual number of vessels offloading garbage increased steadily and significantly during that time period, from 1,937 to 12,518. These data have been interpreted by a USDA official as reflecting increasing levels of compliance with Annex V (Ronald B. Caffey, personal communication to Marine Board staff, August 18, 1992). A similar trend was reported by port authorities in Corpus Christi, Texas, who treated steadily increasing volumes of APHIS waste at their boiler facility between 1989 and 1993.[12]

---

[11]Henceforth, references to the *Federal Register* will be abbreviated using the format 59 Fed. Reg. 18,700 (1994).

[12]The boiler facility treated 30.6 cubic meters (m^3) (40 cubic yards) (yd^3) of APHIS waste in 1989, 79.5 m^3 (104 yd^3) in 1990, 125.4 m^3 (164 yd^3) in 1991, 256.9 m^3 (336 yd^3) in 1992, and 259.9 m^3 (336 yd^3) in 1993 (through August 5 only), port officials reported to the committee. The boiler was shut down in early 1994.

Data from beach cleanups also seem to reflect a slight improvement, although this evidence is soft because the cleanups were not designed to monitor Annex V compliance. Plastic debris is an indicator of Annex V compliance because virtually all overboard discharges of this material are prohibited. Surveys by the Center for Marine Conservation (CMC) found that, in 1989, 60.5 percent of the items found on U.S. coasts were plastic (O'Hara and Younger, 1990); in 1993, the percentage was slightly lower at 53.2 (the rates for individual states from 38.7 percent to 79.7 percent) (Bierce and O'Hara, 1994). On the other hand, a federal beach monitoring project has not detected any improvement. This 5-year beach monitoring pilot program by the National Park Service, which focused on different beaches and employed a different methodology than did CMC, indicated that plastics consistently make up about 90 percent of debris items (Cole et al., 1992).

It is important to recognize that, while beach litter may convince the public that marine debris is a problem, the condition of beaches does not necessarily reflect garbage disposal practices on vessels. Vessels are only part of the marine debris problem. A significant amount of debris originates from land-based sources, including beach goers, wastewater treatment plants, rivers, and combined sewer overflows and storm drains.

Case Histories

Because it is difficult to detect overall trends and progress in controlling vessel garbage, case histories may provide the best portrait of U.S. experiences with Annex V implementation. The selected examples presented here serve not only to illustrate the range and results of past and ongoing efforts, but also to suggest possible model elements of an effective national implementation strategy.

U.S. Navy Compliance

The Navy operates the largest U.S.-flag fleet. The Navy estimates that its ships discharged more than 2,000 metric tons (MT) (4.5 million pounds [lbs.]) of plastic into the oceans each year until 1988. Through leadership and aggressive use of its command organization as well as the willingness of individual crew members, the Navy has made a comprehensive effort to comply with the MPPRCA on its surface combatant fleet. Among its activities, the Navy has established dialogue with outside critics and overseers in the design of its compliance plan, mounted a research and development (R&D) effort to design on-board garbage treatment technology, and instituted a number of progressive policies. Significant progress has been made, but critics note that, even after spending tens of millions of dollars, the Navy still lacks a plan for achieving full compliance (U.S. General Accounting Office, 1994a, 1994b).

To guide the compliance effort, an external advisory committee was created so that Annex V implementation ideas could be discussed in a non-confrontational setting. Participants included senior congressional staff and representatives of environmental groups. The Keystone Center, a dispute resolution organization, was hired to run the committee independent of the Navy. Committee discussions assisted in the initial design of a compliance program that took into account both congressional and environmental concerns, while simultaneously compelling Navy personnel to articulate the challenges involved. The Navy heeded the ad hoc committee's advice and honored the agreements made. Among other things, the decision to reject the use of shipboard incinerators as a permanent solution for disposing of plastics was a result of a consensus-based decision by the committee to avoid combustion technologies that might pollute the air (Ad Hoc Advisory Committee on Plastics, 1988). This decision may warrant reconsideration, however, because it was not based on scientific or engineering investigations and no waste management officials were involved. Moreover, political and technical considerations have changed in the past few years (see Chapters 4 and 5).

Shipboard equipment developed by the Navy beginning in 1979 is expected to enable the fleet to eliminate entirely the discharge of plastics at sea. Heavy-duty solid waste pulpers, small pulpers, metal and glass shredders, and an entirely new device, the plastic waste processor, have been developed. The current focus of the R&D program is the formal testing and evaluation of the plastics processor, so that fleetwide installation can begin in 1995. Testing and evaluation of the pulpers and shredders are in the final stages, but the Navy has no plans to install this equipment because it would not enable compliance with special area mandates. (This issue is discussed further in Chapter 4.) In sum, considerable attention has been devoted to development of on-board technology to support Annex V compliance, although none of the garbage treatment equipment has been installed permanently on ships to date, and the Navy continues to discharge some plastics overboard.

To limit plastics discharges until shipboard equipment can be installed, the Navy invoked several operational changes. Crews now separate out plastic garbage at the source and keep it on board for as long as vessel sanitation and crew habitability can tolerate it. Field trials beginning in 1988 demonstrated that uncompacted, food-contaminated plastic could sit in an unrefrigerated storage locker for a maximum of three days before the stench became intolerable. Clean, uncompacted plastic materials could be collected and retained on board so long as there was storage space anywhere on the ship—about 20 days. When the "3-day/20-day" rule was adopted, dramatic amounts of material piled up on Navy ships, destined for shoreside reception facilities. This simple procedural shift is believed to have reduced overboard discharges of plastic by 70 percent (Chitty, 1989), to 612.4 MT (1.35 million lbs.) per year. Although the 3/20 rule initially was only a recommendation, it is now mandatory,[13] and the Navy is obligated to abide by it until a full suite of shipboard equipment is installed and full fleet compliance is attained.

The naval supply organization also has made a comprehensive effort to support MARPOL compliance. The Plastics Reduction in the Marine Environment (PRIME) program objective is to eliminate plastic packaging and reduce use of disposable packaging in all items in the military supply network. By eliminating unnecessary plastics, using alternative materials, and packing in bulk, an estimated 215.5 MT (475,000 lbs.) of plastic packaging has been eliminated through changes in specifications for more than 350,000 items (Koss, 1994). In addition, nonplastic packaging will be specified in some new contracts (Koss, 1994). Efforts continue to reduce plastic packaging in items used by the Navy but managed by other military services. The long-term focus is on development of alternative materials to replace plastics in some items.

The Navy also adopted an economic incentive suggested in the IMO guidelines for Annex V implementation, by giving to ship crews any income generated from the recycling of garbage materials for their scrap or deposit value. The money is used to purchase amenities, thereby rewarding the crews for their waste reduction efforts. The Navy also has experimented with novel uses for recycled materials. Some 10.4 MT (23,000 lbs.) of plastic wastes from a single ship were transformed into "lumber" for park benches, picnic tables, and other items for use at Navy bases (Middleton et al., 1991). Such efforts can help create new markets and thereby improve the prospects for recycling as a waste management option.

State Initiatives

A number of states have launched initiatives to reduce marine debris and implement Annex V. The effort in Texas has been particularly aggressive and multi-faceted. Even before the federal government ratified Annex V, Texas officials identified marine debris as a serious problem along the states abutting the Gulf of Mexico coast. After the state land commissioner participated in the CMC's first beach cleanup in 1986, the Texas General Land Office took a leadership position in encouraging the U.S. government to ratify Annex V. The office also has worked diligently to implement the agreement, in concert with neighboring state governments along the Gulf of Mexico.

The Texas land commissioner has motivated both the public and Gulf-based industries to understand that Annex V compliance is a serious obligation. The commissioner told the U.S. Congress that state offshore oil inspectors could be of

---

[13]Under the National Defense Authorization Act for Fiscal Year 1993 (P.L. 102-484), ship personnel must store food-contaminated plastic on board for the last three days before entering port, while clean plastic debris must be stored for the last 20 days.

assistance to the federal enforcement program, and that these inspectors pursued violators under existing state laws (Mauro, 1993). In addition, Texas has sponsored the two most extensive surveys of port reception facilities (Hollin and Liffman, 1991, 1993) and is working to further the preparations needed to bring special area status into force for the Wider Caribbean.

Initiatives on the Pacific Coast have stressed education. The Marine Plastic Debris Action Plan for Washington State (Marine Plastic Debris Task Force, 1988; Rose, 1990) identified the types of vessels most common in nearby waters and focused on opportunities for intervening to halt illegal overboard discharges. In addition, noting that prevailing currents would concentrate debris off the Washington coast, the plan focused on education of marine communities, including recreational boaters. Marinas and boaters were targeted in a dedicated program, an unusual undertaking at the time.

The California Marine Debris Action Plan (Kauffman et al., 1990) is the result of a large volunteer effort to establish a continuing program to reduce marine debris. Although federal and state governments participated, responsibility for many of the follow-up activities remains with private and citizen organizations. The plan relies heavily on education to change the habits of marine users.

Pilot Programs by Community and Environmental Groups

Abundant evidence demonstrates the benefits of citizen participation and other private involvement in implementation of Annex V. Community and environmental groups have been highly successful in focusing public attention on marine debris, articulating prevention methods, and convincing citizens to assume responsibility for addressing the problem. Numerous ideas have been tested by these groups, and some of their insights and perspectives have been integrated into government programs.

One popular concept is organized beach cleanups, which not only have cleared unsightly debris, but also have helped document the scope of the problem. The annual CMC beach cleanup began as a project in one state funded by a private contribution. The event quickly grew to international proportions, gaining the support of NOAA, EPA, and the Navy and bringing hundreds of thousands of volunteers to beaches on a regular basis.

Considerable experience also has been acquired in port and marina settings, albeit often in local or short-term projects that ended when initial funding was exhausted. The Coastal Resources Center produced guidelines on how to start a marina recycling program (Kauffman, 1992) and carried out a recycling project at Half Moon Bay, California, that is being duplicated in San Francisco. Such grassroots efforts are an essential means of reaching recreational boaters.

A fishermen's initiative in Oregon, described by Recht (1988), illustrates the effectiveness of integrating vessel and shore garbage disposal. Fishermen using the Port of Newport began a net recycling program in the late 1980s. Initially funded by a grant from MERP, the program included an educational effort targeting fishermen and an agreement with the city to place large dumpsters and storage areas on city-owned piers. Fishermen were encouraged to return to port their netting and cordage formerly discarded at sea; in some cases, they also retrieved netting observed floating at sea. Once on shore, the plastic nets were sorted by type, baled, and transported to recycling centers in Seattle. Although the Newport program has been discontinued, fishermen using various ports in Washington state and Alaska continue to recycle nylon gill-net webbing through a recycling infrastructure established and managed by the Pacific States Marine Fisheries Commission (F.I.S.H. Habitat Education Program, 1994).

Special Situations: The Gulf of Mexico

The Gulf of Mexico is part of the Wider Caribbean special area. Special areas are an important consideration in the development of a U.S. Annex V implementation strategy, for two reasons. First, special areas fall into multiple national jurisdictions, meaning that all nations bordering an area must cooperate to some degree, first to obtain the designation and then to implement and enforce Annex V mandates. The other reason is that Annex V imposes a zero-discharge standard in special areas, and vessels transiting these areas must be able to comply. In most special areas, food waste must be discharged at least 12 nautical miles from shore; in the Wider Caribbean only, comminuted (i.e., ground) food waste may be discharged beyond 3 nautical miles from shore.

The IMO has designated eight special areas under Annex V.[14] The discharge restrictions have gone into force in three areas: the Baltic Sea, the North Sea, and the Antarctic Ocean. The mandates will take effect in the Mediterranean Sea, the Persian Gulf, the Red Sea, the Black Sea, and the Caribbean once IMO determines that sufficient port reception facilities are available bordering the special area. It is important to note that the designation of special areas is a political process, as opposed to an entirely scientific one. The Wider Caribbean was so designated by IMO at the urging of the United States and in consultation with other nations in the region, including Mexico and Cuba, neither of which is a signatory of Annex V.

The special area status of this region will make unique demands on executive agencies of the U.S. government and will require coordination of enforcement and compliance efforts among the countries bordering the Wider Caribbean. The Gulf of Mexico Program (GOMP) is one avenue for such coordination. Organized by the EPA regions[15] spanning the gulf, the GOMP is an interagency effort

---

[14]Different special areas may be designated under other MARPOL annexes. This report addresses only those special areas designated under Annex V.

[15]The EPA divides the United States into 10 regions for administrative purposes. The Gulf of Mexico falls within two jurisdictions, so oversight of the special area requires the cooperation of both the Atlanta and the Dallas EPA headquarters.

that draws heavily on the local expertise of community-based organizations and industries. Programs include educational campaigns and recycling awareness programs. Each decision typically involves the deliberations of most groups that would be concerned about the topic, so the resulting action plan reflects at least some of their needs and objectives. This approach requires a capability to sustain intense participation across a number of organizations that have not interacted previously; success is determined in part by the personal characteristics and skills of the individuals involved.

Although the cooperative decision-making approach is time intensive and demanding, the process has yielded some distinctive results. The program produced the first regional Marine Debris Action Plan (Gulf of Mexico Program, 1991) and provides a forum for sharing the results of local efforts through regional meetings and professional papers. This record demonstrates the utility of a consensus-building approach across several jurisdictions and communities. The consensus-based, open format approach has helped to advance the working-level implementation of Annex V.

THE CHALLENGES AHEAD

The level of independent activity under way to support implementation of Annex V is a positive sign. Clearly this environmental goal has supporters, both within and outside government. However, the many isolated initiatives and current levels of effort do not add up to full compliance, or even a national strategy that will lead to full compliance.

Efforts to improve compliance already are under way at the international level through IMO, which serves as the forum for formally amending MARPOL 73/78 and also offers technical services to help nations overcome obstacles and track compliance. Through its committees, IMO has launched efforts to promote Annex V compliance by further clarifying procedures for port state enforcement with regard to control of pollution from vessels, and to examine the vessel/port interface (including port reception facilities).

The challenge now for the United States is to identify, recruit, organize, integrate, and manage the various elements and resources already in existence that can provide the foundation for a national implementation program. The underpinning of such a program has to be "nuts and bolts" advice of individuals already engaged in the effort, whether at work on the waterfront, volunteering for citizen groups, or holding desk jobs in government. Their observations and experience provide the best evidence on strategies that work, and it is on their shoulders that the ultimate burden for implementation falls. Wide exchange of information about strategies proven to be successful, as well as additional research on and development of promising concepts, clearly could be helpful in implementing Annex V. In addition, common sense suggests that compliance practices ought to be integrated thoroughly into normal vessel and port operations; they should not be disruptive, competitively harmful, or so expensive as to drain the resources of government or private organizations.

While addressing domestic needs and opportunities, U.S. policy also needs to recognize the international aspects of the problem of vessel garbage, which is generated by all maritime nations and taints the environment worldwide. U.S. officials must have a full understanding of both the opportunities and constraints afforded by international law, which provides the context for Annex V implementation. The United States also carries the responsibility of a world leader to provide a model for compliance and promote multilateral cooperation to advance Annex V implementation worldwide.

The following chapter further defines the challenges in the Annex V implementation by examining what is known about the sources, fates, and effects of vessel garbage.

REFERENCES

Ad Hoc Advisory Committee on Plastics. 1988. Reducing Navy Marine Plastic Pollution. A report to the Assistant Secretary of the Navy for Shipbuilding and Logistics. Available from the Office of the Chief of Naval Operations, Washington, D.C. June 28.

Alig, C.S., L. Koss, T. Scarano, and F. Chitty. 1990. Control of plastic wastes aboard naval ships at sea. Pp. 879-894 in Proceedings of the Second International Conference on Marine Debris, 2–7 April 1989, Honolulu, Hawaii (Vol. II), R.S. Shomura and M.L. Godfrey, eds. NOAA-TM-NMFS-SWFSC-154. Available from the Marine Entanglement Research Program of the National Marine Fisheries Service (National Oceanic and Atmospheric Administration), Seattle, Wash. December.

Amos, A.F. 1993. Solid waste pollution of Texas beaches: a Post-MARPOL Annex V study, Vol 1: Narrative. OCS Study MMS 93-0013. Available from the public information unit of the U.S. Department of the Interior, Minerals Management Service, Gulf of Mexico OCS Region, New Orleans, La. July.

Bierce, R. and K. O'Hara, eds. 1994. 1993 National Coastal Cleanup Results. Washington, D.C.: Center for Marine Conservation.

Chitty, F. 1989. Presentation by Fred Chitty, supply officer for the U.S. Navy Atlantic Fleet, to the Ad Hoc Advisory Committee on Plastics, Washington, D.C., May 30, 1989.

Cole, C.A., W.P. Gregg, D.V. Richards, and D.A. Manski. 1992. Annual Report of National Park Marine Debris Monitoring Program, 1991 Marine Debris Surveys with Summary of Data from 1988 to 1991. Tech Rpt. NPS-NRWV/NRT-92/10. Available from the Natural Resources Publications Office of the National Park Service, Denver, Colo.

Debenham, P. and L.K. Younger. 1991. Cleaning North America's Beaches: 1990 Beach Cleanup Results. Washington, D.C.: Center for Marine Conservation. May.

Eastern Research Group, Inc. (ERG). 1992. Report to Congress on Compliance with the Marine Plastic Pollution Research and Control Act of 1987. Report prepared for the U.S. Coast Guard by ERG, Arlington, Mass. (now Lexington, Mass.). June 24.

F.I.S.H. Habitat Education Program. 1994. Net Recycling Program Summary. Fact sheet prepared by the Fishermen Involved in Saving Habitat Education Program, Gladstone, Ore.

Goldberg, E.D. 1976. The Health of the Oceans. Paris: UNESCO Press.

Gulf of Mexico Program. 1991. Marine Debris Action Plan for the Gulf of Mexico. Dallas, Tex.: U.S. Environmental Protection Agency. October.

Hodge, K.L. and J. Glen. 1993. 1992 National Coastal Cleanup Report. Washington, D.C.: Center for Marine Conservation. August.

Hollin, D. and M. Liffman. 1991. Use of MARPOL Annex V Reception Facilities and Disposal Systems at Selected Gulf of Mexico Ports, Private Terminals and Recreational Boating Facilities. Report to the Texas General Land Office by Dewayne Hollin, Texas A&M University Sea Grant College Program, and Michael Liffman, Louisiana State University Sea Grant College Program. September.

Hollin, D. and M. Liffman. 1993. Survey of Gulf of Mexico Marine Operations and Recreational Interests: Monitoring of MARPOL Annex V Compliance Trends. Report to the U.S. Environmental Protection Agency, Region 6, Gulf of Mexico Program by Dewayne Hollin, Texas A&M University Sea Grant College Program, and Michael Liffman, Louisiana State University Sea Grant College Program.

Kauffman, J., M. Brown, and K. O'Hara. 1990. California Marine Debris Action Plan. San Francisco: Center for Marine Conservation.

Kauffman, M. 1992. Launching A Recycling Program at Your Marina. San Francisco: Coastal Resources Center. February.

Koss, L.J. 1994. Dealing With Ship-generated Plastics Waste on Navy Surface Ships. Paper presented at the Third International Conference on Marine Debris, Miami, Fla., May 8–13, 1994. Office of the Chief of Naval Operations, Department of the Navy, Washington, D.C.

Marine Mammal Commission (MMC). 1986. Annual Report of the Marine Mammal Commission, Calendar Year 1985, a report to Congress. Washington, D.C.: MMC. Jan. 31.

Marine Mammal Commission (MMC). 1987. Annual Report of the Marine Mammal Commission, Calendar Year 1986, a report to Congress. Washington, D.C.: MMC. Jan. 31.

Marine Mammal Commission (MMC). 1988. Annual Report of the Marine Mammal Commission, Calendar Year 1987, a report to Congress. Washington, D.C.: MMC. Jan. 31.

Marine Mammal Commission (MMC). 1989. Annual Report of the Marine Mammal Commission, Calendar Year 1988, a report to Congress. Washington, D.C.: MMC. Jan. 31.

Marine Mammal Commission (MMC). 1990. Annual Report of the Marine Mammal Commission, Calendar Year 1989, a report to Congress. January 31.

Marine Mammal Commission (MMC). 1991. Annual Report of the Marine Mammal Commission, Calendar Year 1990, a report to Congress. Washington, D.C.: MMC. Jan. 31.

Marine Mammal Commission (MMC). 1992. Annual Report of the Marine Mammal Commission, Calendar Year 1991, a report to Congress. Washington, D.C.: MMC. Jan. 31.

Marine Mammal Commission (MMC). 1993. Annual Report of the Marine Mammal Commission, Calendar Year 1992, a report to Congress. Washington, D.C.: MMC. Jan. 31.

Marine Plastic Debris Task Force. 1988. Marine Plastic Debris Action Plan for Washington State. Olympia, Wash.: Washington State Department of Natural Resources.

Mauro, G. 1993. Testimony of Garry P. Mauro, commissioner, Texas General Land Office, before the Subcommittee on Superfund, Ocean, and Water Protection of the Committee on Environment and Public Works, U.S. Senate, 102nd Congress, Second Session, Washington, D.C., Sept. 17, 1992. P. 10 in Implementation of the Marine Plastic Pollution Research and Control Act. S. Hrg. 102-984. Washington, D.C.: U.S. Government Printing Office.

Middleton, L., J. Huntley and J. Burgiel. 1991. U.S. Navy Shipboard Generated Plastic Waste Pilot Recycling Program. Washington, D.C.: Council for Solid Waste Solutions of the Society of the Plastics Industry. March.

O'Hara, K.J. and P. Debenham. 1989. Cleaning America's Beaches: 1988 National Beach Cleanup Results. Washington, D.C.: Center for Marine Conservation. September.

O'Hara, K.J. and L.K. Younger. 1990. Cleaning North America's Beaches: 1989 Beach Cleanup Results. Washington, D.C.: Center for Marine Conservation. May.

R.L. Associates. 1988. The Economic Impact of Visitors to the New Jersey shore the summer of 1988. Report prepared for the New Jersey Division of Travel and Tourism by R.L. Associates, Princeton, N.J.

Recht, F. 1988. Report on a Port-Based Project to Reduce Marine Debris. NWAFC Processed Report 88-13. Report prepared for the Northwest and Alaska Fisheries Center of the Marine Entanglement Research Program (National Oceanic and Atmospheric Administration), Seattle, Wash. July.

Roehl, W.S. and R. Ditton. 1993. Impacts of the offshore marine industry on coastal tourism: The case of Padre Island National Seashore. Coastal Management 21:77–89.

Rose, R. 1990. Marine plastic debris: What Washington state has done. Pp. 1020-1028 in Proceedings of the Second International Conference on Marine Debris, 2–7 April 1989, Honolulu, Hawaii (Vol. II), R.S. Shomura and M.L. Godfrey, eds. NOAA-TM-NMFS-SWFSC-154. Available from the Marine Entanglement Research Program of the National Marine Fisheries Service (National Oceanic and Atmospheric Administration), Seattle, Wash. December.

Ryan, P.G. and C.L. Moloney. 1993. Marine litter keeps increasing. Nature 361:23. Jan. 7.

U.S. General Accounting Office (GAO). 1992. International Environment: International Agreements Are Not Well Monitored. GAO/RCED-92-43. Washington, D.C.: GAO Resources, Community, and Economic Development Division. January.

U.S. General Accounting Office (GAO). 1994a. Pollution Prevention: Chronology of Navy Ship Waste Processing Equipment Development. GAO/NSIAD-94-221FS. Washington, D.C.: GAO National Security and International Affairs Division. August.

U.S. General Accounting Office (GAO). 1994b. Pollution Prevention: The Navy Needs Better Plans for Reducing Ship Waste Discharges. GAO/NSIAD-95-38. Washington, D.C.: GAO National Security and International Affairs Division. November.

Younger, L.K. and Hodge, K. 1992. 1991 International Coastal Cleanup Overview. Washington, D.C.: Center for Marine Conservation. May.

2

Sources, Fates, and Effects of Shipborne Garbage

Full implementation of Annex V depends in part on the development of a comprehensive understanding of the sources, fates, and effects of vessel garbage, because this information suggests where interventions are needed. To date, scientific understanding of these phenomena is uneven, and certain aspects have yet to be examined at all. This chapter outlines what is known and identifies important gaps in knowledge.

The chapter opens with an overview of techniques for identifying and monitoring vessel garbage in the marine environment. The heart of the chapter is divided into three sections. The first describes the nine fleets examined by the committee as sources of vessel garbage. The second section outlines what is known about the fate of vessel garbage discarded into the marine environment. The last section and a supporting appendix summarize the effects of vessel garbage and other marine debris on aesthetic enjoyment of oceans and beaches, human health, and the ecology of the marine environment. Although the ill effects of such debris are acknowledged and often visible, they are often difficult to quantify and understand in terms beyond the harm inflicted on marine life. Information on effects is included not only for the sake of completeness, but also because it may be useful in development of educational programs (Chapter 6) and benchmarks for measuring progress in Annex V implementation (Chapter 8).

In tracking vessel garbage, it is important to recognize that an estimate of the quantity of garbage generated is *not* a measure of the amount handled by onboard treatment technologies or port reception facilities. Annex V permits vessel operators to discharge into the oceans non-plastic materials that float, food wastes, and other garbage, so long as the vessel is the prescribed distance from shore (disposal requirements are outlined in Chapter 1, Figure 1-1). Therefore, an unknown amount of garbage continues to be discharged overboard legally, adding to the accumulation of debris already in the marine environment.

IDENTIFYING VESSEL GARBAGE IN THE MARINE ENVIRONMENT

The amounts and precise characteristics of garbage thrown overboard, either before the ratification of Annex V or since, are unknown. Vessel discards are difficult to isolate and identify in the marine environment, due to the littering of coastal waters by land-generated wastes left on beaches, continuing domestic and industrial sewer discharges, and previously discharged waste transported via offshore winds, rivers, and coastal runoff. However, there is at least one way to approximate the level of vessel debris as distinct from other waste—by selecting particular types of sampling sites and then monitoring certain types of debris appearing there.

Plastics, which for all practical purposes are indestructible[1] under marine environmental conditions, may provide a reliable measure of vessel discards if sampled in sediments and on beaches distant from the influences of recreational activities and sewer outfalls. However, because newly discarded plastic items float, they may be transported to locations far from the site of discharge, confounding attempts to identify vessel-generated debris on the basis of location alone. Plastics also may sink over time as they break apart, weather, or accumulate organic coatings, tar, shells, or sand. Sunken items may not be observed. To complicate monitoring efforts further, it is impossible to distinguish plastics tossed overboard lawfully before 1989 from those discarded illegally since then. Still, worldwide, there probably has been a meaningful (albeit unknown) level of compliance with the ban on discharge of plastics.

The types of items discarded from vessels are reflected in beach debris, which encompasses a wide variety of materials. The characteristics of debris items larger than 1 inch (2.5 centimeters) have been summarized from the literature by Ribic et al. (1992). These items include glass, plastic, metal, paper, and a telling variety of fisheries gear, cloth, foodstuffs, wood, rubber, and packaging materials. With the exception of plastics, all these materials may be discharged overboard in certain areas under Annex V. The selection of indicator items for

---

[1]At present, biodegradable plastics are used only on a very limited basis and their ultimate fates in the marine environment are unknown (Palmisano and Pettigrew, 1992). The Environmental Protection Agency (EPA) has published rules setting standards of degradability for plastic six-pack rings (40 C.F.R. §238), and commercial ring carriers appear to meet the standards (Craig Vogt, EPA Oceans and Coastal Protection Division, personal communication to Marine Board staff, July 7, 1994). *Even so, overboard disposal of all plastics, including biodegradable varieties, is prohibited.*

One way to estimate amounts and types of vessel garbage thrown overboard is to examine beach debris, which may vary by geographical area. Milk jugs are a common sight along Gulf of Mexico beaches. Credit: Tony Amos.

measuring vessel discards depends in part on the location of the debris sampling site. Table 2-1 lists items that might be used as indicators for vessel discards washed ashore in the Gulf of Mexico; it should be possible to identify comparable indicator items for vessel garbage in other regions.

To date, the monitoring of debris under the Marine Plastics Pollution Research and Control Act (MPPRCA) has been confined to beach surveys and near-shore urban surveys of harbors, where debris may include materials from waste-water treatment plants and combined sewer overflows and stormwater drains (Trulli et al., 1990). The literature includes occasional reports on underwater surveys or cleanups of sunken debris from harbors or around oil platforms (Debenham and Younger, 1991; Minerals Management Service, 1992).

Sinking debris receives little attention, yet the long-term and perhaps most insidious effects may be upon the benthic biota. Plastics and other wastes are entering the benthos in continuous fluxes. The material may reside for a near-infinite time in the surface sediments. Debris on the coastal sea floor could be monitored by divers or through the use of side-scan sonar imaging, photographic surveys, submersibles, or trawls.

In summary, techniques for monitoring vessel garbage in the marine environment have not been well defined. Improvements are in the offing (Miller, 1993, 1994). Systematic efforts have been made to monitor marine debris[2], but to

---

[2]For example, the National Park Service conducted a five-year sampling program at selected parks (Cole et al., 1990, 1992; Manski et al., 1991; Miller, 1993), and sampling programs have been carried out in Alaska (Merrell, 1980, 1985; Johnson and Merrell, 1988; Johnson, 1990a, 1990b), Hawaii (Henderson et al., 1987), and Texas (Amos, 1993b).

TABLE 2-1   Indicator Items That May be Used to Identify Sources of Beach
Debris in the Gulf of Mexico

SOURCE	ITEMS	OTHER SOURCES
Offshore oil and gas operators	Pipe-thread protectors; 55-gallon drums; 5-gallon pails; large white plastic sheets	Fishing; merchant mariners
Fishing (shrimpers, long-liners)	Rubber gloves; 5-gallon pails, milk jugs; egg cartons; onion sacks; light sticks; plastic sheets	Recreational boaters
Merchant mariners	Galley-waste containers with non-U.S. labels	None
Recreational boaters	Outboard motor oil containers	Fishing; beach goers
Beach goers	Beverage cans; fast food containers	Fishing; recreational boaters; merchant mariners

Source: Amos, 1993a.

date the results have been disappointing in terms of the failure to detect clear
trends. The federal government plans to put a new national monitoring program
in place in 1995. The program will make use of a statistical methodology for
monitoring marine debris that was developed and reviewed by federal agencies
and environmental organizations. Applications for this methodology also are
being studied by Latin American and Caribbean countries.

SOURCES OF SHIPBORNE GARBAGE

Information about sources of shipborne garbage is useful because it can
suggest where Annex V implementation efforts should be directed. The sources
of garbage regulated by Annex V are "all ships," where a ship is defined as ". . .a
vessel of any type whatsoever operating in the marine environment and includes
hydrofoil boats, air-cushion vehicles, submersibles, floating craft and fixed or
floating platforms." (See Appendix B.) Thus, many diverse fleets and vessels are
potential sources of garbage.

The true sources, of course, are the persons aboard these vessels who generate garbage as a normal consequence of all the sundry activities they pursue. The
quantity and nature of vessel discards depend in part on the standards of crew or
passenger accommodations. The amount of garbage is proportional to the
community's standard of living; the higher the standard, the more seafarers are
likely to use packaged prepared foods, supplies, and single-use items rather than
provisions requiring added preparation and cleanup. (Moreover, the use of dis-

posable items and packaging has been encouraged by changes in ship practices, sanitation concerns, and a desire for convenience.) The result is added waste. When an individual is accustomed to a high standard of living on shore, he or she expects similar conveniences on a vessel, despite the cramped living space. Modern vessels are capable of providing many conveniences, even on long voyages.

The task of measuring the amounts of garbage produced during normal voyages is not well supported by present Annex V compliance and enforcement programs. The committee was unable to locate or develop any precise data for any phase of the garbage cycle.[3] There are no reliable data on the characteristics and amounts of vessel garbage generated by all the maritime sectors to which Annex V applies. Nevertheless, drawing on numerous sources, the committee sought to characterize as completely as possible the various fleets and the garbage they generate. Nine major maritime sectors are addressed in this report.[4] The information presented in this chapter is deliberately brief; additional details about each fleet and its garbage management practices are provided in Chapter 4.

The only all-inclusive estimates of amounts of garbage generated by U.S. maritime sectors were developed in support of MARPOL/MPPRCA rule making for the Department of Transportation by the Eastern Research Group (1988) and later revised (Cantin et al., 1990). (See Table 2-2.) These estimates, while based on some flawed assumptions, provide an initial perspective on sources of vessel garbage. The Cantin data identified recreational boaters as generating the largest amount of garbage (by weight), more than 50 percent of the total. Day boats and fishing vessels each were thought to contribute close to 20 percent of the total.

The Cantin data must be employed carefully because they are based on some fleet-specific assumptions that are either outdated or, in the committee's judgment, questionable. The former problem is obvious with regard to the merchant marine, for example. The maritime industry has changed considerably in recent years. Environmental awareness has increased within the industry, while the continued depression in worldwide shipping has spurred operators to reduce crew sizes, change organizational structures and voyage patterns, and expand shoreside responsibilities for vessel garbage management. These factors can influence the amounts of garbage generated. An example of a questionable assumption may be found in the Cantin calculations for the recreational boating sector, in which per-person garbage generation was presumed to be similar to that for cargo ships. This correlation seems doubtful, considering that boaters generally eat only one meal per voyage, while merchant mariners may consume three meals daily and generate additional garbage from food preparation. Thus, the Cantin estimate for

---

[3]A now-outdated study by the National Research Council (1975) estimated that ocean-going vessels discard 635,000 MT (14 billion pounds) of wastes every year.

[4]Each sector reflects a general type of vessel; most surface vessels would fit into one of the nine categories (the committee did not examine submarines). Any omissions of specific sectors or vessels are due only to limits on the committee's time and resources.

TABLE 2-2 Annual Garbage Generation by U.S. Maritime Sectors[a,b]

Sector	Garbage Generated (MT)	Percent of Total
Recreational Boats	636,055	51.4
Day Boats	245,108	19.8
Fishing Vessels	233,177	18.8
Small Public Vessels		3.2
U.S. Navy	34,611	
U.S. Coast Guard	4,317	
U.S. Army	490	
Schools	266	
Cargo Ships	30,949	2.5
Navy Surface Combatant Vessels	21,968	1.8
Offshore Industry		1.4
Platforms	14,721	
Service	1,989	
Passenger Cruise Ships	13,347	1.1
Miscellaneous Vessels	1,161	0.1
Research Vessels		<0.1
NOAA	317	
Other	213	
Total	1,238,689	99.99

[a]This garbage is not necessarily discharged overboard.

[b]The original presentation of the data has been revised to conform with the committee's maritime sectors.

Source: Cantin et al., 1990.

boaters' garbage seems high. Other salient observations on the Cantin data may be found in the forthcoming descriptions of each sector.

The data presented in Table 2-2 reflect garbage *generation*. The Cantin study also estimated amounts of garbage discharged ashore and overboard by each maritime sector, both before and after ratification of Annex V. These estimates were incorporated into a congressionally mandated study of plastic waste materials, including marine debris (U.S. Environmental Protection Agency, 1990). In the committee's judgment, neither the Cantin nor the EPA results with respect to garbage discharged overboard can be relied on, even to gain an initial perspective on disposal practices. The committee's misgivings are due primarily to the absence of any way to know whether the estimates are even reasonable. Indeed, little is known about the amounts of garbage discarded at sea, or, correspondingly, whether these disposal levels are environmentally acceptable. Examination of these issues is beyond the scope of the present report.

However, recognizing the shortcomings of available data, the committee developed its own estimates of vessel garbage generation based on weighting factors obtained from a variety of sources (see Table 2-3). These rough approxi-

TABLE 2-3   Characterization of Vessel Garbage Generated in U.S. Maritime Sectors[a]

Estimate of Annual

	Number of Vessels	Average Crew/Passengers			Vessel Utilization[c]
		Low	Average	High	
Recreational Boats	7,300,000	1	2	6	0.06
Fishing Vessels	129,000	1	4	200	0.66
Cargo Ships	7,800	17	20	25	0.96
Day Boats	5,200	6	46	330	0.66
Small Public Vessels	3,194				
U.S. Navy	284	25	150	300	0.33
U.S. Coast Guard	2,316	5	8	140	0.3
U.S. Army	580	5	6	40	0.2
Schools	14	50	100	150	0.35
Offshore Industry	2625				
Platforms	1125	15	22	40	1
Service Vessels	1500	3	7	20	1
Navy Combatant					
Surface Vessels	360	200	436	5900	0.33
Passenger Cruise Ships	128	125	2,250	3,300	0.96
Research Vessels	125				
NOAA	25	10	90	110	0.75
Other[e]	100	10	30	50	0.5
Miscellaneous Vessels[f]	85	7	23	30	1
Total					

[a]U.S. maritime sectors include foreign-flag vessels that call at U.S. ports as well as all U.S.-flag vessels.

[b]Domestic garbage includes food waste and personal care items; operational/maintenance wastes include fuel oil and fishing wastes; cargo-related garbage includes packaging materials and dunnage.

[c]Vessel utilization is an estimate of the number of days per year vessels are used (1.00 = 365 days).

[d]Day use is an estimate of how long vessels operate during a day of use (1 = 24 hours).

[e]Other research vessels include those operated by private institutions or by federal agencies other than NOAA (e.g., EPA).

[f]Miscellaneous vessels include those operated by private industry.

Sources: All figures are based on the best information available to the Committee on Shipborne Wastes. Estimates of garbage generation (shown in the column entitled "Total [metric tons]") were derived by multiplying together all the preceding figures in each row (using only the average number of crew/passengers). The committee relied on the following sources in developing the table: *Recreational Boats*: Cantin et al., 1990; American Red Cross, 1991; U.S. Coast Guard, 1992a. (The total

Garbage Generation

Day Use[d]	Per Capita Garbage (kilograms/ person/day)	Total (metric tons)	Character of Typical Voyage A= Nearshore B= Offshore	Types of Garbage[b] Operational/ Domestic	Maintenance	Cargo Related
1	0.5	159,900	A	x		
1	1.85	230,500	A, B	x	x	
1	2	111,700	B	x		x
0.5	2	57,623	A, B	x		
		14,932	A, B	x	x	x
1	2	10,262				
1	2	4,058				
0.5	2	254				
1	2	358				
		25,733		x	x	x
1	2	18,068	B			
1	2	7,665	A, B			
1	2	37,812	B	x	x	
1	2	201,830	A	x		
		1,779	A, B	x	x	
1	2	1,232				
0.5	2	548				
1	2	1,427	A, B	x		
		843,236				

number of vessels is all boats registered in coastal states or in states bordering the Great Lakes. *Fishing Vessels:* Cantin et al., 1990; National Research Council, 1991. *Cargo Ships*: U.S. Maritime Administration, 1992a, 1992b; 1992 data obtained from the Maritime Administration's Office of Trade Statistics and Insurance, Washington, D.C.; 1993 data obtained from the U.S. Coast Guard's Marine Information Management System database. (The total number of cargo ships is the number of different ships of all flags calling at U.S. ports annually.) *Day Boats:* U.S. Coast Guard, 1994a. *Small Public Vessels:* U.S. Coast Guard, 1992b. *Offshore Oil Industry:* U.S. Coast Guard, 1994b; Minerals Management Service, 1992; 1994 data obtained from Offshore Marine Services Association, New Orleans, La. *U.S. Navy Surface Combatant Vessels:* Cantin et al., 1990; Polmar, 1992; Forecast International, 1992; 1994 data obtained from U.S. Navy International Programs Office, Washington, D.C. *Passenger Cruise Ships:* Cantin et al., 1990; Cruise Lines International Association, 1994. *Research Vessels:* Cantin et al., 1990; National Research Council, 1994. *Miscellaneous Vessels:* Cantin et al., 1990.

mations, which seem reasonable to the committee, suggest that fishing vessels produce the most garbage (by weight), followed by passenger cruise ships and then recreational boaters. The major differences between the committee's and Cantin's estimates illustrate how seemingly minor changes in assumptions can skew the data, thereby casting doubt on the utility of such exercises. It must be emphasized, once again, that the data provide only an initial perspective on where Annex V implementation problems may lie. In any case, the precise numbers become less important in light of the committee's determination that amount is only one of several factors related to garbage sources that are significant from the standpoint of implementing Annex V. Several other key factors are reflected in Table 2-3.

Of special interest are the numbers of vessels in each sector, the duration of voyages, and the nature of the garbage generated. The number of vessels reflects the quantity of isolated points at which garbage is generated and must be handled properly. The huge number of recreational boats poses a unique challenge in this respect. Voyage duration is also a significant factor. Some Navy ships face extreme challenges in managing garbage because they remain at sea for weeks or even months, so shoreside disposal is a rare option. The problems are fewer on day boats, which easily can store garbage for the duration of their brief voyages. And the nature of the garbage is important because some materials can be disposed of more easily than can others. Vessels that produce multiple types of garbage (especially when many different materials are involved) may require unusually involved Annex V compliance strategies.

Thus, a complex of factors must be considered in identifying which fleets pose the greatest challenges in terms of garbage management. None of the key factors—amounts of garbage, numbers of vessels, duration of voyages, or types of garbage—can be defined with precision across all sectors, because reliable data are scarce and vessel characteristics, even within a single sector, vary widely. Each sector presents unique issues and must be examined individually. The following presentation is organized according to the number of vessels in each sector (the most objective factor), beginning with the largest fleet.

Recreational Boats

Recreational boats produce relatively small amounts of garbage per person and per vessel, due to the short duration of voyages. However, there are an estimated 7.3 million recreational boats in the United States, far more than in any other sector. This sector therefore poses unique challenges in Annex V implementation. Still, the total amount of garbage generated is probably lower than the Cantin estimate, which the committee believes was based on inflated assumptions for numbers of passengers per vessel and, as noted earlier, per-person garbage generation.

Recreational boats produce mainly domestic garbage. Most of these boats

Although many recreational boaters are concerned about the marine environment, there is visible evidence that Annex V compliance levels need to be improved. Even when marinas have reception facilities, garbage still may end up in the water or on the shore. Credit: Coastal Resources Center.

operate within 3 nautical miles of shore, so they are supposed to store all garbage for disposal ashore. Actual disposal practices are difficult to ascertain; the discards are virtually indistinguishable from those of land-side sources, and boats may use innumerable docks and launch ramps that are exempt from requirements for port reception facilities. Nevertheless, there is some evidence that less garbage is thrown overboard today than was in the past. Mudar (1991) has documented a 91 percent Annex V compliance rate among Nantucket boaters using a public marina.

Commercial Fisheries

The United States supports a large and diverse fishing industry that makes a significant contribution to the national and regional economies.[5] The fleets are unique to each catch, and a wide variety of gear is employed. Vessels range from small powered craft and row boats to those over 1,000 tons. In 1990, some 30,000 fishing vessels over 5 net tons were documented by the federal government and

---

[5]In 1992, the United States ranked sixth in total world harvest behind China, Japan, Peru, Chile, and the former Soviet Union. U.S. fishermen landed 4.7 million metric tons (MT) (10.5 billion pounds [lbs.]) of fish valued at $3.5 billion, and the U.S. vessels transferred to foreign ports or vessels an additional 216,000 MT (476.8 million lbs.) valued at $195.4 million (National Marine Fisheries Service, 1994).

TABLE 2-4   Estimated Number of Fishing Industry Vessels Active During 1987 (by Region Fished)[a]

Region	Documented Vessels	State-Numbered Vessels
North Atlantic		
New England	1,800	16,500
Mid-Atlantic	800	5,500
Chesapeake Bay	2,500	3,500+[b]
South Atlantic	2,700	13,500
Gulf/Caribbean		
Gulf Coast	10,000	26,500[c]
Caribbean	[d]	1,500
Great Lakes	[e]	[e]
West Coast	5,000	6,000
Alaska	8,000	9,000
Hawaii/Southwest Pacific	200	200
Total	31,000±	80,000±[f]

[a]Numbers are composite estimates from regional sources. Principal sources include records of fish landings maintained by National Marine Fisheries Service regional offices, permit data maintained by the Commercial Fishing Entry Commission in Juneau, Alaska, and regional assessments commissioned for this study, and economic analyses available for some fisheries.

[b]Based on 1986 estimate of Chesapeake Bay oyster fishery (Sutinen, 1986).

[c]Includes a large number of small boats engaged in shrimp fisheries in bays, sounds, and estuaries.

[d]Negligible.

[e]Current information is not available.

[f]The number of commercial fishing vessels bearing state numbers is not known. West Coast and Alaska figures are close approximations. All other data presented are general estimates.

Source: National Research Council, 1991.

about 80,000 vessels were registered by the states (a breakdown by region is provided in Table 2-4). Trip length varies; most smaller craft take day trips, but the largest vessels, such as tuna seiners, may be at sea for two or three months at a time.

In addition to vessels used by professionals who catch fish for sale as food, there are significant numbers of smaller craft known as charter or head boats, which carry recreational fishermen offshore. Most operate as day boats, rarely venturing beyond 12 miles from shore. A few operate up to 2,500 nautical miles from port for 15 to 20 days. Most use marinas or small docks (as opposed to fishing piers) to support operations. Garbage generally is stored on board and disposed of ashore.

Fishing vessels generate significant amounts of garbage, as is evident from both Cantin's and the committee's estimates (they are nearly identical). Equally notable is the type of garbage, which can include fishing nets, monofilament lines, hooks, traps, and packing bands and containers for frozen bait. The preferred materials for nets are polyethylenes and polypropylenes, which float, so nets lost or discarded overboard are likely to drift and wash ashore, or sink as they accumulate denser materials. Monofilament lines typically are made of nylon, which sinks.

Substantial amounts of gear may be lost due to storms, entanglements on reefs and rocks, and other mishaps. Lost or discarded fishing equipment causes significant harm to wildlife, as documented later in this chapter. Provided that "all reasonable precautions" are taken to prevent such mishaps, accidental losses are not violations of Annex V (which addresses deliberate discharges only), but the IMO implementation guidelines encourage measures to prevent and recover lost gear.

Fishermen have economic motivations to comply with Annex V. More than other seafarers, fishermen rely on the well-being of the oceans for their livelihood and enjoyment; moreover, the capture or entanglement of plastic debris on hooks, in trawls, and in propellers is a costly nuisance. Some fisheries communities are striving to implement the mandate. For instance, in a new program for the Gulf of Maine and Cape Cod Bay, fishermen, fishing ports, and state governments are working with local governments to encourage all vessel operators to return their garbage ashore, and to establish appropriate reception facilities in fishing ports. Within the recreational fishing community, a number of tackle manufacturers have initiated recycling programs for monofilament fishing lines. However, much remains to be accomplished in fisheries communities. Only the very largest fishing ports, such as Seattle, can accommodate all the garbage generated by the commercial fishing fleet; obsolete and worn-out gear and plastic food containers continue to end up in the water and on shorelines.

Cargo Ships

The U.S. merchant fleet operating along the coasts includes U.S.-flag vessels in domestic or international trade, and an international trade fleet of foreign-flag vessels that call at U.S. ports.[6] Some 7,800 different cargo ships call at U.S. ports each year; this figure reflects the true size of this sector more accurately than does the small domestic fleet. The U.S.-flag fleet is just over 400 ships, ranging from breakbulk carriers that may operate anywhere in the world to liquified natural gas tankers that may serve only specialized ports. The average crew complement on

---

[6]The domestic fleet of U.S.-flag vessels, including coastal tows, moves cargoes originating in domestic ports to other U.S. ports. An example would be a chemical tanker moving small lots of specialty chemicals from a Gulf of Mexico port to ports along the Atlantic coast. The international trade fleet of U.S.-flag vessels moves cargoes between the United States and other nations but may call at several U.S. ports on any given voyage. The international trade fleet of foreign-flag vessels moves cargoes to and from the United States, often calling at more than one U.S. port.

a merchant ship in the U.S. fleet is about 21, after falling from as high as 50 in the 1960s (Landsburg et al., 1990).

This sector is a greater Annex V implementation challenge than is suggested by the Cantin study, which significantly underestimated the number of U.S. port calls by foreign-flag ships. (Another major difficulty in this sector, addressed in Chapter 7, is securing compliance by foreign-flag ships.) On the positive side, the trend in crew size suggests that amounts of domestic garbage generated have been declining. Regarding types of Annex V garbage generated, the most noteworthy factor is the need to contend with bulky dunnage, which must be retained on board under some circumstances.[7]

Passenger Day Boats and Ferries

Coastal day boats include many specialty craft, usually carrying passengers on leisure excursions such as sport fishing, whale watching, bird watching, or touring of the coastal waters. Small freighters are included in this category due to the short duration of their voyages. Casino ships, which may remain in port most of the time, are also included.

The Cantin estimate for garbage generated by this sector seems much too high, apparently because day boats were considered to be in full-time service, taking overnight trips (like cruise ships). As defined by the committee, day boats are in service part-time and take brief voyages; they should not be a major challenge in Annex V implementation. Most passenger ferries in the United States transport a large number of persons on relatively brief voyages, during which perhaps one meal and snacks may be eaten. Garbage, consisting primarily of leftover food and packaging, is off-loaded to port reception facilities at the beginning and end of each trip. Most ferries are regulated under domestic laws enacted prior to the ratification of Annex V and long have operated on a zero-discharge standard.

Small Public Vessels

The committee examined Coast Guard, naval auxiliary vessels, and other public vessels (such as those operated by the U.S. Army and military academies) as a single category. This sector includes more than 3,100 vessels, all small in size compared to the Navy's warships. The Cantin data appear to overestimate the amounts of garbage generated by this sector, in part because the assumptions for crew sizes on naval auxiliaries were high. In the committee's judgment, these fleets do not generate huge amounts of garbage but do face special challenges in

---

[7]Dunnage may be discharged overboard beyond 25 nautical miles from shore, but it must be retained on board for shoreside disposal inside 25 nautical miles and in special areas.

managing it, due to the severe shortage of on-board space and the vessels' varied missions and operating profiles.

The Coast Guard fleet includes large and small cutters, several ice breakers, and numerous patrol boats and other small craft. Crew size and voyage length vary, but garbage handling problems can become severe; a cutter may have as many as 140 officers and crew on board, and a voyage often lasts more than 10 days. Domestic and operational wastes are generated on board, but most maintenance is performed while vessels are in port, so repair wastes are a lesser concern.

Each naval auxiliary vessel is designed and constructed uniquely to accomplish some special task, whether it is to support larger Navy vessels, move troops and materiel, maintain station off the coast, or some other specialized duty. Because auxiliary vessels often must carry large amounts of heavy equipment, meet very high survivability standards, and pursue missions that are incompatible with certain garbage treatment options (e.g., compactors pose a magnetic problem on minesweepers), Annex V compliance is a unique challenge for this fleet.

Offshore Industry Rigs, Platforms, and Supply Vessels

Nearly all offshore oil and gas exploration occurs in the Gulf of Mexico.[8] Some 10,000 persons work offshore every day. An offshore rig or platform typically is operated through a contractual arrangement involving the leaseholder, who owns the platform and shoreside base terminal that serves as a "port"; the drilling contractor; and the offshore vessel operator, who transports personnel, supplies, and garbage between the platform and the shore. This sector is subject to assorted domestic laws and regulations that impose discharge restrictions independent of but consistent with Annex V, which prohibits the disposal of any garbage except comminuted food waste from fixed or floating platforms (and from all vessels within 500 meters [547 yards] of such platforms).

This sector encompasses 1,125 manned platforms and 1,500 supply boats.[9] Hazardous wastes hauled from platforms are documented, but few data are available on operational and domestic wastes. Mineral Management Service regulations require platform operators to record and report accidental overboard losses of materials, but these data are not collected centrally. The EPA is gathering information on operational wastes to support development of regulations.

The industry has made an effort to comply with the various discharge restrictions, but overboard loss of equipment and materials due to less-than-exemplary

---

[8]Because of this concentration of activity and the designation of the Gulf of Mexico as part of an Annex V special area, the committee treated offshore oil and gas platforms and vessels as a regional fleet.

[9]These figures reflect the size of the offshore sector in 1994.

handling and management remains a major issue in this sector.[10] Petroleum product containers, at least some of which can be traced to offshore operations, continue to wash ashore on Gulf of Mexico beaches.

Navy Surface Combatant Vessels

The Navy operates the U.S. government's largest and most varied fleet, some 360 surface combatant vessels in addition to the auxiliary fleet addressed earlier. Navy vessels typically carry much larger crews than do merchant ships (Navy ships also may carry troops), so substantial amounts of plastics and other garbage are generated and there is little extra space available to store it. Although the Navy apparently generates far less garbage by weight than do some other fleets (according to both Cantin's and the committee's estimates), this is the only sector with a single owner and operator, which therefore faces singular burdens. Furthermore, the Navy has been singled out by the Congress for particular attention, not only in the MPPRCA, but also in a recent congressional investigation of its Annex V compliance plans, equipment, and expenditures (U.S. General Accounting Office, 1994a, 1994b).

The length of time spent at sea varies, depending on a ship's mission and degree of self-sufficiency in terms of fuel capacity and food storage capability. A ship may remain at sea for several days to several weeks. Naval vessels have dedicated home ports, but they also may visit foreign ports routinely, and they may operate in special areas, where no garbage except food waste may be discharged.

Crews on naval ships generate not only domestic and operational waste but also, in some cases, unique wastes, such as those from amphibious and aircraft operations, troop transport, or document shredding. The exact composition of garbage generated on Navy ships has been documented: 1.36 kilograms (kg) (3 lbs.) of solid waste per person per day. Forty-one percent of the garbage (by weight) is food wastes; 35 percent is paper and cardboard; 17 percent is metal, glass, and "other"; and 7 percent is plastic (Alig et al., 1990). Overall, the amount of garbage generated has been declining because the Navy has cut back on waste, such as packaging (Schultz and Upton, 1988).

Passenger Cruise Ships

The vast majority of the world's 128 passenger cruise ships are foreign flag, but the United States has an interest in assuring the fleet's compliance with

---

[10]Although accidents are not violations of Annex V, the treaty stipulates that "all reasonable precautions" must taken to avoid loss of fishing gear, which can be equated with equipment and materials—particularly plastics—used by the offshore industry.

Annex V because most passengers are American (4 million took vacation cruises in 1991), and 6 of the world's 8 leading cruise markets[11] are in or adjacent to U.S. waters. Voyages typically consist of fairly short ocean passages punctuated by visits to one or more tourist ports of call. Garbage consists primarily of food and other domestic wastes.

Far more garbage probably is generated by this sector than is suggested by the Cantin data, which underestimated the numbers of persons on board and have been superseded by the expansion of the fleet. A large cruise ship in today's market can carry 2,500 passengers and employ a crew of 800, so a single vessel can have as many as 3,300 persons on board (Florida-Caribbean Cruise Association, 1993). The garbage from normal operations on one ship exceeds 1 ton a day. More significant than the weight is the volume of the accumulated garbage, which demands that ship spaces and procedures be well planned and organized. Meanwhile, the size of the fleet has grown considerably since the Cantin study was conducted and continues to expand. The number of passengers taking cruises is expected to double before the turn of the century. Forty-eight new cruise ships have been built since the U.S. ratification of Annex V in 1987, and 21 more are scheduled for delivery by the end of 1998.

Most garbage is treated on board. When garbage must be off-loaded, cruise ships can put a strain on port reception facilities, due to the volumes of materials landed, the short port times and congested schedules of many cruise itineraries, and the minimal landfill capacities or shoreside treatment capabilities in many tourist ports. (Many tourist destinations, particularly those in the Caribbean and Mexico, are finding it increasingly difficult to manage land-generated waste, let alone ships' garbage.) New investment and construction in North American cruise ports has been substantial, but in other itinerary ports, selected initially for their pristine or exotic ambience, the logistical challenges of handling vessel garbage pose a chronic impediment to Annex V implementation.

The cruise industry may be able to accommodate that challenge, because its revenues are sufficient to cover the cost of new ships incorporating state-of-the-art equipment. Working with shipbuilders and equipment manufacturers, the cruise industry has developed and equipped its fleet with the latest in garbage treatment systems—including compactors, incinerators, pulpers, and shredders—not only to maintain compliance with Annex V, but also to reduce reliance on port reception facilities.

Research Vessels

Dozens of public and private marine research vessels operate in the United States. Most are small ships operated by private universities and research organi-

---

[11]The eight markets, in order of prominence, are the Caribbean, Western Mexico, Mediterranean, TransPanama Canal, Europe, Alaska, Bermuda, and Hawaii (Maritime Reporter, 1993).

zations. In addition, the University National Oceanographic Laboratory Systems fleet has some two dozen ships, while the National Oceanic and Atmospheric Administration (NOAA) has 18 in active service, the largest government research fleet. The EPA conducts most of its oceanographic research from a single vessel.

Much of the information available about this fleet comes from NOAA. Typically, a NOAA vessel spends perhaps 250 to 295 days a year away from port. A single voyage may last two or three weeks and may linger in remote locations, including special areas, where no garbage except food may be discharged overboard. These vessels operate out of home ports, although on long voyages they may visit civilian or foreign ports. No data have been collected on the garbage generated by these ships or discarded in ports.

Personnel aboard NOAA research ships include uniformed officers, merchant mariners, and visiting scientists. Approximately two-thirds of the garbage generated is thought to be food and other domestic waste (Art Anderson Associates, 1993); much of it is similar to that found in landfills, because research vessels are provisioned individually for each voyage, and consumable items are purchased at local markets. Other garbage, including used scientific instruments and their packing materials, may result from research activities.

FATES OF SHIPBORNE GARBAGE

Before the 1987 ratification of Annex V, it was assumed widely that the garbage generated aboard vessels was tossed into the sea. In 1975, a National Research Council committee prepared an initial approximation of all potential ocean pollutants, including marine litter, based on the assumption that all vessel wastes were discarded at sea (National Research Council, 1975). Clearly, this assumption is no longer valid.

Vessel garbage generally is disposed of in one of two places: at sea, or in port reception facilities. (Recreational boaters may take their garbage home, for disposal with municipal solid waste.) This section describes what is known about the fates of garbage discarded at sea. The use of port reception facilities is addressed in Chapter 5.

General Observations

The fate of garbage after it is discharged overboard depends on a number of factors, including whether it is loose or bagged and the physical and chemical characteristics—particularly the density—of the solids (Swanson et al., 1994). Large, dense particles, such as ground glass and shredded metal, quickly sink. In areas where there is a strong pycnocline,[12] small particles tend to disperse in the

---

[12]A pycnocline is a region of rapidly increasing density in the ocean caused by a decrease in temperature or an increase in salinity. It is a stable layer, usually found beneath the well-mixed, neutrally stable surface layer.

surface layer, while emulsified particles may remain in the water column for long periods of time. Pulped paper may tend to settle, and unpulped paper may float for a time. Little research has been completed on the degradation of paper in the marine environment (Swanson et al., 1994). Organic material may or may not sink; garbage and sewage-related discharges have been observed in windrows up to 5 kilometers long in the coastal ocean and often "wash ashore as waves of debris" (Swanson et al., 1994).

Regardless of how materials are predicted to flow in the ocean, wash ups of some types of debris on beaches (e.g., World War II munitions that would be expected to sink to the bottom and remain there) seem to defy logical explanation. A member of the Committee on Shipborne Wastes has found numerous containers and appliances on beaches that had been weighted or holed expressly to ensure sinking. Clearly there are forces, such as oceanic currents, that influence the fate of debris in ways that have yet to be explained fully.

Many materials in vessel garbage are persistent. This is obviously the case for glass, cement, brick, metals, and rigid and film plastics, but even timber, hemp, sisal, and cloth can persist for a long time. It is well known that cigarette butts persist; they have been used for decades as markers for sewage sludge deposited in sediments (Swanson et al., 1994). But plastic is the primary concern, as is reflected in Annex V regulations. This material is not only persistent, but also abundant. Plastics dominate the debris found on beaches (see Table 2-5) and in sediments. It is appropriate, therefore, to focus here on the fate of plastics.

According to results of a study carried out in Panama, the time frames for the movements of plastics into and out of beach areas appear to be on the order of months or a year (Garrity and Levings, 1993). The residence time of marked items on beaches appeared to be about a year. The marked items were replaced at the same place where they were found; there was little evidence of down-shore or

TABLE 2-5   Plastic Contributions to Beach Debris (% of Total Items Found)[a]

Location	Plastic
Olympic National Park	98% (1,350/1,385 items)
Cape Cod National Seashore	95% (1,322/1,396)
Channel Islands National Park	94% (953/1,013)
Canaveral National Seashore	92% (1,095/1,192)
Assateague Island National Seashore	86% (395/458)
Gulf Islands National Seashore	85% (681/803)
Cape Hatteras National Seashore	75% (165/220)

[a]These data were collected in 1990-1991 by the National Park Service in quarterly surveys at 36 segments of beaches in seven parks and seashores. (Data also were collected at Padre Island in Texas, but those results are not included here because a different methodology was used.)

Source: Cole et al., 1992.

lateral movement along the beaches. Thus, items leaving the beach areas probably floated away in surface waters or sank to the sea floor following accumulation of dense debris. Beaches that were cleared completely accumulated about 50 percent of the original amount of plastic in about three months, and 61 percent of the original amount in six months.

On the other hand, the persistence of debris on beaches of Padre Island, off the southeastern coast of Texas, appears to be on the order of days, judging by daily collections and observations of marked debris (Miller, 1993). The area is subject to intense tidal movements.

Thus, the time frames of debris retention on beaches, before removal to the adjacent waters or sediments, ranges from days to months. Debris also collects in other places, such as in "backshore" areas behind sand dunes on undeveloped barrier beaches. Debris buried in such areas on Padre Island includes materials deposited well over a decade ago (Miller, 1993).

Coastal sediments constitute an important long-term sink for litter, but few investigations have explored the quantitative aspects of submerged debris. To date, concern about this issue has been limited primarily to fishermen, who often catch debris in their nets. This was evident in a recent British study of the Swansea Bay area, which receives discards from commercial fishing as well as wastes from four rivers and 99 stormwater outfalls (Williams et al., 1993). Macerated industrial discharges and screened sewage also are introduced into the bay. Litter was collected from 30 static gill nets, which varied in length from 57 to 732 meters (62 to 801 yards), with a mesh size of 15 to 20 centimeters (6 to 8 inches). Plastics accounted for 66 percent of the articles. Twenty-four percent of the 3,670 litter items were found to be of sewage origin, a finding attributed to ineffective screening and to inputs from stormwater outfalls. Sources of the remainder of the debris could not be identified.

Implications for Special Areas

The fate of garbage discharged in or near special areas is of particular concern, because these areas need extra protection. The behavior of waste materials may vary significantly among these areas; factors that warrant study include the proximity to shore of discharges, winds and currents, water column stratification, biological production, oxygen levels at depth, and whether a seasonal pycnocline develops. Some areas, such as parts of the Gulf of Mexico, have low oxygen levels at great depth, a situation that generally reduces the rate of decomposition of garbage on the bottom (Swanson et al., 1994). Key characteristics of each special area are summarized in Appendix E.

Regardless of the features of a special area, the zero-discharge rules do not protect these areas fully, because marine debris (including vessel garbage) can be transported over long distances. For example, debris originating from the Hudson/Raritan Estuary has been stranded nearly 100 kilometers (more than 50 miles) from the mouth of the estuary on New Jersey and Long island beaches (Swanson and Zimmer, 1990). A member of the Committee on Shipborne Wastes has found messages in bottles in the Gulf of Mexico that were sent from as far away as Australia. Such findings clearly imply that garbage discharged legally at sea could drift into a special area.

Implications for Implementation of Annex V

When fully and successfully implemented, Annex V will eliminate the disposal of plastics and shift the disposal of other vessel garbage away from the coast, into the open ocean. This change, it is hoped, may reduce to a meaningful degree the most visible pollution problems caused by vessel garbage. However, recently acquired knowledge concerning the fates of marine debris calls into question the ultimate effectiveness of Annex V as currently conceived.

Oceanographic and satellite data gathered during the past 20 years have improved understanding of ocean circulation and marine water dynamics, which influence the fate of materials thrown off ships. In some waters, the net transport is offshore, while elsewhere the net transport is onshore, and in still other areas, such as the Gulf of Mexico or the Atlantic waters bordering the Gulf Stream, the motion is essentially cyclical. For example, tar balls created by the oily residues discharged from ships have been "tracked" in the ocean, and the tracks indicate that the same persistent tar ball can move with the water body, returning to the same spot more than once (Butler et al., 1973). Drifting bits of debris, litter, or garbage are affected by small-scale water movements as well as by larger-scale currents and seasonal changes.

In 1990, approximately 80,000 athletic shoes in containers were lost overboard in a storm in the north Pacific Ocean; at least 1,300 shoes were transported more than 2,000 kilometers (km) (more than 1,240 miles) in seven to nine months, washing shore in bunches along roughly 1,000 km (approximately 620 miles) of coastline across Oregon, Washington, and Vancouver Island (Ebbesmeyer and Ingraham, 1992; Swanson et al., 1994). Because previous satellite studies suggested that such objects would not be distributed so widely, processes other than oceanic dispersion, probably coastal currents, appeared to be at work (Ebbesmeyer and Ingraham, 1992). Furthermore, although the initial drift apparently was due east, shoes were reported washing ashore as far away as Hawaii to the south, and researchers expected some to reach Asia and Japan eventually after drifting west (Ebbesmeyer and Ingraham, 1992). The same researchers also have reported that a broken container from a ship released about 29,000 bath toys into the middle of the Pacific Ocean in January 1992, and that hundreds of the plastic animals appeared along the shores of southeastern Alaska in the fall of that year (Ebbesmeyer and Ingraham, 1994). The scientists predicted that some of the toys would float through the Arctic Ocean, past Greenland, and into the Atlantic Ocean.

These types of studies suggest that the arbitrary Annex V disposal demarcations of 12 and even 25 miles from land may not protect coastal areas fully against pollution from vessel garbage. Initial research in the Gulf of Mexico indicates that floating plastic sheets (used by the offshore oil industry to cover materials in transit) persist and remain a nuisance as long as they remain anywhere in the water (Lecke-Mitchell and Mullin, 1992).

ENVIRONMENTAL AND PHYSICAL EFFECTS
OF MARINE DEBRIS

Marine debris may accumulate on beaches, on the surface waters, and in the benthos. The potential environmental and physical effects[13] of this debris, whether from vessel or land-based sources, include

- aesthetic degradation of surface sea waters and beach areas;
- physical injuries to humans and life-threatening interference with their activities;
- ecological damage caused by the interference of plastics with gas exchange between overlying waters and those in the benthos;
- alterations in the composition of ecosystems caused by debris that provides habitats for opportunistic organisms;
- entanglements of birds, fish, turtles, and cetaceans in lost or discarded nets, fishing gear, and packing materials; and
- ingestion of plastic particles by marine animals.

The aesthetic problems are obvious to anyone who has visited a debris-littered beach or observed garbage floating in the sea. Indeed, the aesthetic degradation that is evident when a beach is littered may be more compelling to the public and to policymakers than is any number of numerical analyses of debris levels, animal mortality, or other effects. Yet these other effects are significant. Following is a summary of what is known about the health and ecological effects of marine debris, including vessel garbage. Additional details concerning ecological effects may be found in Appendix F.

Human Health Problems

Aside from the potential for beach goers to step on or in some other way be injured by pieces of glass, metal, or other sharp objects, the most widely perceived threat has been from the fear of contamination by medical waste washed

---

[13]Marine debris also has economic effects, as noted in Chapter 1 and implied in the forthcoming discussion of ghost fishing. The committee did not examine this aspect of the problem in detail.

up on beaches[14]. Public health officials believe the risk of contracting blood-borne diseases from exposure to medical wastes found on beaches is low, but the EPA has asserted that "inadvertent exposure is publicly unacceptable and should be prevented" (U.S. Environmental Protection Agency, 1989).

A similar threat is posed by debris items containing hazardous waste. At Padre Island National Seashore in Texas, hazardous wastes such as acids have been found in bottles and other containers washed up on beaches. Although no serious incidents have occurred, National Park Service employees consider themselves lucky (John Miller, National Park Service, personal communication to member of the Committee on Shipborne Wastes, November 1, 1993).

Marine debris also has been known to disable divers and vessels, with potentially life-threatening results. Divers sometimes become entangled in pieces of monofilament fishing line that have snagged on reefs or other underwater structures. Fishing has been banned from some oil platforms in the North Sea because of related problems experienced by divers (Bourne, 1990). In addition, large debris items have caused boat collisions, while smaller items have been reported to wrap around propellers or clog cooling water intakes, causing engine failure. These problems have not been studied in detail. To improve understanding of the magnitude of the problem, incidents involving debris could be coded and recorded in the Coast Guard's accident database; insurance agencies might be another source of information.

Ecological Effects

Little scientific information is available concerning how debris may affect marine invertebrate species, plant life, or marine habitats in general, aside from observations that debris damages coral reefs, is ingested by squid (Araya, 1983; Machida, 1983), and may offer a new habitat niche for encrusting marine species (Winston, 1982). Concern has been expressed about the biological uptake of minute suspended particles possibly contaminated with heavy metals or other toxic substances. Such particles may result from the degradation of large plastic items, cosmetic additives (minute plastics are added as abrasives), and aero-blasting (use of plastic "sand" to remove paint from ship hulls) (Gregory, 1994). Also, concern has been expressed that floating plastics may facilitate the

---

[14]During the summers of 1987 and 1988, medical wastes appearing on beaches in the Northeast raised concerns over the potential threat of exposure to diseases such as Acquired Immune Deficiency Syndrome (AIDS). In fact, several syringes, needles, and blood vials that were found tested positive for the AIDS antibodies and the hepatitis B virus, and there were reports of persons being punctured by these items. But the majority of the items reported were syringes generated by land-based sources, not ships (ICF, 1989).

transoceanic or regional introduction of aggressive alien taxa into new areas (Winston et al., 1994).

Many questions remain concerning effects of plastics and other debris on the benthos. When they accumulate, these materials can interfere with dissolved gas exchange between the pore waters of the sediment and the overlying waters, potentially leading to hypoxic or anoxic environments[15] that can kill some organisms. Community structure may be altered further by opportunistic organisms that may colonize plastic debris. There has yet to be any systematic and continuous surveillance to determine how the increasing coverage of the sea floor with plastics and other indestructible materials affects the functioning of ecosystems.

Entanglement of Marine Animals

Plastic debris causes considerable mortality of marine wildlife. Entangled animals may be unable to breathe, swim, feed, or care for their young properly (Laist, 1987, 1994). Studies have indicated that each year as many as 50,000 northern fur seals were becoming entangled and dying in plastic debris, primarily fishing nets and strapping bands (Fowler, 1982). Indeed, marine debris is blamed for a significant decline in the fur seal population (Laist, 1994). Research continues to show that plastic also causes widespread mortality among other marine mammals, turtles, birds, and fish, either through entanglement or ingestion (Laist, 1987, 1994). Even land-based creatures, including foxes and rabbits, become ensnared in plastic debris on coastlines (Fowler and Merrell, 1986; O'Hara and Younger, 1990).

The National Marine Fisheries Service (NMFS) is a key resource for biologists and others documenting wildlife interactions with debris; NMFS workshops and the resulting proceedings are important mechanisms for exchange of information on the subject among researchers and agencies.[16] For example, the NMFS collected most of the information available on pinniped interactions with debris. The agency also conducted the first comprehensive assessment of turtle entanglement, compiling a list of 60 cases of sea turtle entanglements worldwide involving green, loggerhead, hawksbill, olive ridley, and leatherback turtles (Balazs, 1985).

While pinniped and sea turtle entanglement in plastic debris has been documented, no agency has collected extensive data on bird mortality due to entanglement in debris, even though entanglements have been reported for at least 51 (16 percent) of the world's 312 seabird species. Likewise, little is known about the

---

[15]An hypoxic environment is oxygen deficient; anoxia results when oxygen is absent entirely.

[16]Other resources include a growing body of papers in journals such as the *Marine Pollution Bulletin* and conferences such as the North Pacific Rim Fishermen's Conference on Marine Debris (Alverson and June, 1988).

extent of entanglement among cetaceans, perhaps because these animals are found only on occasions when they wash ashore, and necropsies are done only when adequate expertise and funding are available. Entanglements have been reported for 10 (13 percent) of the 75 cetacean species (Laist, 1994). Information on the entanglement of fish in marine debris is also largely anecdotal.

Ingestion of Plastics by Marine Species

The most highly publicized example of plastic ingestion may be the consumption of plastic bags or sheeting by sea turtles, which are thought to mistake these items for jellyfish, squid, and other prey. Turtles, especially hawksbills, also eat encrusting organisms that grow on floating plastic and ingest plastic pieces as a "by-product" (Plotkin and Amos, 1988). The effect of plastics ingestion on sea turtle longevity and reproductive potential is unknown. It is thought that ingested plastics may cause mechanical blockage of the digestive tract, starvation, reduced absorption of nutrients, and ulceration. Buoyancy caused by plastics also could inhibit diving activities needed for pursuit of prey and escape from predators (Balazs, 1985; Lutz, 1990).

Birds and fish also ingest plastics. At least 108 of the world's 312 seabird species are known to ingest plastic debris (Laist, 1994). Individuals from 33 fish species have been reported to ingest plastics (Laist, 1994); a list compiled by Hoss and Settle (1990) included larva, juvenile, and adults from benthic to pelagic habitats.

Limited information is available concerning ingestion of plastic debris by marine mammals, although information from marine parks and zoos suggests that debris ingestion has the potential to be a direct cause of mortality (Walker and Coe, 1990). A dying pygmy sperm whale rescued by the National Aquarium in Baltimore had ingested several pieces of plastic bags and balloons (Craig Vogt, EPA, personal communication to Marine Board staff, August 4, 1994). The Texas Marine Mammal Stranding Network has records of necropsies revealing debris ingestion by several cetaceans. In one highly publicized case, a rough-toothed dolphin died of peritonitis (inflammation of the abdominal lining) attributed to ingestion of a plastic snack food bag, while a 4-ton minke whale died with a plastic bag in its stomach. This information demonstrates that data on the effects of marine debris can be obtained through existing research mechanisms designed to achieve other goals. It might be feasible to expand other research projects focusing on non-Annex V topics, such as fish feeding behavior, to record any ingestion of plastics and other debris. Data also could be gathered by conducting regular necropsies on dead, stranded marine mammals and other animals.

The value of using existing procedures to compile and maintain a database on debris interactions with wildlife is demonstrated by a report on plastic ingestion by the West Indian manatee, an endangered species. In the southeastern United States, manatee carcasses routinely are salvaged to determine cause of death and collect biological information. In Florida, personnel from the U.S. Fish and Wildlife Service and the University of Miami have performed systematic necropsies on dead manatees. Based on this information, Beck and Barros (1991) found that of 439 manatees necropsied between 1978 and 1986, 63 (14.4 percent) had ingested debris. Pieces of monofilament fishing line were the most common debris items ingested. (Marine debris is not, however, the leading cause of manatee deaths, which are attributed most often to collisions with vessel propellers.)

Ghost Fishing

Ghost fishing—a term referring to lost or discarded fishing gear that continues to catch finfish and shellfish species indefinitely—may significantly reduce some commercial stocks and ultimately could affect marine ecosystems. This is a difficult problem to study. Few data are available on the number of gear units deployed in various fisheries, the number lost, or the capability of various types of gear to ghost fish (Natural Resources Consultants, 1990).

Nevertheless, available estimates suggest that ghost fishing could be a significant problem. Lobster and crab traps and gillnets have been found to have a significant potential to ghost fish. For the inshore lobster fishery of Maine, it has been estimated that 25 percent of all traps are lost each year, and that each lost trap can continue to catch lobsters up to 1.1 kg (2.5 lbs.) (Smolowitz, 1978). An estimated 10 to 20 percent of traps used in the coastal Dungeness crab and American lobster fisheries are lost each year; many crab and pot fisheries now are required to mark traps and use timed-release devices on panels to minimize ghost fishing (Breen, 1990).

Lost gillnets can capture many fish and shellfish over long periods of time. According to one estimate, lost gillnets can fish at a 15 percent effectiveness rate for up to eight years (Natural Resources Consultants, 1990). A 24-day gillnet retrieval project in 1976 recovered 176 nets containing 4,813 kg (10,611 lbs.) of groundfish and 2,593 kg (5,717 lbs.) of crab (Brothers, 1992). A report of 10 lost nets found in 37.5 hours of searching off Massachusetts (Carr, 1986) suggests there may be numerous lost nets in some sink gillnet fishing areas in the northeastern United States (Laist, 1994).

SUMMARY

This chapter yields four basic findings, which provide the foundation for the remainder of the report. The first three findings are straightforward. First, considerable amounts of garbage are generated by seafarers in most if not all maritime communities. Second, garbage discarded into the sea can be transported far from the point of discharge. Third, the disposal of plastics in the marine environment is causing considerable harm, including mortality among marine mammals, turtles, birds, and fish, either through entanglement or ingestion.

The fourth finding, which is multifaceted, is that available data on the sources, fates, and effects of marine debris—particularly vessel-generated debris—are often of poor quality, incomplete, and out of date. Because this problem may be too diffuse to be obvious, the specific deficiencies are enumerated here:

Sources: Detailed, comprehensive data on garbage generation have been collected only for the Navy, and neither U.S. nor international Annex V compliance and enforcement regimes support the gathering of such data for other maritime sectors. Amounts of garbage generated and discarded overboard by the various fleets can only be estimated.

Fates: Knowledge concerning the fates of vessel garbage is derived primarily from beach surveys, and the percentage of beach debris that comes from vessels is unknown. Few data are available on debris that ends up in sediments or the benthos.

Effects: Although the harm to individual animals is apparent, the ecological effects of marine debris (including frequency of harm to wildlife and population impacts) cannot be established on the basis of surveys and other information gathered to date. Even for endangered species subject to continuous monitoring, cause-effect relationships have yet to be established.

Part of the problem is that data on effects of marine debris have been gathered largely by individual researchers, working without an overall program of data collection. There is little centralized data analysis, and reporting on wildlife interactions with debris is not standardized. The value of systematically compiling and maintaining a database on debris interactions with wildlife is demonstrated by the information collected on West Indian manatees. Another problem is that no systematic effort coordinates the exchange of information on wildlife interactions with marine debris, other than through NMFS workshops and proceedings, and published literature on the topic remains scarce.

There are fundamental barriers to the development of comprehensive knowledge about the effects of marine debris on wildlife. It is and will remain difficult to detect entangled and dead animals at sea and to distinguish the effects of marine debris from other impacts. Indeed, the true magnitude of the effects of marine debris on wildlife may never be defined absolutely.

REFERENCES

Alig, C.S., L. Koss, T. Scarano, and F. Chitty. 1990. Control of plastic wastes aboard naval ships at sea. Pp. 879-894 in Proceedings of the Second International Conference on Marine Debris, 2–7 April 1989, Honolulu, Hawaii (Vol. II), R.S. Shomura and M.L. Godfrey, eds. NOAA-TM-NMFS-SWFSC-154. Available from the Marine Entanglement Research Program of the National Marine Fisheries Service (National Oceanic and Atmospheric Administration), Seattle, Wash.

Alverson, D. and J.A. June, eds. 1988. Proceedings of the North Pacific Rim Fishermen's Conference on Marine Debris. Seattle, Wash.: Natural Resources Consultants.

American Red Cross. 1991. American Red Cross National Boating Survey. Washington, D.C.: American Red Cross.

Amos, A.F. 1993a. Technical Assistance for the Development of Beach Debris Data Collection Methods. Final Report submitted to U.S. Environmental Protection Agency, Gulf of Mexico Program, Dallas, Tex. TR/93-002. May 31.

Amos, A.F. 1993b. Solid waste pollution of Texas beaches: a Post-MARPOL Annex V study, Vol 1: Narrative. OCS Study MMS 93-0013. Available from the public information unit of the U.S. Department of the Interior, Minerals Management Service, Gulf of Mexico OCS Region, New Orleans, La. July.

Araya, H. 1983. Fishery biology and stock assessment of Ommastrephes bartrami in the North Pacific Ocean. Mem. of the National Museum in Victoria (Australia). 44:269-283.

Art Anderson Associates. 1993. NOAA Fleetwide Shipboard Waste Management. Report prepared for the National Oceanic and Atmospheric Administration by Art Anderson Associates, Bremerton, Wash.

Balazs, G.H. 1985. Impacts of ocean debris on marine turtles: entanglement and ingestion. Pp. 387-429 in Proceedings of the Workshop on the Fate and Impact of Marine Debris, 27–29 November 1984, Honolulu, Hawaii, R.S. Shomura and H.O. Yoshida, eds. NMFS NOAA-TM-NMFS-SWFC-54. Available from the Marine Entanglement Research Program of the National Marine Fisheries Service (National Oceanic and Atmospheric Administration), Seattle, Wash.

Beck, C.A. and N.B. Barros. 1991. The impact of debris on the Florida manatee. Marine Pollution Bulletin 22(10):508-510. October.

Bourne, W.R.P. (chair). 1990. Report of the working group on entanglement of marine life. Pp. 1207-1215 in Proc. of the Second International Conference on Marine Debris, 2–7 April 1989, Honolulu, Hawaii (Vol. II), R.S. Shomura and M.L. Godfrey, eds. NOAA-TM-NMFS-SWFSC-154. Available from the Marine Entanglement Research Program of the National Marine Fisheries Service (National Oceanic and Atmospheric Administration), Seattle, Wash.

Breen, P.A. 1990. A review of ghost fishing by traps and gillnets. Pp. 571-599 in Proc. of the Second International Conference on Marine Debris, 2–7 April 1989, Honolulu, Hawaii (Vol. I), R.S. Shomura and M.L. Godfrey, eds. NOAA-TM-NMFS-SWFSC-154. Available from the Marine Entanglement Research Program of the National Marine Fisheries Service (National Oceanic and Atmospheric Administration), Seattle, Wash.

Brothers, G. 1992. Lost or abandoned fishing gear in the Newfoundland aquatic environment. Paper presented at the C-Merits Symposium: Marine Stewardship in the Northwest Atlantic. Department of Fisheries and Oceans, St Johns, Newfoundland, Canada. November. Cited in. Laist, D.W. 1994. Entanglement of Marine Life in Marine Debris (draft). Paper prepared for the Third International Conference on Marine Debris, Miami, Fla., May 8–13, 1994. Marine Mammal Commission, Washington, D.C.

Butler, J.N., B.F. Morris, and J. Sass. 1973. Pelagic Tar from Bermuda and the Sargasso Sea. Bermuda Biological Station Report No. 10. Available from the librarian, Bermuda Biological Station for Research, St. George's West, Bermuda.

Cantin, J., J. Eyraud, and C. Fenton. 1990. Quantitative estimates of garbage generation and disposal in the U.S. maritime sectors before and after MARPOL Annex V. Pp. 119-181 in Proceedings of the Second International Conference on Marine Debris, 2–7 April 1989, Honolulu, Hawaii (Vol. I), R.S. Shomura and M.L. Godfrey, eds. NOAA-TM-NMFS-SWFSC-154. Available from the Marine Entanglement Research Program of the National Marine Fisheries Service (National Oceanic and Atmospheric Administration), Seattle, Wash. December.

Carr, H.A. 1986. Observation on the occurrence of impacts of ghost gillnets on Jeffrey's Ledge (abstract). Pp. 134-135 in Program and Abstracts, Sixth International Ocean Disposal Symposium, April 21–25, 1986, Pacific Grove, Calif. Washington, D.C.: National Oceanic and Atmospheric Administration. Cited in. Laist, D.W. 1994. Entanglement of Marine Life in Marine Debris (draft). Paper prepared for the Third International Conference on Marine Debris, May 8–13, 1994, Miami, Fla. Marine Mammal Commission, Washington, D.C.

Cole, C.A., J.P. Kumer, D.A Manski, and D.V. Richards. 1990. Annual Report of National Park Marine Debris Monitoring Program: 1989 Marine Debris Survey. Tech Rpt. NPS/NRWV/NRTR-90/4. Available from the Natural Resources Publications Office of the National Park Service, Denver, Colo.

Cole, C.A., W.P. Gregg, D.V. Richards, and D.A. Manski. 1992. Annual Report of National Park Marine Debris Monitoring Program: 1991 Marine Debris Surveys with Summary of Data from 1988 to 1991. Tech. Rpt. NPS/NRWV/NRT-92/10. Available from the Natural Resources Publications Office of the National Park Service, Denver, Colo.

Cruise Lines International Association (CLIA). 1994. The CLIA Fleet: Post 1987. New York: CLIA.

Debenham, P. and L.K. Younger. 1991. Cleaning North Anerica's Beaches: 1990 Beach Cleanup Results. Washington, D.C.: Center for Marine Conservation. May.

Eastern Research Group. 1988. Development of Estimates of Garbage Disposal in the Maritime Sectors. Final Report prepared for the Transportation Systems Center, Research and Special Programs Administration, U.S. Department of Transportation by ERG, Arlington, Mass. (now Lexington, Mass). December.

Ebbesmeyer, C.C. and W.J. Ingraham Jr. 1992. Shoe spill in the North Pacific. EOS, Transactions American Geophysical Union 73(34):361,365. Aug. 25.

Ebbesmeyer, C.C. and W.J. Ingraham Jr. 1994. Pacific toy spill fuels ocean current pathways research. EOS, Transactions American Geophysical Union 75(37):425,427,430. Sept. 13.

Florida-Caribbean Cruise Association. 1993. The Cruise Industry's Role in Waste Management. Paper prepared by the association, Miami, Fla. April.

Forecast International. 1992. Warships Forecast, Appendix V: U.S. Navy Force Levels. Newtown, Conn.: Forecast International/DMS Market Intelligence Report. August.

Fowler, C.W. 1982. Interactions of northern fur seals and commercial fisheries. Pp. 278-292 in Transactions of the 47th North American Wildlife and Natural Resources Conference. Washington, D.C.: Wildlife Management Institute. Cited in. Laist, D.W. 1994. Entanglement of Marine Life in Marine Debris (draft). Paper prepared for the Third International Conference on Marine Debris, May 8–13, 1994, Miami, Fla. Marine Mammal Commission, Washington, D.C.

Fowler, C.W. and T.R. Merrell. 1986. Victims of plastic technology. Alaska Fish and Game 18(2):34-37.

Garrity, S.D. and S.C. Levings. 1993. Marine debris along the Caribbean coast of Panama. Marine Pollution Bulletin. 26(6):317-324. June.

Gregory, M.R. 1994. Plastic micro-litter: an underestimated contaminant of global oceanic waters. Paper prepared for the Third International Conference on Marine Debris, Miami, Fla., May 8–13, 1994.

Henderson, J.R., S.L. Austin, and M.B. Pillos. 1987. Summary of webbing and net fragments found on Northwestern Hawaiian Islands beaches, 1982–1986. Rpt. H-87-11. Southwest Fisheries Science Center of the National Marine Fisheries Service (NMFS). Honolulu Lab. Long Beach, Calif.: NMFS.

Hoss, D.E. and L.R. Settle. 1990. Ingestion of plastics by fishes. Pp. 693-709 in Proceedings of the Second International Conference on Marine Debris, 2–7 April 1989, Honolulu, Hawaii (Vol. I), R.S. Shomura and M.L. Godfrey, eds. NOAA-TM-NMFS-SWFSC-154. Available from the Marine Entanglement Research Program of the National Marine Fisheries Service (National Oceanic and Atmospheric Administration), Seattle, Wash.

ICF, Inc. 1989. Inventory of Medical Waste Beach Washups, June–October 1988. Report prepared for the U.S. Environmental Protection Agency, Office of Policy, Planning and Evaluation, Washington, D.C. March 13.

Johnson, S.W. and T.R. Merrell Jr. 1988. Entanglement debris on Alaskan beaches, 1986. NOAA Tech. Memo. NMFS F/NEC-126. Auk Bay, Alaska: National Marine Fisheries Service, Northwest and Alaska Fisheries Center. Available from the Marine Entanglement Research Program of the National Marine Fisheries Service (National Oceanic and Atmospheric Administration), Seattle, Wash.

Johnson, S.W. 1990a. Entanglement Debris on Alaskan beaches, 1989. NWAFC Processed Report 90-10. Auk Bay, Alaska: Alaska Fisheries Science Center. Available from the Marine Entanglement Research Program of the National Marine Fisheries Service (National Oceanic and Atmospheric Administration), Seattle, Wash.

Johnson, S.W. 1990b. Distribution, abundance, and source of entanglement debris and other plastics on Alaskan beaches, 1982–1988. Pp. 331 in Proceedings of the Second International Conference on Marine Debris, 2–7 April 1989, Honolulu, Hawaii (Vol. I), R.S. Shomura and M.L. Godfrey, eds. NOAA-TM-NMFS-SWFSC-154. Available from the Marine Entanglement Research Program of the National Marine Fisheries Service (National Oceanic and Atmospheric Administration), Seattle, Wash. December.

Laist, D.W. 1987. Overview of the biological effects of lost and discarded plastic debris in the marine environment. Marine Pollution Bulletin 18 (6B):319-326.

Laist, D.W. 1994. Entanglement of Marine Life in Marine Debris (draft). Paper prepared for the Third International Conference on Marine Debris, Miami, Fla., May 8–13, 1994. Marine Mammal Commission, Washington, D.C.

Landsburg, A.C., E. Gabler, G. Levine, R. Sonnenschein, and E. Simmons. 1990. U.S. commercial ships for tomorrow. Marine Technology. 27(3):129-152. May.

Lecke-Mitchell, K.M. and K. Mullin. 1992. Distribution and abundance of large floating plastic in the north-central Gulf of Mexico. Marine Pollution Bulletin 24(12):598-601. December.

Lutz, P.L. 1990. Studies on the ingestion of plastic and latex by sea turtles. Pp. 719-735 in Proc. of the Second International Conference on Marine Debris, 2–7 April 1989, Honolulu, Hawaii (Vol. I), R.S. Shomura and M.L. Godfrey, eds. NOAA-TM-NMFS-SWFSC-154. Available from the Marine Entanglement Research Program of the National Marine Fisheries Service (National Oceanic and Atmospheric Administration), Seattle, Wash. December.

Machida, S. 1983. A brief review of the squid fishery by Hoyo Maru No. 67 in southeast Australian waters in 1979/80. Mem. of the National Museum in Victoria (Australia). 44:291-295. Cited in. Walker, W.A. and J.M. Coe. 1990. Survey of marine debris ingestion by odontocete cetaceans. Pp. 747-774 in Proc. of the Second International Conference on Marine Debris, 2–7 April 1989, Honolulu, Hawaii (Vol. I), R.S. Shomura and M.L. Godfrey, eds. NOAA-TM-NMFS-SWFSC-154. Available from the Marine Entanglement Research Program of the National Marine Fisheries Service (National Oceanic and Atmospheric Administration), Seattle, Wash.

Manski, D.A., W.P. Gregg, C.A. Cole, and D.V. Richards. 1991. Annual Report of the National Park Marine Debris Monitoring Program: 1990 Marine Debris Surveys. Tech. Rpt. NPS/NRWW/NRTR-91/07. Available from the Natural Resources Publications Office of the National Park Service, Denver, Colo. September.

Maritime Reporter. June 1993. Cruise Shipping—Recent Orders Help to Buoy Market. 31.

Merrell, T.R. Jr. 1980. Accumulation of plastic litter on beaches of Amchitka Island, Alaska. Mar. Env. Res. 3:171-184.

Merrell, T.R. Jr. 1985. Fish nets and other plastic litter on Alaskan beaches. Pp. 160-182 in Proceedings of a Workshop on the Fate and Impact of Marine Debris, 27–29 November 1984, Honolulu, Hawaii, R. Shomura and H. Yoshida, eds. NOAA-TM-NMFS-SWFC-54. Available from the Marine Entanglement Research Program of the National Marine Fisheries Service (National Oceanic and Atmospheric Administration), Seattle, Wash.

Miller, J.E. 1993. Marine Debris Investigation: Padre Island National Seashore, Texas. Corpus Christi, Tex.: National Park Service. December.

Miller, J.E. 1994. Marine Debris Point Source Investigation: Padre Island National Seashore, Texas. Paper prepared for the Third International Conference on Marine Debris, Miami, Fla., May 8–13, 1994. Resource Management Division, Padre Island National Seashore, Corpus Christi, Tex.

Minerals Management Service. 1992. Federal Offshore Statistics. Washington, D.C.: Department of the Interior.

Minerals Management Service (MMS). 1992. Proceedings: Twelfth Annual Gulf of Mexico Information Transfer Meeting, Nov. 5–7, 1991, New Orleans, La., compiled by Geo-Marine Inc. OCS Study MMS 92-0027. New Orleans: U.S. Department of the Interior, MMS, Gulf of Mexico OCS Region. December.

Mudar, M.J. 1991. Reducing Plastic Contamination of the Marine Environment under MARPOL Annex V: A Model for Recreational Harbors and Ports. Ph.D. dissertation. Rensselaer Polytechnic Institute, Troy, New York. February.

National Marine Fisheries Service. 1994. Fisheries of the United States: Current Fisheries Statistics 9,300. Washington, D.C.: U.S. Government Printing Office. May.

National Research Council (NRC). 1975. Assessing Potential Ocean Pollutants. Ocean Affairs Board, NRC. Washington, D.C.: National Academy of Sciences Printing and Publishing Office (now National Academy Press).

National Research Council (NRC). 1991. Fishing Vessel Safety: Blueprint for a National Program. Marine Board, NRC. Washington, D.C.: National Academy Press.

National Research Council (NRC). 1994. Review of NOAA's Fleet Replacement and Modernization Plan. Marine Board, NRC. Washington, D.C.: National Academy Press.

Natural Resources Consultants. 1990. Survey and Evaluation of Fishing Gear Loss in Marine and Great Lakes Fisheries of the United States. Report prepared for the National Marine Fisheries Service, Marine Entanglement Research Program, Seattle, Washington.

O'Hara, K.J. and L.K. Younger. 1990. Cleaning North America's Beaches: 1989 Beach Cleanup Results. Washington, D.C.: Center for Marine Conservation. May.

Palmisano, A.C. and C.A. Pettigrew. 1992. Biodegradability of plastics. BioScience 42(9):680-685. October.

Plotkin, P.E. and A.F. Amos. 1988. Entanglement in and ingestion of marine debris by sea turtles stranded along the south Texas Coast. Pp. 79-82 in Proc. of the Eighth Annual Workshop on Sea Turtle Biology and Conservation. NOAA-TM-NMFS-SEFC-214. Available from the Southeast Fisheries Center of the National Marine Fisheries Service, Miami, Fla.

Polmar, N. 1992. Ships and Aircraft of the U.S. Fleet, 15th ed. Annapolis, Md.: Naval Institute Press.

Ribic, C.A., T.R. Dixon and I. Vining. 1992. Marine Debris Survey Manual. NOAA Tech. Rpt. NMFS 108. Available from the Marine Entanglement Research Program of the National Oceanic and Atmospheric Administration, Seattle, Wash.

Schultz, J.P. and W.K. Upton, III. 1988. Solid Waste Generation Aboard USS *O'Bannon* (DD 987). DTRC/SME-87/92. Bethesda, Md.: U.S. Navy, David W. Taylor Naval Ship Research and Development Center.

Smolowitz, R.J. 1978. Trap design and ghost fishing: Discussion. Marine Fisheries Review 40(5-6):59-67.

Sutinen, J.G. 1986. Enforcement Issues in the Oyster Fishery of Chesapeake Bay. Paper presented at the Conference on Economics of Chesapeake Bay Management, Annapolis, Md., May 28–29, 1986.

Swanson, R.L. and R.L. Zimmer. 1990. Meteorological conditions leading to the 1987 and 1988 washups of floatable wastes on New York and New Jersey beaches and comparison of these conditions with the historical record. Estuarine, Coastal and Shelf Science (U.K.) 30:59–78. Cited in. Swanson, R.L., R.R. Young, and S.S. Ross. 1994. An Analysis of Proposed Shipborne Waste Handling Practices Aboard United States Navy Vessels. Paper prepared for the Committee on Shipborne Wastes, Marine Board, National Research Council, Washington, D.C.

Swanson, R.L., R.R. Young, and S.S. Ross. 1994. An Analysis of Proposed Shipborne Waste Handling Practices Aboard United States Navy Vessels. Paper prepared for the Committee on Shipborne Wastes, Marine Board, National Research Council, Washington, D.C.

Trulli, W.R., H.K. Trulli, and D.P. Redford. 1990. Characterization of marine debris in selected harbors of the United States. Pp. 309-324 in Proceedings of the Second International Conference on Marine Debris, 2–7 April 1989, Honolulu, Hawaii (Vol. I), R.S. Shomura and M.L. Godfrey, eds. NOAA-TM-NMFS-SWFSC-154. Available from the Marine Entanglement Research Program of the National Marine Fisheries Service (National Oceanic and Atmospheric Administration), Seattle, Wash. December.

U.S. Coast Guard (USCG). 1992a. Boating Statistics. Washington, D.C.: Consumer and Public Affairs Division of the USCG Office of the Boating Safety.

U.S. Coast Guard (USCG). 1992b. Register of Cutters of the U.S. Coast Guard. Washington, D.C.: USCG.

U.S. Coast Guard (USCG). 1994a. Draft Regulatory Evaluation for Revision of 46 CFR Subchapter T. Washington, D.C.: USCG Vessel Documentation Branch.

U.S. Coast Guard (USCG). 1994b. Offshore Statistics. Washington, D.C.: USCG Office of Fishing Vessel Safety and Offshore Operations.

U.S. Environmental Protection Agency (EPA). 1989. Medical Waste: Environmental Backgrounder. Washington, D.C.: EPA Office of Public Affairs (A-107). March.

U.S. Environmental Protection Agency (EPA). 1990. Methods to Manage and Control Plastic Wastes—Report to Congress. EPA/530-SW-89-051. Washington, D.C.: Office of Solid Waste (OS-300) and Office of Water (WH-556). February.

U.S. General Accounting Office (GAO). 1994a. Pollution Prevention: Chronology of Navy Ship Waste Processing Equipment Development. GAO/NSIAD-94-221FS. Washington, D.C.: GAO National Security and International Affairs Division. August.

U.S. General Accounting Office (GAO). 1994b. Pollution Prevention: The Navy Needs Better Plans for Reducing Ship Waste Discharges. GAO/NSIAD-95-38. Washington, D.C.: GAO National Security and International Affairs Division. November.

U.S. Maritime Administration (MARAD). 1992a. Annual Report. Washington, D.C.: MARAD.

U.S. Maritime Administration (MARAD). 1992b. Merchant Marine Data Sheet. Aug. 14. Washington, D.C.: MARAD Office of Public Affairs.

Walker, W.A. and J.M. Coe. 1990. Survey of marine debris ingestion by odontocete cetaceans. Pp. 747-774 in Proceedings of the Second International Conference on Marine Debris, 2–7 April 1989, Honolulu, Hawaii (Vol. I), R.S. Shomura and M.L. Godfrey, eds. NOAA-TM-NMFS-SWFSC-154. Available from the Marine Entanglement Research Program of the National Marine Fisheries Service (National Oceanic and Atmospheric Administration), Seattle, Wash. December.

Williams, A.T., S.L. Simmons, and A. Fricker. 1993. Offshore sinks of marine litter: A new problem. Marine Pollution Bulletin 26(7):404-405. July.

Winston, J.E. 1982. Drift plastic—an expanding niche for a marine invertebrate? Marine Pollution Bulletin 13(10):348-351.

Winston, J.E., M.R. Gregory, and L.M. Stevens. 1994. Encrusters, epibionts, and other biota associated with pelagic plastics: a review of biographical, environmental, and conservation issues. Paper prepared for the Third International Conference on Marine Debris, Miami, Fla., May 8–13, 1994.

3

Implementation

Implementation of MARPOL Annex V has been and continues to be problematic. The two greatest obstacles are the difficulties experienced by mariners seeking to comply with Annex V and the ease with which violators can escape detection. While both problems could be mitigated to some degree, the former—difficulty with compliance—probably is easier to resolve. This judgment is based on the assumption that human beings generally want to be helpful and can change if the barriers to voluntary compliance can be overcome. By contrast, increasing direct surveillance of mariners to identify Annex V violations would be difficult and in many cases impossible.

Logic dictates, therefore, that an Annex V implementation strategy should focus on fostering voluntary compliance, while also ensuring robust enforcement capabilities. But the specific elements of an effective strategy are more difficult to determine. To pinpoint the opportunities to improve Annex V implementation, a systematic approach is needed. To that end, this chapter examines Annex V and the hazards it targets from a *comprehensive hazard management* perspective.

The chapter opens with a description of a generic hazard evolution model. Then the model is adapted to the problem of vessel garbage, and opportunities for intervening in the evolution of the hazard (marine debris) are identified within this framework. Throughout the discussion, behavioral and organizational principles are introduced that must be considered in developing mechanisms for successful implementation of Annex V. The analysis also underscores how the provisions of the Annex welcome and support a very broad range of methods for facilitating compliance.

HAZARD EVOLUTION MODEL

A framework for considering the problem of vessel garbage may be found in the work of social geographer Roger Kasperson and public administration specialist David Pijawka (Kasperson and Pijawka, 1985), who considered the ways in which technological hazards pose different and more challenging problems to the public and the government than do natural hazards such as hurricanes, tornadoes, or earthquakes. This work draws upon an overall conceptual framework of analysis developed at Clark University (Hohenemser et al., 1985; Kasperson et al., 1985).

Kasperson and Pijawka proposed that technological hazards challenge institutions and communities for a number of reasons: (1) These hazards are new and unfamiliar; (2) there is a lack of accumulated experience with control or coping measures; (3) there is a lack of full appreciation of the hazard chain; (4) the broad opportunities for control mechanisms make these hazards seem more controllable than natural hazards are; and (5) there is a perception that technological hazards can be corrected with technical solutions, regardless of the social context or the significance of social costs. In sum, technological hazards are viewed erroneously as easily "fixable," with the result that less attention and effort are devoted to them than is warranted.

In an effort to adjust perceptions to reality, Kasperson and Pijawka proposed using a hazard evolution model to clarify analysis of unfamiliar technological hazards. This comprehensive yet simple model examines the ways in which a society generates technological hazards and deals with the resulting impacts (see Figure 3-1). Human needs result in human wants, which are satisfied by a choice of technology. The selected technology can produce a product or byproduct (waste) that poses a hazard. For example, the production technology may create air, water, or ground pollution that requires constant controls. In addition, the product or its packaging can create hazards after its intended use if disposal is not controlled (e.g., discarded plastic six-pack rings may ensnare small animals). Once the material is in use or released into the environment, humans or other organisms can be exposed to and be harmed by the hazard.

Using this flow diagram, Kasperson and Pijawka identified general types of interventions at each stage of the hazard-generating process that could prevent the hazard and its concomitant risks. The model provides an organizing framework for confronting a technological hazard and permits a full appreciation of the intervention opportunities available, both "upstream" and "downstream" of the initiating events. Upstream (toward the left), human wants can be modified, or the technology used to address the wants can be altered, or an initiating event during use of the material can be prevented, or release of the materials can be prevented. Downstream (toward the right), once release or use occurs, exposure of organisms to the hazard can be blocked, or the negative consequences of

FIGURE 3-1   The Chain of Technological Hazard Evolution.   Source: Kasperson and Pijawka, 1985.

exposure can be prevented. As a last resort, intervention after exposure to the hazard may be able to mitigate the harm done.

In this manner, the flow diagram facilitates an examination of the changes needed and how they may be accomplished, by suggesting (1) the location of the required change within the social organization, (2) the costs of the change to society and how those costs are spread or concentrated, (3) the array of segments of society involved, and (4) ways to facilitate the change. Consideration of these four issues assists in determining how far the benefit(s) of a particular change or effort would go toward mitigating the targeted hazard.

ADAPTING THE MODEL TO VESSEL GARBAGE MANAGEMENT

The Kasperson and Pijawka model has considerable application to the problems of managing vessel garbage and facilitating the implementation of MARPOL Annex V. However, in adapting the model for its own analysis, the committee found it appropriate to make two modifications. First, the committee eliminated the first intervention option ("modifying human needs") because, while this action may be possible, it is very difficult to accomplish and likely is not an intervention that marine user groups currently have the capacity to accomplish. It is important, however, to recognize the human needs that draw individuals to the marine environment: the need to earn a living, to engage in recreation, and to transport resources that enhance quality of life at the destination. It is also important to recognize that human needs might be modified, especially in the sense of altering perceptions about what needs and behaviors are regarded as appropriate in the marine environment. The committee's flow diagram, which identifies both the stages of hazard evolution and the possible interventions, is shown in Figure 3-2.

The second change made by the committee was to re-label the second box ("human wants") to focus on behavior. (This change is not reflected in Figure 3-2 due to the simplicity of the diagram but appears in matrixes presented later in this chapter and in Chapter 4.) Because Annex V establishes new performance standards, it is appropriate to focus here on the changes in behavior that must be achieved to accomplish the mandated level of performance. No change in "human wants" is required, although this might help, indirectly.

Interventions to Remedy the Hazard

The committee recognized that a comprehensive set of approaches to implementation of Annex V was essential, based on the durable nature of the hazard created by garbage thrown overboard, and the need for a broad-based effort to halt permanently many longstanding practices of all sectors of the maritime community. Therefore, in addition to modifying the Kasperson and Pijawka hazard evolution model to focus on vessel garbage, the committee further set the stage

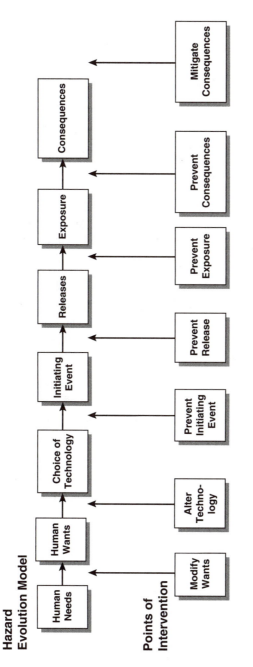

FIGURE 3-2 Intervention opportunities in hazard management. Source: Adapted from Kasperson and Pijawka, 1985.

for its analysis by identifying five general approaches that might yield successful interventions and that, taken together, would cover all aspects of maritime activities. These five approaches were selected based on the committee's judgment, but, fortuitously, they also are suggested by Annex V. Specifically, the committee proposes that potent interventions to support successful implementation of Annex V can come from

- technological innovations,
- organizational and operational changes,
- educational communications,
- government and private regulation and enforcement, and
- economic incentives.

These five approaches were incorporated, along with the elements identified by Kasperson and Pijawka, into a generic matrix designed by the committee. In the matrix, the rows represent the five general intervention approaches, and the columns represent the boxes (as modified) from the Kasperson and Pijawka flow diagram.

To illustrate the strength of the hazard evolution concept in clarifying the unfamiliar, the committee used its generic matrix to analyze the intervention options permitted and encouraged by Annex V regulations and the International Maritime Organization (IMO) *Guidelines for the Implementation of Annex V (Garbage)*, which propose an integrated garbage management regime that combines waste reduction, treatment, and disposal strategies. (Table 3-1 illustrates this application of the model.) The following commentary on this application of the matrix also describes the general intervention approaches and establishes why the committee found them so appropriate to the hazard in question.

Technological Innovations

While most pollution-control technology to date has been designed to minimize the release of waste into the environment, technology also can assist in reducing both the amount of waste generated and exposure to the waste once it is introduced into the environment. Clearly, technology could be a useful intervention at numerous stages in the evolution of the hazard posed by vessel garbage. Table 3-1 shows how both Annex V and the IMO implementation guidelines encourage the use of technologies to intervene against the hazard at every step.

In some cases, new technology may be needed, or existing technology may require further development to make it suitable for use on vessels or in port reception facilities. Research by the U.S. Navy has demonstrated some of the possibilities in garbage treatment equipment for military use. According to the four criteria established by Navy developers, shipboard systems should (1) be appropriate to handle the garbage generated by ships with different populations and missions, (2) be sized to the space available on boats and ships, (3) be reliable and cost effective and (4) produce processed garbage in appropriate form for safe disposal as defined by Annex V (Smookler and Alig, 1992). Technologies tailored for use in port reception facilities also would support Annex V; existing technologies used to manage other waste streams need to be adapted for use in ports.

As valuable as technology may be, it is not a panacea for environmental problems. Research suggests there is a maximum 35 percent ongoing implementation level for federal environmental regulations that include technological applications (Burby and Patterson, 1993).[1] Therefore, supplementary interventions are required.

Organizational and Operational Changes

Organizational and operational changes are crucial to the Annex V regime, which attempts to change some very old practices. Responsible handling of vessel garbage has not been emphasized in business or government until recently, and there is much to modify. Necessary organizational and operational changes that have been identified include (1) consistent articulation of commitment to comply with Annex V by top executives of corporations involved in marine activities, (2) standardization and closer integration of vessel and port garbage handling practices so that vessel operators know what to expect, and (3) improved coordination among the various federal agencies responsible for implementation of Annex V. The first of these factors is addressed here. The integration of garbage handling practices is addressed in Chapter 5, and the issue of coordinating federal agency activities is addressed in Chapter 7.

Bassow (1992) emphasizes that implementation of environmental regulations requires combining appropriate changes in technology with changes in corporate culture. He explains:

> In the last 20 years, many U.S. companies have adopted comprehensive environmental policies. They have introduced new procedures and technologies to reduce and eliminate harmful impacts on the environment and human health. But these are technological fixes, engineering fixes. The much more difficult challenge is to change the way people within a company think about the company's environmental policies, to change their attitudes and their mind set, in effect, to change their collective beliefs about the way the company does business. We're now talking about changing the **corporate culture.**

---

[1]The 35 percent figure was derived as follows: 70 percent of the requisite technology was installed, and adequate maintenance to permit the technology to function was provided 50 percent (or half) of the time; half of 70 percent equals 35 percent total ongoing implementation (Burby and Patterson, 1993).

TABLE 3-1   Applying the Hazard Evolution and Intervention Model to
MARPOL Annex V Provisions

Hazard Evolution Model	Behavior that Encourages Generating Garbage	On-board Generation of Garbage	Breakdown in Compliance
Intervention Model	Modify Behavior that Encourages Generating Garbage	Reduce Garbage Generation during Voyage	Prevent Breakdown in Compliance
Technological	Behavior modification is encouraged throughout guidelines but not mandated by Annex V regulations.	While not mandated by Annex V regulations, waste reduction is encouraged explicitly by Guideline 3 (Minimizing the amount of potential garbage).	
Organizational and Operational	No restrictions are imposed. As long as the garbage generated is disposed of properly, no on-board activities need be constrained.	Waste reduction is encouraged but not required.	

Discharge of Garbage into Sea	Exposure to Discharged Garbage	Consequence of Discharged Garbage
Block Discharge of Garbage into Sea	Block Exposure to Discharged Garbage	Diminish Consequences of Discharged Garbage
Guideline 4 (Shipboard garbage handling and storage procedures) and Guideline 5 (Shipboard equipment for processing garbage) address means for meeting the need to retain garbage on board for disposal in port reception facilities (addressed in Guideline 6).	Pretreatment prior to release apparently is intended to minimize, although not block, the exposure to the garbage. Both food and nonfood garbage may be discharged after comminution to particles less than 25 mm in diameter.	Guideline 4.3.5 encourages recovery of garbage at sea, but retrieval is not mandated.
Discharge is only partially blocked. Overboard disposal of plastics is prohibited but many other items may be so discharged (Regulation 3). Pretreatment (i.e., grinding) is required in some cases.	Restrictions vary by the location of discharge. Annex V provides maximum protection to coastal sea within 25 miles of shore. Floating nonplastic garbage may be discharged beyond 25 miles (Regulation 3.1.b.i). "Sinkable" garbage may be discharged beyond 12 miles (Regulation 3.1.b.ii). No discharges except ground food waste are permitted from fixed or floating structures.	Annex V emphasizes the elimination of plastic discharges, which are judged most harmful. IMO guidelines encourage prevention and retrieval of lost fishing gear, even though such loss does not violate Annex V.

TABLE 3-1 Continued

Hazard Evolution Model	Behavior that Encourages Generating Garbage	On-board Generation of Garbage	Breakdown in Compliance
Intervention Model	Modify Behavior that Encourages Generating Garbage	Reduce Garbage Generation during Voyage	Prevent Breakdown in Compliance
Educational (Target Population/ Content)	Guideline 2 calls on governments to "develop and undertake training, education and public information programmes suited for all seafaring communities under their jurisdictions." Guideline 2.2 encourages exchange of information on compliance strategies.		Placards and notices must be provided to crews of vessels over a certain size (under U.S. law). Annex V can be used as a tool in fostering public support for and raising mariners' environmental awareness.
Regulation and Enforcement (by governments and private organizations in signatory nations, as required by the treaty and international law)			Guideline 7.3 recommends that national governments assist and recognize compliance initiatives by private and professional organizations.

Discharge of Garbage into Sea	Exposure to Discharged Garbage	Consequence of Discharged Garbage
Block Discharge of Garbage into Sea	Block Exposure to Discharged Garbage	Diminish Consequences of Discharged Garbage

To effect the changes mandated by Annex V, the guidelines encourage provision of both general information and specific education about means to comply. Guidelines also encourage technical exchange concerning improvements achieved in performance and equipment used for complying with garbage restrictions.

Annex V requires signatory nations to provide "adequate" reception facilities. Guideline 7.1 (Enforcement) suggests means to organize national authorities, record use of port reception facilities, and verify vessel operators' activities.

Guideline 1.3 encourages the maximum use of port reception facilities rather than continued discharges at sea, even where legal.

TABLE 3-1 Continued

Hazard Evolution Model	Behavior that Encourages Generating Garbage	On-board Generation of Garbage	Breakdown in Compliance
Intervention Model	Modify Behavior that Encourages Generating Garbage	Reduce Garbage Generation during Voyage	Prevent Breakdown in Compliance
Economic (Market Forces)			Guidelines include directions for estimating the required capacity of reception facilities but many uncertainties remain. Guideline 7.2 addresses compliance incentives, such as funding for capital investment in port facilities or garbage hauling infrastructure.

It is likely that changing the orientation of personnel at all levels of an organization (i.e., changing the corporate culture) becomes more important when compliance with a regulation is very challenging, as in the case of Annex V.

Indeed, organizational and operational changes may be essential in order to engage personnel and other resources in the effort to comply with Annex V. Such changes can range from modifying a procurement officer's job description to specifying that suppliers use reusable packaging, to reorganizing a port's waste management operations so that services are integrated. Organizational activities also can include development of company rules for handling garbage and internal penalties for violations of the rules, up to and including dismissal (Estes, 1993). These approaches must be supported by training, to prepare the organization for

Discharge of Garbage into Sea	Exposure to Discharged Garbage	Consequence of Discharged Garbage
Block Discharge of Garbage into Sea	Block Exposure to Discharged Garbage	Diminish Consequences of Discharged Garbage
Annex V does not establish cost criteria for reception facilities but acknowledges implicit costs (delay to ships). Garbage hauling fees add to ship operating expenses. Annex V does not require ports to charge fees or ships to land garbage		

Annex V compliance, and commitment of organizational resources to develop new internal garbage management plans.

Again, Table 3-1 makes it clear that the drafters of Annex V expected seafarers to include organizational and operational changes in their compliance plans. No specific changes are mandated, however. By establishing performance standards, the drafters left managers and operators the flexibility to devise a compliance program that best suits their circumstances.

To effect a change in corporate culture, according to Bassow, there must be communication, involvement of all managers and employees, training and support, and system alignment to the new goals. "The experience of large corporations shows that synergy between technological change and a responsive corporate culture that supports an environmental vision is essential to achieving environmental goals" (Bassow, 1992).

The committee heard from commercial ship operators who discussed the way in which Total Quality Management (TQM) (Barkley and Saylor, 1993) is being adopted in the industry. The TQM approach may enable the commercial marine culture to incorporate new objectives more rapidly than would have been feasible through other management practices. Globally, the maritime industries have been trying to articulate the role management has to play in meeting the ambitious goals set by IMO and individual governments. Members of IMO have tried to unify years of isolated efforts into a comprehensive document that speaks to shoreside executives as well as watchstanding crew members. Guidelines for ship management and quality assurance have been developed (International Maritime Organization, 1993).

While vessel crew members carry out garbage discharges, a corporate culture perspective would suggest that crew behavior reflects the values of their employer corporations and their professional membership organizations. These corporations and organizations may not have fully developed cultures committed to reducing environmental hazards. This lack of commitment may be influenced by the intractability of the marine waste management problem as well as the threat it poses to other corporate values, such as profit.

It is important to remember that integrated waste management is more than technology—it is an organizational concept that employs technology. Organizational changes are possible in all phases identified in the hazard evolution model for vessel garbage, but particularly "upstream," such as in modifying seafarers' behaviors, preventing initiating events, and blocking the discharge of garbage into the sea. A supportive organizational culture can make it easier to introduce new operating practices through measures such as restricting the distribution of supplies (Gallop, undated) and eliminating packaging that seafarers are accustomed to having ashore but do not need at sea. An example of such an intervention is to switch from small to large ketchup bottles in a dining room on an offshore platform. An oil company reported that it expected to eliminate 4,800 bottles from its annual waste stream through this change, without asking anyone to give up ketchup (Babin and Toll, 1992).

Educational Communication

Education has been shown to be an effective intervention against the problem of vessel garbage. Education often is employed in the movement to combat environmental hazards; such approaches have been used effectively, for example, in lobbying the U.S. Congress to ratify Annex V and enact the Marine Plastics Pollution Research and Control Act (O'Hara et al., 1988). But in the present context, persuasive public information is not enough. To achieve its potential in supporting Annex V, education must target specific users in each maritime sector. Within each sector, certain individuals and groups responsible for different aspects of Annex V compliance will have distinct requirements for education, training, and information exchange. Table 3-1 outlines how the drafters of Annex V viewed education and information exchange.

Educational messages should be structured to build the seafarer's self-image as someone who engages in environmentally sound behavior (i.e., MARPOL compliance) voluntarily rather waiting for imposition of external controls (i.e., prosecution for a violation). Messages not only should persuade users to comply and provide information about legal responsibilities, but also should describe compliance methods, because a mariner needs tools to make it possible to follow the rules.

Educational interventions are possible throughout the hazard evolution model. Mariners can be made aware of the way in which their needs are manifested in behaviors. They also can be taught alternative ways to satisfy their needs or even persuaded to constrain their needs while at sea because of the difficulties with garbage disposal. Simultaneously, it may be necessary to confront lingering, outdated attitudes among those who still view the ocean as a garbage receptacle. The educational message might be that it is inappropriate to continue garbage-generating activities at sea, unless one is willing to take responsibility for proper disposal.

There are also promising possibilities for educational intervention at the waste generation and waste release stages. Objectives could include stimulating recognition and modification of behaviors that result in garbage being carried or generated on board vessels or disposed of improperly. If mariners were made aware of the ways in which marine organisms are exposed to garbage and the impact of the hazard, then they might recognize the need to modify their behaviors "upstream" in the hazard evolution model. Once informed about the legal restrictions, many mariners will modify their behavior in order to avoid sanctions. Multilingual and cross-cultural educational efforts are needed due to the international character of many crews.

Educational communication can target either the individual user or those in authority who can change an organization to improve implementation of Annex V. Either way, it is important to recognize that no individual decides in isolation whether to comply. Norms of behavior exist for all subcultures in the marine community, whether the groups encompass specific types of users, particular regions of the country, or specific communities within a region. Therefore, the educational process should reinforce the message through group dynamics (Laska, 1990). To ignore the power of a group in influencing its members is to lose an important opportunity to alter behavior. User organizations need to be employed as much as possible as vehicles for communicating the importance of compliance with Annex V and for altering behavior throughout the intervention.

Government and Private Regulation and Enforcement

Environmental policies and enabling legislation must be accompanied by a commitment to enforcement if the regulatory process is to be successful. It is important to punish violators of environmental laws, which are designed to protect the commons. Annex V, like most environmental regulations, gives new enforcement authority to governments. The matrix details how the drafters of Annex V established both new standards and means to enforce them.

Compliance rates are likely to fall if agencies charged with enforcement are not adequately funded or committed to enforcement. If no one ever is punished, then many seafarers will feel no pressure to comply. If no one is even caught, then more seafarers will disregard the law and continue to toss garbage overboard. Clearly, enforcement must serve as a visible deterrent to potential violators. But it is important to acknowledge that enforcement is only one way for a government to intervene. Overemphasizing this responsibility as the major obligation of government could skew implementation of Annex V toward a small group of violators, leaving unmet the needs of seafarers who are trying to comply. In other words, government needs to address the entire hazard evolution process.

Regulations have focused on later stages of the hazard evolution process, by punishing illegal garbage discharges and requiring ports to provide reception facilities. Of course, these strategies may affect the upstream points of intervention indirectly, because good enforcement encourages modifications early in the process that reduce the costs of compliance later.

The federal government is not the only U.S. regulatory body involved in control of vessel garbage. Some state and local jurisdictions have established regulations that differ from and sometimes are more stringent than federal requirements. (An example is California's quarantine requirements, which are more stringent than federal standards [Mendel, 1992]). This situation may exacerbate the difficulty of complying with Annex V and other related regulations and thereby increase the incidence of violations. The private sector also regulates the garbage disposal practices of employees, clients, and others. Private firms and other organizations can establish internal systems of penalties for violations of Annex V or for policy infractions that could lead to illegal garbage discharge.

Economic Incentives

If compliance is cheaper than committing a violation, then economic theory holds that seafarers will tend to comply. Several types of interventions encourage compliance by offering an economic benefit or opportunity. Whenever possible, it is important to facilitate compliance in a cost-effective manner. For example, the cost of disposing of vessel garbage in ports varies widely, even for the same service. Furthermore, the basis for the pricing is so variable that it is difficult for vessel operators to assess which ports offer the best value. Costs may be quoted in cubic meters, kilograms, bins, dumpsters, skips, truckloads, barges, or other measures. If there were uniform and affordable garbage disposal rates, then levels of compliance probably would rise.

Another possible economic incentive lies in the development of on-board garbage processing equipment that is reasonably priced, reliable, and effective. The government could be involved in this effort. For example, existing mechanisms could be used to disseminate to the private sector any relevant technical developments by government engineering facilities. Another option would be to assist the private sector (e.g., through loans) with research to improve on-board garbage treatment methods.

Economic incentives are powerful means of encouraging compliance. In fact, for commercial marine users, these likely are the most important incentives. Because regulation and enforcement requires these users to bear the cost of business externalities such as garbage disposal, the cost of noncompliance (e.g., fines, bad publicity, reduced product and/or service demand) is weighed against the cost of compliance. Actions taken at each stage of the hazard evolution model may be considered means of reducing compliance costs.

Application of the Model to the Seafarer Communities

The foregoing overview of the two-dimensional hazard evolution model illuminates the range of interventions that could facilitate implementation of Annex V. The drafters of Annex V clearly intended that signatory nations would use many methods to encourage compliance or enable enforcement.

There is an additional dimension of the problem of Annex V implementation that the model does not address adequately in its present form—the breadth and diversity of the regulated maritime fleets and ports. To remedy that shortcoming and assure that recommendations based on the model contain sufficient detail to be useful to policymakers, the committee decided to develop a separate matrix for each of the nine maritime sectors addressed in this report. This approach enabled the committee to consider specific interventions by type of action and phase of the process for each user group. In doing so, commonalities across user groups became evident, thereby suggesting where combined efforts might provide economy of scale. The examination of the nine maritime sectors may be found in Chapter 4.

In filling in the cells of each matrix, two related approaches were employed. The activities of the user group were considered with respect to the actual and the potential waste they generated; the committee determined how the activities could be modified through various types of efforts. Concurrently, the committee focused on the different types of garbage generated by each group and how each type might best be controlled.

A Final Modification to the Model

In adapting the matrix to each maritime sector, the committee omitted the final column ("Consequences") because it is usually beyond the ability of most seafarers to intervene once garbage is loose in the ocean. This is, of course, a critical point in the hazard evolution model, where the peril evolves from a possibility into reality. (The committee's review of the ecological and social consequences of discharging garbage from ships is summarized in Chapter 2 and Appendix F.) From the perspective of planning for Annex V implementation, however, the source of discharged waste is not a factor in intervening against the consequences. No matter what types of mariners discharge garbage into the sea, their capability to intervene is limited.

It is possible for others to take effective action once debris is discharged into the marine environment. What is important in the present context is the issue of whether and how to integrate these types of interventions into the Annex V implementation strategy. Beach cleanups, for instance, are readily identifiable as interventions, however modest, against the consequences of a hazard. Each volunteer who bags a piece of marine debris is helping to reduce the hazard to wildlife, lessen aesthetic degradation of the beach, and reroute the pollutant into the shoreside waste management system. For materials still in the water, retrieval serves the same purpose. Fishermen who capture debris in their nets and bring it back to shore for disposal are helping to mitigate the consequences of someone else's discards.

Such efforts are neither encouraged nor rewarded in the present Annex V implementation regime. Rewards could be offered to encourage seafarers to retrieve marine debris; this approach has been employed in fishing tournaments in the Gulf of Mexico (Louisiana State University Sea Grant Program, 1989).

SUMMARY

A systematic approach to Annex V implementation can help government authorities and regulated seafarers take full advantage of all options available to address the challenges posed by a potent pollutant—vessel garbage. The hazard evolution model described in this chapter is an example of such a systematic approach.

An important feature of the model is the inclusion of waste reduction as a garbage management option. The committee's analysis demonstrates that, to date, most efforts to reduce the hazard—whether economic incentives, educational programs, or enforcement of the law—have been "downstream." That is, most interventions are carried out after packaging and other items made of nondegradable materials are brought on board. Recently, the Environmental Protection Agency revised its hazardous waste policy, which formerly emphasized *waste* management, to focus "upstream" on reducing waste generation (U.S. Environmental Protection Agency, 1990). In seeking ways to improve implementation of Annex V, the federal government might benefit from emulating the shift in EPA's waste management policy.

REFERENCES

Babin, D.A. and B. Toll. 1992. One company's response to offshore waste management. Pp. 310-317 in Proceedings: Twelfth Annual Gulf of Mexico Information Transfer Meeting, Nov. 5–7, 1991, New Orleans, La., compiled by Geo-Marine, Inc. New Orleans: U.S. Department of the Interior, Minerals Management Service, Gulf of Mexico OCS Region. December.

Barkley, B. and J.H. Saylor. 1993. Customer-Driven Project Management: A New Paradigm in Total Quality Implementation. New York: McGraw Hill.

Bassow, W. 1992. Environmental policy, corporate culture. Environmental Protection 3:10-13.

Burby, R. and R.G. Patterson. 1993. Improving compliance with state environmental regulations. Journal of Policy Analysis and Management 12(4):753-772.

Estes, J.T. 1993. Testimony of John T. Estes, president, International Council of Cruise Lines, before the Subcommittee on Superfund, Ocean, and Water Protection of the Committee on Environment and Public Works, U.S. Senate, 102nd Congress, Second Session, Washington, D.C., Sept. 17, 1992. Pp. 18 in Implementation of the Marine Plastic Pollution Research and Control Act. S. Hrg. 102-984. Washington, D.C.: U.S. Government Printing Office.

Gallop, M. Undated. USS Theodore Roosevelt Environmental Compliance Program Cookbook. Available from the commanding officer, USS Theodore Roosevelt, homeport in Norfolk, Va.

Hohenemser, C., R.E. Kasperson, and R.W. Kates. 1985. Casual Structure. Pp. 25-42 in Perilous Progress: Managing the Hazards of Technology, R.W. Kates, C. Hohenemser, and J.X. Kasperson, eds. Boulder, Colo.: Westview Press.

International Maritime Organization (IMO). 1993. International Management Code for the Safe Operation of Ships and For Pollution Prevention (International Safety Management [ISM] Code). Resolution A.741(18). Adopted November 4, 1993. Available from IMO, 4 Albert Embankment, London, SE1 7SR.

Kasperson, R.E. and K.D. Pijawka. 1985. Societal response to hazards and major hazard events: Comparing natural and technological hazards. Public Administration Review 45:7-18. Special issue.

Kasperson, R.E., R.W. Kates, and C. Hohenemser. 1985. Hazard Management. Pp. 43-66 in Perilous Progress: Managing the Hazards of Technology, R.W. Kates, C. Hohenemser, and J.X. Kasperson, eds. Boulder, Colo.: Westview Press.

Laska, S. 1990. Designing effective educational programs: The attitudinal basis of marine littering. Pp. 1179-1190 in Proceedings of the Second International Conference on Marine Debris, 2–7 April 1989, Honolulu, Hawaii (Vol. II), R.S. Shomura and M.L. Godfrey, eds. NOAA-TM-NMFS-SWFSC-154. Available from the Marine Entanglement Research Program of the National Marine Fisheries Service (National Oceanic and Atmospheric Administration), Seattle, Wash. December.

Louisiana State University Sea Grant Program. 1989. Saltwater Anglers and Research. Aquanotes 18(spring):1-3.

Mendel, N. 1992. Presentation by Neil Mendel, Animal and Plant Health Inspection Service supervisor, Port of Oakland, to the Committee on Shipborne Wastes of the National Research Council, at Coast Guard Island, Alameda, Calif., Oct. 15–17, 1992.

O'Hara, K.J., S. Iudicello, and R. Bierce. 1988. A Citizens Guide to Plastics in the Ocean: More Than A Litter Problem. Washington, D.C.: Center for Environmental Education (now the Center for Marine Conservation).

Smookler, A. and C. Alig. 1992. The Navy's shipboard waste management research and development program. Naval Engineers Journal 104(3):89-97. May.

U.S. Environmental Protection Agency (EPA). 1990. Reducing Risk: Setting Priorities and Strategies for Environmental Management. SAB-EC-90-021. Washington, D.C.: EPA Science Advisory Board.

4

Elements of an Implementation Strategy

G iven the diversity among vessels passing through U.S. waters and the ports they visit, it is clear that no single Annex V implementation approach will work across the board. No one reward or punishment will bring all mariners and ports into compliance with Annex V. The interventions chosen must be appropriate to the targeted maritime sector and sustainable within resource limitations. At the same time, the various interventions need to be integrated into a coherent national strategy, to conform with U.S. policy calling for the establishment of integrated waste management practices wherever possible.

Using the hazard evolution matrix described in Chapter 3 and drawing on first-hand observations and research, the committee considered how Annex V compliance could be achieved within each sector of vessels and ports. To assess barriers and opportunities, the committee sought input from each community. Levels of preparedness and capabilities varied widely among the various groups as well as the government agencies tasked to enforce the rules. It became clear that many different individuals, not just vessel masters or port managers, can influence compliance levels.

This chapter provides an initial assessment of promising intervention points and implementation methods for each maritime sector. (Later chapters offer a national perspective on how these elements could be woven together into a national strategy.) Key to the committee's assessment is the analytic approach of Kasperson and Pijawka (1985), who focused on intelligence gathering and control capabilities as the basis for selection of an effective management strategy. The chapter opens with a brief description of this approach.

INTRODUCTION

According to the Kasperson and Pijawka model, the selection of an effective management strategy depends on an assessment of the amount of intelligence (i.e., first-hand information) that can be collected to support interventions, and the degree of control—direct or indirect—that can be exercised over the target community. The same approach can be used to identify elements of an effective Annex V implementation strategy.

There are obvious limits on the federal government's information-gathering and control capabilities. The mandates of Annex V are difficult to enforce directly, and MARPOL depends on seafarers to continue to comply even when beyond sight of land. In the United States, federal enforcement depends on reporting of incidents and vessel boardings[1] in port to a far greater extent than on surveillance at sea. Furthermore, because vessel operators may select from a range of compliance options, no single indicator can serve as proof that a vessel has complied with or violated the law. Individual infractions at sea are almost impossible to detect, and violators are difficult to prosecute unless witnesses come forth.

Even when garbage washes ashore that may have been discharged from a vessel, the burden remains on the enforcement agency to prove which mariner is the violator—often an impossible task. Thus, implementation of Annex V cannot rely solely on the government's ability to identify violators and enforce the law. Fortunately, the government is not the only party capable of gathering intelligence. As will become evident in this chapter, private managers, vessel operators and passengers, or other members of the maritime community may be in better positions to monitor practices than are government officials.

In ensuring compliance with pollution laws, the first line of control is direct government regulation. In some maritime sectors, government licenses, certificates, or other approvals may be withheld if a mariner, vessel, or port fails to comply with the law. Even when that authority is absent, there may be opportunities to exert indirect control if a fleet is subject to federal regulation for another purpose directly tied to mariners' livelihood.

Federal control capabilities vary by sector. Some fleets, notably cargo vessels and passenger cruise lines, are regulated directly by the Coast Guard. Military and public fleets, as arms of government, also are subject to direct control. Many commercial fishing vessels are regulated indirectly by the National Marine Fisheries Service (NMFS) through fisheries management and, on matters of

---

[1]MARPOL inspections are conducted as a component of port safety boardings. Coast Guard inspectors use a checklist. Garbage logs are required on many U.S.-flag vessels. Foreign-flag vessels are not required to keep written records, so the inspector interviews crews and officers (language barriers can be a problem).

safety, by the Coast Guard. Offshore oil and gas platforms in federal waters are inspected for pollution compliance as part of Minerals Management Service (MMS) regulation of the industry. The recreational boating community is subject to little direct federal control, because relevant authorities have been delegated in large part to the states.

In sum, government capacity for intelligence gathering and control is uneven and limited, but creative strategies may be devised to capitalize on any opportunities that exist. Identification and analysis of the opportunities could serve to stimulate their use. For example, forms are available from the International Maritime Organization (IMO) for reporting inadequate port reception facilities, but mariners rarely fill them out. This is a potential source of intelligence that has not been exploited. An examination of why this mechanism is ignored, and how this situation might be reversed, could suggest ways of improving Annex V compliance.

In the forthcoming analysis, a matrix is presented for each fleet containing a range of intervention options. (As in Chapter 3, the columns are the headings from the modified Kasperson and Pijawka model and the rows are the five types of intervention options.) Some of the measures suggested have been tried—albeit usually in isolated locations—while others were conceived by the committee. There has been some pre-screening, to the extent that all the options listed are plausible and worthy of serious consideration; however, practical considerations may argue against or eliminate some of the ideas.[2] The committee's views concerning the various intervention options will become evident in the commentary on each matrix and in later chapters. The final screening criteria and recommendations may be found in Chapter 9.

ANALYSIS OF INTERVENTIONS

Recreational Boats and Their Marinas

Intelligence

There is no formal intelligence-gathering network for recreational boaters, but the community is monitored by private groups and some research has been conducted. Available information suggests that recreational boaters are very concerned about the marine environment and many want to comply with Annex V, but that awareness of the mandate is far from universal and educational informa-

---

[2]Interventions actually fall into five groups: (1) activities now conducted effectively that should be encouraged further, (2) activities currently under way that require improvement, (3) activities currently under way that should cease, (4) activities not being conducted that should be, and (5) activities that might be useful but are considered too costly or impractical.

tion is inadequate (Boat Owners Association of the United States, 1990; Wallace, 1990).

Most recreational boaters, because they make short trips, simply hold garbage on board until they return to land. But anecdotal reports suggest that compliance within this sector needs to be improved. One member of the Committee on Shipborne Wastes observed that his port has continuing problems with recreational boaters, especially sport fishermen, who dump refuse into the harbor in full view of the shore.

Control

Little direct control can be exercised over recreational boaters, because vessels are privately owned and management of marinas and other portside facilities is highly decentralized. Coast Guard and customs officials and state marine police occasionally interact with recreational boaters, but the only routine government contact occurs through state boat registration for tax collection purposes, and programs such as the courtesy motorboat examinations offered by the Coast Guard Auxiliary, a volunteer organization that supports the agency's efforts. Moreover, because recreational boaters are so diverse, there is no single way of reaching them, even indirectly. Approximately 38 percent of these boats are used for fishing (American Red Cross, 1991), but many recreational fishermen do not consider themselves boaters and therefore may not, for example, read boating magazines.

Persuasion and peer pressure are viewed as the most effective management tools in this community. An example of an ongoing initiative of this type is the Boater's Pledge Program, an effort to persuade boaters to promise to stop discharging garbage in the Gulf of Mexico. Established educational programs, such as boating safety courses taught by the Coast Guard Auxiliary and the nonprofit U.S. Power Squadron, also can be avenues for dissemination of Annex V information.

Analysis of Interventions

Table 4-1 suggests interventions that might improve Annex V implementation in the recreational boating sector. Technological options include development of food and fishing equipment to permit safe and efficient storage of supplies in bulk. While on-board space is especially constrained in this sector, installation of garbage treatment equipment may be appropriate, especially for boats taking extended voyages.

Among organizational and operational interventions, the distribution of Annex V information through licensing and registration processes could be a straightforward way to reach many boaters. Use of disposable items clearly could be reduced through careful purchasing. Beach cleanups could be held more often and in more places. Another approach would be to promote the retrieval of debris observed while on the water; this has been done in the Gulf of Mexico through the offering of rewards in fishing tournaments.

Education is a critical tool, due to the poor intelligence and minimal control capabilities in this sector. Information about Annex V and compliance strategies can be distributed through existing channels, such as boating safety courses and the Sea Grant Marine Advisory Service (described in Chapter 6), and new activities, such as volunteer efforts by boating groups. International channels, such as racing associations, could be employed as well. Instructors can exploit group dynamics (i.e., peer pressure and the desire of individuals to conform with group behavior). In addition, it might be useful to train Coast Guard and customs officials and state marine police in techniques for persuading boaters to comply.

Selected regulatory and enforcement interventions might be effective. For example, boat racers must comply with racing rules, which could be amended to mandate Annex V compliance and disqualify violators. Such a measure would affect only a small segment of the boating community, however. To reach more boaters, state boating and marine officials might be authorized to assess fines for Annex V violations. Peer reporting could be a useful supplementary tool; the Coast Guard plans to publicize the telephone number for reporting violations to the National Response Center (1-800-424-8802).

Economic interventions include several that might promote recycling—offering boaters credits on marina fees for return of recyclables, holding deposits for return of garbage to shore, and charging extra for return of unsorted garbage. While such schemes might be complicated to implement, recycling merits promotion because it reduces amounts of garbage (which may be discharged overboard, legally or otherwise) and has become a standard component of integrated land-based waste management (see Chapter 5). Other options include imposing surcharges on disposable items sold at marina stores, and increasing and publicizing fines for Annex V violations.

Commercial Fisheries and Their Fleet Ports

Intelligence

The federal government has scrutinized the practices of U.S. commercial fisheries for decades, but the focus has been on ensuring the strength of biological stocks rather than reviewing garbage disposal practices. Some information is available on numbers of vessels and their general operations while at sea, but reports of garbage management practices are largely anecdotal (see sidebar). Until recently, neither vessels nor operators were regulated directly by the Coast Guard, and the fishing community argued strenuously against government oversight of vessel conditions and operations. It is only since 1989 that the Coast Guard has had congressional authority to oversee the safety of fishing vessel

TABLE 4-1   Applying the Hazard Evolution and Intervention Model to Recreational Boats and Their Marinas and Waterfront Facilities

Hazard Evolution Model	Behavior that Encourages Generating Garbage	On-board Generation of Garbage
Intervention Model	Modify Behavior that Encourages Generating Garbage	Reduce Garbage Generation during Voyage
Technological	Create products that require little or no packaging.	Develop food and fishing equipment that permit use of bulk items.
Organizational and Operational	Choose bulk liquids and beverages. Choose food with few byproducts. Prepare foods ashore. Choose recyclable, compactible, and reusable containers. Repackage condiments in small reusable containers.	Remove equipment and replacement parts from packaging and dispose of the wrapping ashore. Cut back on purchases of items that can be discarded. Encourage sale of items with minimal packaging at convenience stores near marinas.
Educational (Target Population/ Content)	Instill respect for clean environment. Make boaters aware of alternative ways to satisfy their needs. Address behavior change in ecotourism presentations.	Select bulk and repackage in reusable containers. Use "retensiles"—cloth napkins, cotton dish towels, sponges, reusable cutlery, mugs, and drinking glasses. Avoid disposable eating materials. Buy resealable packages to hold food waste that may spoil. Buy recyclable, compactible, packaging.

Breakdown in Compliance	Discharge of Garbage into Sea	Exposure to Discharged Garbage
Prevent Breakdown in Compliance	Block Discharge of Garbage into Sea	Block Exposure to Discharged Garbage

Build garbage storage areas into new boats.	Develop and install appropriate on-board garbage handling equipment.	
Include Annex V information in boating license and registration packets.	Return all materials for shoreside disposal. Provide waste management at marinas to encourage boaters to return their garbage.	Retrieve debris observed while on the water. Hold beach cleanups.
Serve meals in individual reusable lunch kits that also can hold garbage. Encourage volunteer groups to implement Annex V educational programs. Distribute Annex V information through boating safety courses, registration Sea Grant agents, and international channels. Train officials how to persuade boaters to comply.		

TABLE 4-1   Continued

Hazard Evolution Model	Behavior that Encourages Generating Garbage	On-board Generation of Garbage
Intervention Model	Modify Behavior that Encourages Generating Garbage	Reduce Garbage Generation during Voyage
Government and Private Regulation and Enforcement	Require recycling in municipal laws and permits for marinas.	
Economic (Market Forces)	Encourage boaters to buy items that can be reused, recycled, or compacted; buy in bulk; and avoid foamed plastic and other disposables. Impose surcharge on disposables sold at marina stores.	Encourage marina recycling programs with incentives (e.g., offer credits on marina fees). Marine stores and chandleries could stock reusable products. Encourage equipment manufacturers to recycle or offer credit for returned (used) equipment.

construction and operation (National Research Council, 1991), so the agency has had little time to become familiar with the diverse operations of fishing fleets.

Fisheries employ a wide variety of gear and methods and therefore produce assorted wastes. But the vast majority of fishing vessels take short trips, so most should be able to refrain from discharging any garbage at sea. Exceptions to this rule include the vessels in some fleets that eviscerate or process the catch and discard the processing waste at sea. On some vessels, the combined fishing/processing waste can far outweigh the garbage generated by the crew.

Breakdown in Compliance	Discharge of Garbage into Sea	Exposure to Discharged Garbage
Prevent Breakdown in Compliance	Block Discharge of Garbage into Sea	Block Exposure to Discharged Garbage
Establish citizen patrols to monitor Annex V compliance and report violations. Publicize the toll-free telephone number for reporting violations to the Coast Guard.	Amend racing and association rules to mandate compliance with Annex V and disqualify violators. Require waste management plans in event permits and licenses. Extend authority to levy fines to state boating and marine authorities.	
Increase and publicize rewards for reporting violations. Publicize fines levied against violators. Make boaters aware of costs of damage to boats by debris.	Greatly increase fines for Annex V violations. Post cleanup costs and pass them on to marina tenants. Hold deposits for return of garbage to shore. Charge extra for unsorted garbage returned to shore.	Promote compliance as a means of reducing boat maintenance costs (by keeping water clean). Offer rewards for recovered debris.

Despite the shortage of official intelligence, informal communications net-works proliferate in this sector. Commercial fisheries typically require that a catch be landed at a fishing port rather than a general-purpose waterfront. A sense of community can develop among fishermen working out of local ports, and vessel operators using the same facility usually become well acquainted. This community often is extended, because fishing can be a family business. In addi-tion, the harbor master or other individual acting as a port authority often is

FISHERIES GARBAGE DISPOSAL PRACTICES

Commercial fisheries have employed various strategies to comply with Annex V, some by installing shipboard trash compactors or incinerators, others by retaining garbage on board until they reach port. The biggest problem is handling of garbage in port. In some remote ports, there is no landfill space for vessel garbage, and waste hauling from fishing piers is generally irregular across the nation. Disposal of nets is a major problem, in that there is no national infrastructure for recycling them. However, a regional infrastructure has been established in the Pacific Northwest; fishermen in Alaska and Washington are recycling about 680.4 metric tons (150,000 pounds) annually of nylon gill-net webbing, which is marketed to Taiwan and Hong Kong for use in bicycle seats, electronics and appliance parts, kitchen utensils, and other items (F.I.S.H. Habitat Education Program, 1994).

Most fishing vessels operating in the coastal ocean, Great Lakes, and other inland waters have little extra storage space, so discharge of garbage ashore depends on the availability of adequate reception facilities. Because many of these vessels are operated from remote ports in Alaska, Maine, and Southern Louisiana, and along inland waterways, vessel-generated garbage frequently accumulates on shore. Fishing gear is retrieved each day to extract the catch, or, if large numbers of traps are used, at the end of the season. Inevitably, some gear is lost.

An unusual case among coastal fisheries is the menhaden fleet operating from Maine to Texas. These large ships have extra storage space, in part because crew accommodations are provided aboard carrier vessels. Garbage is stored on board for disposal in port, where the vessel owners maintain sophisticated facilities not only for processing the catch but also for handling garbage. Some haul their own garbage, while others contract for waste disposal.

Among the near-coastal fisheries, the shrimp fleet is alleged by the National Park Service to be a major contributor to the debris in the Gulf of Mexico. Empty food containers and other wrapping from ship suppliers frequently are found on beaches during routine cleanups. Shrimp vessel operations also may contribute pieces of netting and cordage discarded during repairs to damaged shrimp trawls.

Vessels in the Alaskan Pacific groundfish fishery can be very large (up to 300 feet long) and may sail for weeks at a time, and fish-processing ships must carry all packaging materials as well as substantial stores of food and spare parts. As a result, these vessels must manage considerable amounts of garbage. On some ships, waste materials are burned using "burn barrel" technology (Chang, 1990).

familiar with the operations of boat owners. Thus, there are many informal sources and conduits of information among fishermen.

A potential official intelligence-gathering capability may be found in the complicated NMFS regulatory regime, which establishes fishing seasons and catch allocations designed to permit the maximum allowable harvest of the standing stock, now solely reserved for U.S.-based fisheries. The legal framework for fisheries management within the 200-nautical-mile-wide Exclusive Economic Zone (EEZ), the Fisheries Conservation and Management Act of 1976 (P.L. 94-265), was developed in the mid-1970s to control access to U.S. fishing stocks, especially by foreign commercial fleets and fish processors. What has evolved is a regulatory system that emphasizes annual stock estimates, catch quotas, and seasons and perpetuates some operational inefficiencies or creates new ones. But at least fishing operations are monitored. In some fisheries, management plans call for on-board observers who remain on vessels as long as they are at sea. In other fisheries, the catch is assessed when landed at the pier. In both cases, a survey program is in place that could be a mechanism for providing information on net and gear disposal alternatives. In practice, however, this may not be feasible.

Control

Control of fishing vessels is decentralized among private owners, who are difficult to reach for the purpose of persuading them to comply with Annex V. Neither Coast Guard nor Animal and Plant Health Inspection Service (APHIS) boarding parties routinely inspect fishing vessels. Some degree of public control can be exerted, however, because these vessels typically must operate within federally managed fisheries in accordance with plans created by public agencies, and NMFS agents routinely meet arriving vessels to verify the weight and type of fish caught. In addition, the United States can exert some control over nearby foreign fisheries through joint fishing or scientific agreements.

The present fisheries management regime is not highly effective, in that compliance has been difficult to achieve (Sutinen et al., 1990). More to the point, the regime is not designed to support implementation of Annex V; in fact, it is obstructive. The regime has been criticized widely for establishing gear practices that encourage fishermen to disregard safety and environmental protection in pursuit of the catch. Some regional fisheries management plans create situations in which it may be to a fisherman's advantage to deliberately cut away and discard any remaining gear at the end of the season (even though such discards are prohibited by Annex V). But these regulatory practices may be ending as a result of severe economic dislocation among fishermen and the collapse of fish stocks, and implementation of Annex V is proceeding. The first major Annex V enforcement action in this fleet was taken in April 1993 against a fishing vessel operator based in Seattle; the operator was fined $150,000 for 85 counts of instructing crew members to throw all garbage over the side (Weikart, 1993). The incident, first reported by several disgruntled fisheries employees, attracted considerable attention on the Pacific Coast.

The NMFS also has taken selective action to increase its control. In the summer groundfish fisheries off the Pacific Coast, fisheries observers sail with the larger processing vessels for the entire season, to witness the operations and verify that the operators catch fish in accordance with the law. Authorities had so mistrusted this fleet that fishermen agreed to the surveillance so they could continue fishing. However, such direct federal presence is costly and therefore rare.

The Coast Guard can exert some control over fisheries. A Certificate of Adequacy (COA) must be obtained for piers serving vessels that off-load more than 500,000 pounds of commercial fishery products annually. In addition, Coast Guard regulations effective May 19, 1994, require that U.S.-flag, ocean-going commercial vessels over 12.2 meters (about 40 feet) keep records of garbage discharges.[3] Commercial fisheries are among the fleets affected.[4] By promoting knowledge of regulations and awareness of garbage handling practices, as well as means of verifying that responsibilities are being carried out, the use of such records is expected to "promote compliance, facilitate enforcement, and reduce the amount of plastics discharged into the marine environment" (59 Fed. Reg. 18,700 [1994]).

Fortunately, the government is not the only source of control. Experience has shown that indirect control can be exerted through employee complaints to law enforcement authorities and peer pressure (Alverson and June, 1988; Recht, 1988; Buxton, 1989; DPA Group, 1989). Attempts are being made to harness these tools to influence fisheries behavior (Center for Marine Conservation, 1989). Increasingly, regional councils focusing on the prevention of marine debris are enlisting the active support of fishermen to encourage voluntary change (Buxton, 1989; Gulf of Mexico Program, 1991; Pearce, 1992).

Some form of influence clearly is needed to improve port reception facilities, which (as in most maritime sectors) are considered inadequate for handling all the garbage generated by fishing vessels. Fishing ports are owned and managed by a variety of government organizations, city docks, and commercial enterprises. As with any new standard that imposes changes in waste handling, complying with the mandate for port reception facilities can be prohibitively expensive for a small harbor, pier, or terminal. The government may be able to exert some influence in this area by offering to subsidize modification costs, guarantee loans for facility construction, or classify costs of port reception facilities as pollution-control devices for bond underwriting purposes.

Analysis of Interventions

Table 4-2 indicates options for intervening to improve Annex V implementa-

---

[3]Under 33 C.F.R. §151, garbage logs must show when and where garbage is incinerated or discharged (overboard, to another ship, or to a port reception facility); the date, time, location, and volume of the discharge; and the specific contents of garbage discharged overboard. The final rule was published in 59 Fed. Reg. 18,700 (1994).

[4]The regulations also affect the limited number of U.S.-flag cargo ships, all manned offshore platforms, some research vessels, and the few U.S.-flag cruise ships. Public vessels and foreign-flag vessels are not required to comply, although proposed amendments to the Marine Plastics Pollution Research and Control Act would allow the regulations to be extended to any ship of a size and use specified by the Secretary of Transportation.

tion in the fisheries sector. In general, it is important to take into account regional differences, to use whatever intelligence is available, and to capitalize on the existing government control structure established by NMFS oversight of fishing activities and Coast Guard regulation of vessels and operators.

Technological interventions need to be tailored to the conditions on fisheries vessels. Trash compactors, for example, need to be the right size. Special storage procedures may be needed depending on the size and condition of waste materials. Measures also could be taken to reduce gear losses, as encouraged by the International Maritime Organization (IMO) guidelines for Annex V implementation.

A key organizational intervention, suggested by the preceding discussion of control, would be to modify criteria of restricted fishing seasons to enable retrieval of gear left in the water. Another promising approach would be for fishing cooperatives and other organizations to obtain advice and support from federal and state agencies to help establish port reception facilities tailored to local needs. In addition, fishermen could be encouraged to return to shore any debris recovered in nets or other gear.

Education to encourage voluntary compliance with Annex V must continue to consolidate some of the early success in this community. Annex V information could be disseminated through existing channels, such as fishing license renewal and boat registration processes as well as the Sea Grant Marine Advisory Service. Sea Grant agents might be able to provide the necessary technical assistance as well. Another promising educational strategy would be to distribute data on lost gear and its possible effects on the marine environment, including fish stocks. In addition, fisheries management councils could be educated in how to encourage Annex V compliance in their planning.

Annex V enforcement, including vigorous prosecution of violators and imposition of significant penalties, is important in this sector. Debris from fishing activities—net fragments, monofilament lines, broken traps, and other gear—is associated consistently with injuries to wildlife and damage to vessels. In some regions of the United States, debris originating from fishing vessels dominates the garbage washing ashore; where this occurs, securing compliance from local fishing fleets could yield significant environmental benefits. If the objectives of Annex V cannot be met through voluntary compliance (and the work of Sutinen et al. [1990] points out how the fisheries regime struggles to achieve compliance), then federal authorities should focus their limited enforcement resources on the most effective strategies. Options include expanding the duties of NMFS on-board observers to include monitoring for Annex V violations, and requiring the reporting of gear losses (not covered by the Coast Guard record-keeping regulations). In addition, international agreements could encourage or require Annex V compliance by participating nations; this approach might be valuable, for example, in fostering compliance by the Mexican shrimp industry, which is blamed in part for debris in the Gulf of Mexico (Boudreaux, 1993).

TABLE 4-2   Applying the Hazard Evolution and Intervention Model to Commercial Fisheries and Their Fleet Ports

Hazard Evolution Model	Behavior that Encourages Generating Garbage	On-board Generation of Garbage
Intervention Model	Modify Behavior that Encourages Generating Garbage	Reduce Garbage Generation during Voyage
Technological		Reduce use of discardable material.
Organizational and Operational	Repair nets ashore. Modify criteria of restricted fishing season to enable retrieval of gear left in water.	Examine materials now in use to identify where use of substitute materials can reduce waste generation. Sort garbage at site of generation. Use only vendors committed to packaging and storage techniques that minimize waste.
Educational (Target Population/ Content)	Educate vessel operators about alternate processing methods that generate less waste than conventional approaches. Communicate that cleaner water may increase value of fish and minimize damage to vessel and gear.	Examine methods now in use to identify where alternative methods would generate less waste.

Breakdown in Compliance	Discharge of Garbage into Sea	Exposure to Discharged Garbage
Prevent Breakdown in Compliance	Block Discharge of Garbage into Sea	Block Exposure to Discharged Garbage
Build garbage storage space and processing equipment into new vessels and retrofit where feasible. Keep shipboard systems well maintained.	Develop and install appropriate garbage handling equipment. Try to prevent storms and vessels from dislocating set fishing gear.	Use products made of biodegradable materials (except plastic). Tag gear with pingers or other devices to help relocate it.
Provide reminders via posters and placards on vessels. Audit practices regularly. Keep records on gear losses and disposal.	Establish port reception facilities tailored to local needs. Establish an incentive for manufacturers to buy back nets.	Encourage crews and captains to bring to shore any debris recovered in gear.
Circulate data on lost or discarded gear and effects on wildlife. Distribute Annex V information via Sea Grant agents and fishing license and boat registration processes.	Train crews to hold garbage (including items often discharged) for shoreside recycling. Educate fisheries management councils to incorporate Annex V compliance into fisheries management planning.	

TABLE 4-2   Continued

Hazard Evolution Model	Behavior that Encourages Generating Garbage	On-board Generation of Garbage
Intervention Model	Modify Behavior that Encourages Generating Garbage	Reduce Garbage Generation during Voyage
Government or Private Regulation and Enforcement		Prohibit use of certain plastic materials in the manufacture of gear. Prohibit fishing methods that promote setting of excess gear or wasteful discards.
Economic (Market Forces)	Develop equivalent products using alternative materials. Establish an incentive for manufacturers to buy back nets. Improve remanufacturing of old nets.	Determine overall costs (throughout product life cycle) of using discardable materials. Create market demand for recycled nets (intact and fragments) and materials.

Financial incentives may be particularly useful. Canadian interviews[5] reported by Buxton (1989) suggest that economic incentives will drive compliance in some circumstances. Buxton reports that " . . . it makes business sense to change present disposal practices. This may relate to quality issues, real or perceived, or avoiding losing fish." Interviewees expressed concerns about the cost of garbage handling equipment and even greater anxiety about the high fines for illegal discharges (Buxton, 1989). Interventions to encourage the return of used

---

[5]Canada is not a signatory to Annex V but has strict domestic regulations that parallel the mandates of Annex V.

Breakdown in Compliance	Discharge of Garbage into Sea	Exposure to Discharged Garbage
Prevent Breakdown in Compliance	Block Discharge of Garbage into Sea	Block Exposure to Discharged Garbage
Enforce and publicize enforcement of ban on plastics discharges. Make penalties larger than the gains achieved through violations. Require reporting of gear losses.	Require vessel operators to educate crews about discharge restrictions and compliance strategies. Incorporate Annex V compliance provisions into fisheries management criteria and international agreements. Train on-board fisheries observers to recognize Annex V violations and instruct fishermen in compliance.	Create incentives and remove disincentives for returning to shore any debris recovered in nets or other gear.
Require deposits on nets and lines to encourage return of gear after use or recovery. Offer small ports financial aid to provide reception facilities.	Retrieve fishing lines. Require deposits on lines or offer rebates for returned (used) lines. Encourage waste exchanges.	

fishing gear to shore include requiring deposits on nets and lines and promoting recycling of fishing gear.

Cargo Ships and Their Itinerary Ports

Intelligence

Although cargo vessels are boarded routinely by inspectors from the Coast Guard and APHIS, few overall data are available concerning garbage handling practices in this sector. Record keeping can be expected to improve for U.S.-flag cargo ships as a result of the Coast Guard requirements for garbage logs.

Garbage transactions seldom are recorded by ports. Waste haulers may record the weight or volume of discards, but this information is not logged in a consistent manner and it may reflect the total volume in the hauler's container rather than a specific ship's garbage. Data on quarantined waste are sometimes available, but they reflect numbers of containers rather than weight of the garbage, and APHIS inspectors do not monitor off-loading of all of this waste. Furthermore, there is no way to determine what proportion of the waste stream APHIS garbage constitutes. The COA regime does not require any tally of garbage.

Control

The potential for federal control over this sector is significant, in that the Coast Guard determines the professional qualifications of U.S. mariners, monitors vessel construction and operation, and has authority to board or inspect all vessels in U.S. waters to assure safety and environmental protection. The Coast Guard can discipline mariners either through fines or by suspending or removing their licenses. In practice, however, control is inconsistent. Some fleets are managed centrally and effectively by their operators and flag states, while others are under little or no control.

Most of the cargo ships entering U.S. ports fly a foreign flag, a factor that has hindered enforcement of Annex V. U.S. port state authorities under international law provide a basis for enforcing Annex V but are not unlimited (see Appendix C for a discussion of these authorities). Recently, however, the Coast Guard decided to change the way it exercises port state enforcement authorities so that direct U.S. action would be taken against increased numbers of foreign-flag vessels that violate MARPOL, and fewer cases would be referred to flag states (see Chapter 7). Proposed amendments to the Marine Plastics Pollution Research and Control Act (MPPRCA) would provide for even greater federal control in this sector by allowing the requirement for garbage logs to be extended to foreign-flag vessels that make U.S. port calls.

Another potential source of control in this sector would be an international requirement that flag states issue certificates confirming that a ship's waste management system meets or exceeds some minimum criteria. This approach, which would have to be instituted through IMO, would be analogous to the International Oil Pollution Prevention Certificate issued to confirm that a ship has been surveyed and that its structure, equipment, systems, fittings, arrangements, and materials comply with requirements of MARPOL Annex I. (Annex I requires such surveys.) A waste management certificate could be issued or renewed based on an audit of a ship by either the flag state or a classification society.[6] Failure to

---

[6]Classification societies establish standards, guidelines, and rules for the design, construction, and survey of ships.

obtain such a certificate would be considered a violation of Annex V. (Apart from providing some international control over shipboard garbage management, the certificate approach also could be a mechanism for confirming whether a ship has a comprehensive capability to manage all its Annex V garbage and APHIS wastes on board. Such a capability could exempt a ship from any requirement to off-load garbage at U.S. ports [an option discussed in Chapter 5].)

The Animal and Plant Health Inspection Service exerts fairly tight control over cargo ships. Inspectors board many arriving vessels; for violations of APHIS disposal regulations, penalties may be assessed and fines must be paid within 72 hours. A "blacklist" is maintained of vessels with recent violations, and these vessels are monitored closely. During boardings, in addition to checking for compliance with quarantine regulations, APHIS inspectors also ask four questions concerning Annex V.[7] However, any Annex V violations discovered must be referred to the Coast Guard, and a decision may not be rendered for months.

Shoreside garbage disposal can be a problem for cargo ship operators, because disposal costs often are perceived as too high (see Chapter 5) and port reception facilities may be inconveniently located or their use may be denied. The Coast Guard exerts some control over U.S. public ports and operators of large private terminals through the COA program, but cost and convenience levels are not regulated. Some cargo ships, such as bulk carriers and chemical tankers, never call at a public port; instead, they go directly to the private waterfront terminals of the cargo owner. Some private terminals have been reported to turn cargo ships away when they attempt to off-load garbage, while other facilities, notably refineries, are so remote that it is difficult to arrange for services, such as off-loading of food-contaminated plastics and other garbage that must be quarantined.

Analysis of Interventions

Table 4-3 outlines possible interventions to improve Annex V implementation in the cargo ship sector. Due to the international profile of this sector, the most useful options are those that can improve compliance by foreign-flag as well as U.S.-flag ships.

Technological innovations can be adopted by any ship operator. But it is clear that experts outside the merchant marine—designers, vendors, engineers—

---

[7]The four questions, all requiring "yes" or "no" responses, are included as items 23–26 on APHIS Form 288, Ship Inspection Report. They are: (23) Plastic materials requiring disposal are used aboard the vessel. (24) There are waste plastics in the vessel's trash for disposal ashore. (25) There is a functional incinerator or other disposal method aboard. (26)(a) Responsible vessel operator was requested to show garbage pickup receipt or other evidence of lawful disposal of plastics ashore. (b) Responsible vessel operator produced garbage pickup receipt or other evidence of lawful disposal of plastics ashore.

TABLE 4-3   Applying the Hazard Evolution and Intervention Model to Cargo Ships and Their Itinerary Ports

Hazard Evolution Model	Behavior that Encourages Generating Garbage	On-board Generation of Garbage
Intervention Model	Modify Behavior that Encourages Generating Garbage	Reduce Garbage Generation during Voyage
Technological		Reduce use of discardable packaging. Design packaging techniques and storage systems that minimize need for plastic wrappings and bindings as well as packing materials.
Organizational and Operational	Assure that organizational culture encourages commitment to proper garbage management at all levels, using TQM methods and expediting implementation of ISM.	Use only vendors committed to packaging and storage techniques that minimize waste. Sort garbage at the site of generation.
Educational (Target Population/ Content)	Modify crews' comfort expectations and attitudes about waste management. Encourage acceptance of need to avoid individually packaged items. Train shoreside personnel, vessel operators, and crews in TQM/ISM principles. Train regulatory authorities at federal, state, and port levels in TQM principles to break down barriers and achieve regulatory synergy.	Inform management about packaging alternatives. Encourage vendors to develop alternate packaging. Encourage packaging manufacturers to develop affordable, reusable containers.

Breakdown in Compliance	Discharge of Garbage into Sea	Exposure to Discharged Garbage
Prevent Breakdown in Compliance	Block Discharge of Garbage into Sea	Block Exposure to Discharged Garbage
Design garbage storage space into ships. Keep shipboard systems well maintained. Establish system for garbage pickup at ports that meets Annex V and APHIS requirements.	Develop and install appropriate garbage handling equipment, such as efficient, safe incinerators and reliable shredders and compactors.	Promote affordable compactors that create non-buoyant waste slugs (with no plastics).
Establish internal company penalties for noncompliance. Encourage commitment to garbage management at the level of the individual.	Establish clear policies and procedures for a comprehensive garbage management system. Standardize port disposal services.	
Provide constant reminders via posters and placards aboard ships. Educate vessel operators and crews about the types of garbage subject to Annex V versus APHIS regulations. Require crew education for entry into U.S. waters or ports.	Inform crews of compliance requirements and methods and the harm caused by improper discharges. Inform managers of compliance methods, both organizational and technological. Inform regulators about ways to improve integration of Annex V and quarantine regimes. Develop recycling programs for items (cans) often discarded overboard.	

TABLE 4-3   Continued

Hazard Evolution Model	Behavior that Encourages Generating Garbage	On-board Generation of Garbage
Intervention Model	Modify Behavior that Encourages Generating Garbage	Reduce Garbage Generation during Voyage
Government or Private Regulation and Enforcement	Change regulatory balance to emphasize cooperation rather than control. Clarify Annex V roles and relationships of federal, state, and port agencies. Coordinate efforts at ship/port interface.	Restrict use of certain materials on ships.
Economic (Market Forces)	Include environmental impacts in cost-benefit analyses of garbage management systems (typically rated on profitability and efficiency). Establish cost benefits for all possible solutions (i.e., conduct impact analysis); identify optimal solution from cost-benefit standpoint.	Develop reusable packaging that is more cost effective than traditional materials or has a life-cycle cost benefit.

are essential to technological advancement in this sector. For example, alternative packaging and storage systems need to be developed that minimize use of plastics. Appropriate garbage treatment equipment needs to be designed into new ships and, where necessary and feasible, purchased or developed and retrofitted on existing ships.

Because this is an industrial community, organizational interventions are important. Garbage management strategies must be integrated into standard and emerging industrial practices, such as Total Quality Management (TQM) and the International Safety Management (ISM) Code adopted recently by the Interna-

Breakdown in Compliance	Discharge of Garbage into Sea	Exposure to Discharged Garbage

Prevent Breakdown in Compliance	Block Discharge of Garbage into Sea	Block Exposure to Discharged Garbage
Audit practices to ensure full compliance. Tighten inspection of port reception facilities. Require flag states to issue waste management certificates.	Require logs of waste handling transactions. Tighten port state controls and inspections. Require off-loading of Annex V (and APHIS) garbage at port calls.	
Bounty provision in U.S. law may encourage peer surveillance and discourage violators. Return monies from recycling to vessel crew for their discretionary use.	Make on-board waste treatment equipment and use of port reception facilities affordable. Incorporate disposal costs into port user fees/tariffs. Spread cost across entire port user base.	

tional Maritime Organization (1993). The ISM lays the foundation for a new organizational and cultural framework for ship management, requiring that policies and actions be consistent within an organization and focusing attention on human factors. Shipping company operators can establish an organizational culture that supports proper garbage management by using only vendors that minimize waste, establishing clear and effective policies and procedures, and imposing internal penalties for infractions of the rules.

Educational interventions must target not only vessel crews but also shipping company managers, vessel agents and brokers, suppliers, and government regulators. Managers need to be informed about effective compliance strategies. Vessel agents and brokers are responsible for knowing local conditions and are the primary conduits of information about services between ship and shore. Suppliers can be encouraged to develop alternative packaging that minimizes waste. Vessel operators and crews need to be educated about the types of garbage subject to Annex V and APHIS regulations, and government officials need to be encouraged to improve integration of these two regimes (an issue addressed in Chapter 5).

Regulatory interventions include requiring foreign-flag ships that call at U.S. ports to keep logs of garbage transactions (now mandated for U.S.-flag ships only). To reach foreign-flag cargo ships, the Coast Guard prepared a small, well-illustrated book aimed at providing ships' agents and other shoreside personnel with the information needed to help arrange garbage reception facilities and services for arriving ships, especially those discharging quarantined garbage (Kearney/Centaur, 1994). Another alternative would be to tighten control over port reception facilities. For example, the Coast Guard has proposed MPPRCA amendments that would require inspections of port reception facilities (including those not covered by the COA program) under certain conditions. A possible intervention at the international level would be to require that flag states audit shipboard garbage management systems and issue certificates confirming that they meet or exceed minimum standards.

Finally, because cost is a driving force in cargo operations, it is important to offer economic incentives to vessel owners and operators. Promising options include returning monies from recycling programs to vessel crews and revamping the highly inconsistent fee structures for garbage disposal (the latter issue is addressed in Chapter 5).

Passenger Day Boats, Ferries, and Their Terminals

Intelligence

Few data are available on garbage disposal by passenger day boats and ferries, but direct observation is relatively simple as the trips are short and predictable (and some casino ships don't move at all). In any case, there is minimal concern about Annex V implementation and enforcement in this sector, because voyages tend to be brief, and port calls are frequent and usually at dedicated facilities controlled by the vessel operator (Eric Scharf, National Association of Passenger Vessel Owners, personal communication to Marine Board staff, July 11, 1991).

Control

Control of passenger day boats and ferries is fairly stringent, in that all vessels are U.S. flag, and most terminals are owned by the vessel operators or are under long-term contracts to vessels. The Coast Guard regulates the construction and operation of these vessels (46 C.F.R. Subchapter T). The operators of these vessels must have Coast Guard documentation. In general, garbage management is not a problem in this sector because vessels operate regularly out of the same terminals and have standard waterfront garbage-hauling contracts. However, vessel operators report that disposing of garbage in shoreside facilities is becoming more expensive.

Analysis of Interventions

Table 4-4 lists possible interventions to improve Annex V implementation in the day boat sector. Although it appears that minimal assistance is needed with Annex V compliance, this is a significant maritime sector contributing to coastal traffic, the fastest growing segment of maritime transportation today. There are probably ways of improving Annex V implementation, notably through waste minimization, passenger education, crew training, and improvements in shoreside disposal systems.

Operational interventions might include offering passengers drinks in paper cups from large dispensers[8] rather than individual cans. Because passengers come and go quickly and may remain in a limited area, ample Annex V information must be provided through posters, placards, and public address announcements, throughout both vessels and terminals. Regulatory interventions include auditing of shipboard practices and requiring Annex V compliance on ferries with international routes as a condition of joint agreements with the other nation involved (e.g., Canada).

Because an individual vessel typically uses the same piers repeatedly, it should not be difficult to integrate the garbage disposal needs of vessels into waste management planning for ports. Simple improvements could yield a high level of compliance.

Small Public Vessels

Intelligence

At one time, operators of Coast Guard and small naval auxiliary vessels expected to base their Annex V compliance strategies on the Navy's compliance

---

[8]It is important that such dispensers not leak or attract insects (Emshwiller and McCarthy, 1993).

TABLE 4-4   Applying the Hazard Evolution and Intervention Model to
Passenger Day Boats, Ferries, and Waterfront Facilities

Hazard Evolution Model	Human Behavior that Encourages Generating Garbage	On-board Generation of Garbage
Interventions	Modify Behavior that Encourages Generating Garbage	Reduce Garbage Generation during Voyage
Technological		Use alternate packaging materials.
Organizational and Operational	Employ Total Quality Management principles. Provide high standard of service with new materials.	Use only vendors committed to packaging and storage techniques that minimize waste.
Educational (Target Population Content)	Encourage passengers to respect clean oceans and support tenets of Annex V. Train crews to provide same service with new materials.	
Government and Private Regulation and Enforcement	Control activities of vessel operators.	Prohibit use of certain materials.
Economic (Market Forces)	Make vessel operators aware that clean water may encourage increased business.	

Breakdown in Compliance	Discharge of Garbage (already prohibited by national laws)	Exposure to Discharged Garbage (not applicable)
Prevent Breakdown in Compliance	Block Discharge of Garbage	Block Exposure to Discharged Garbage
Provide ample on-board storage capacity.	Make room to store garbage in places other than weather deck.	
Provide Annex V posters, placards, and public address announcements and many trash cans on board vessels and in terminals. Audit shipboard practices.	Establish garbage sorting system. Establish integrated garbage management (coordinated with shoreside recycling programs).	
Train crews in Annex V compliance procedures. Foster peer enforcement among passengers. Educate vessel operators through literature directed at this sector.		
Audit vessel operations to assure compliance. Require compliance on international routes as condition of joint agreements.	Ensure adequacy of port reception facilities.	Grind garbage before discharge.
Encourage peer enforcement through bounty provisions of U.S. law.	Assure that port reception facilities are cost effective.	Increase fees for receiving unsorted wastes. Pay premium for recyclables returned to port.

program for warships. But it has become clear that the Navy's management strategies and technologies are ill-suited to the distinct operational needs and close quarters of small vessels.[9] The Navy's strategic plan (U.S. Navy, 1993) concedes this point. However, the Navy does not consider garbage management a major challenge for most auxiliary vessels because trips tend to be short; the problems arise on a small number of vessels, such as minesweepers, that remain at sea for longer periods and may not be able to use compactors due to their magnetic effects (Larry Koss, U.S. Navy, personal communication to Marine Board staff, August 12, 1994). An effective strategy, apart from addressing the unique problems of small vessels, would have to assure zero-discharge capability to permit operations in special areas.

The Coast Guard has a strong tradition of decentralized management of vessel operations, so information about garbage disposal practices is difficult to gather. However, it is clear that attempts to comply with Annex V have created unpleasant conditions for crews. Icebreakers on patrol, for example, can become clogged with plastic debris in every available space. In one instance observed by the committee, a cutter crew retained all plastic garbage, both clean and food-contaminated, and hung it in a large net on the weather deck. While malodorous and unpleasant, this solution was tolerated to ensure compliance with Annex V.

Realizing that centralized technical support and decision making were needed to alleviate these problems, Coast Guard senior management has developed plans to retrofit on-board garbage treatment equipment (Bunch, 1994). The plans call for polar icebreakers to be fitted with systems consisting of an incinerator, a trash compactor, and a pulper. On cutters with more than 50 crew members and endurance[10] of five days or more, commercial-grade compactors and possibly incinerators and small pulpers will be installed. Small cutters will be equipped with household compactors for treating plastics and other garbage. Numerous compactors and a prototype incinerator have been installed; the key fleetwide issue to be resolved is whether vessels have sufficient space to accommodate the requisite equipment (Sara Ju, U.S. Coast Guard, personal communication to Marine Board staff, August 18, 1994).

Control

Although all public vessels are under direct federal command and control, the effectiveness of garbage disposal procedures is limited in practice by the nature of service management structures and the slim margin for operational

---

[9]Even application of the Navy's 3-day/20-day rule for plastic wastes severely degrades living conditions; for example, a troop transport vessel can be so loaded with personnel and supplies that individuals must squeeze past each other under ordinary circumstances, so there is literally no room for the garbage generated during even a single day.

[10]Endurance refers to the length of time a vessel may remain at sea without returning to port.

changes on some military vessels. The Navy has yet to develop a complete compliance solution for auxiliary vessels, and it is not yet certain that the Coast Guard's plans can and will be implemented.

Analysis of Interventions

Table 4-5 lists possible interventions to improve Annex V implementation in this sector. In addition to retrofitting garbage treatment equipment on the Coast Guard fleet as planned, technological options include adapting these strategies for use on other small public vessels. Continued development of alternative packaging strategies and biodegradable materials is likely to be useful as well. For example, vessel operators might be able to use only paper packaging and then install pulpers to dispose of this waste.

Among organizational strategies, each service would do well to foster fleet-wide support for ending temporary coping mechanisms in favor of permanent compliance strategies. Other alternatives include development of recycling programs for items, such as cans, now thrown overboard (where permitted). There may be less need for new educational programs in this sector than in some others, because Coast Guard and Navy personnel are well aware of and willing to comply with Annex V. Still, there is room for improvement, such as with standardized training in compliance strategies.

Possible regulatory interventions include a ban on use of certain disposable items, and extending to public vessels the requirement for garbage logs (now applied to U.S.-flag commercial vessels). The latter option might not accomplish much in terms of raising compliance levels, considering that uniformed personnel generally want to comply but face technical obstacles.

Offshore Platforms, Rigs, Supply Vessels, and Base Terminals

Intelligence

The government's capacity for gathering information about the offshore oil and gas industry is significant, although the system is not geared to Annex V. The MMS collects data on outer continental shelf activities, but little of it relates to garbage. Platforms are inspected at least once a year for compliance with operating rules. In reviewing possible sources of pollution, inspectors focus on oil leaks rather than garbage but may check for compliance with equipment handling regulations designed to minimize overboard losses. Platform operators are required to mark equipment, tools, and containers weighing over 40 pounds for purposes of identification and to report equipment losses to MMS as well as record them in daily operations reports.

Some information on disposal practices has been obtained from beach surveys. At Padre Island National Seashore in Texas, for example, National Park

TABLE 4-5   Applying the Hazard Evolution and Intervention Model to Small
Public Vessels and Their Home Ports

Hazard Evolution Model	Human Behavior that Encourages Generating Garbage	On-board Generation of Garbage
Intervention Model	Modify Behavior that Encourages Generating Garbage	Reduce Garbage Generation
Technological	Reduce or eliminate convenience packaging of supplies and foods.	Provide alternate packaging when feasible (given packaging standards for electronic equipment).
Organizational and Operational	Centralize or oversee provisioning to foster widespread innovation.	Use only vendors committed to minimizing waste. Sort garbage at site of generation. Hold garbage on board for shoreside recycling. Coordinate or review provisioning to extend innovation through fleets.
Educational (Target Population/ Content)	Provide standard training in compliance methods for officers and crews.	
Government and Private Regulation and Enforcement	Modify fleet supply contracts for provisions that trigger garbage generation.	Prohibit use of certain disposable items (e.g., plastics). Impose mandatory sorting and holding of garbage for shoreside recycling.

Breakdown in Compliance	Discharge of Garbage into Sea	Exposure to Discharged Garbage
Prevent Breakdown in Compliance	Block Discharge of Garbage into Sea	Block Exposure to Discharged Garbage
Promote development of improved on-board garbage management equipment for small vessels. Keep shipboard systems well maintained.	Install on-board garbage treatment equipment on Coast Guard vessels. Develop appropriate units for other small military vessels. Make room to store garbage in places other than weather deck.	Promote compacting of legal discards to minimize garbage in water column, avoid blanketing ocean bottom, and minimize harm to wildlife. Use biodegradable materials (except plastics).
Provide reminders for crew with posters and placards.	Foster fleet support for permanent compliance procedures and equipment. Establish on-board recycling programs for items (such as cans) that otherwise would be discharged overboard legally.	
Encourage peer enforcement of internal guidelines.	Require internal records of legal discharges at sea. Keep receipts issued by port reception facilities. Establish and enforce internal penalties (fleet policies).	Develop in-house guidelines and directives.

TABLE 4-5   Continued

Hazard Evolution Model	Human Behavior that Encourages Generating Garbage	On-board Generation of Garbage
Intervention Model	Modify Behavior that Encourages Generating Garbage	Reduce Garbage Generation
Economic (Market Forces)	Provide budgets for shipboard compliance to avoid conflicts with operating, maintenance, and repair budgets.	Demonstrate any cost benefits from switch to reusable items.

Service employees report constant washups of items ranging from 55-gallon drums to small plastic bottles containing waste oil, acids, and a variety of other hazardous chemicals. A related problem is the significant expense associated with removing containers that have washed ashore and are suspected of containing hazardous substances. The equipment identification system helps in identifying owners, who are expected to cover removal costs ($1,700 per 55-gallon drum in 1993).

Little information is available on garbage reception facilities at supply boat terminals. Because most offshore service vessels weigh less than 400 gross tons, the base terminals are not required to obtain COAs, and the Coast Guard has no other reason to visit the terminals or the vessels that call there (Green, 1993). Amendments to the MPPRCA have been proposed that would require inspection of non-COA garbage holding facilities.

Control

The federal government wields considerable power over this sector through an array of laws and regulations. All vessels are U.S. flag, and platforms in federal waters operate under direct permit from the MMS, which regulates equipment handling and overboard discharges under the Outer Continental Shelf Lands Act (P.L. 83-212), as amended. In addition, permits issued by the Environmental

Breakdown in Compliance	Discharge of Garbage into Sea	Exposure to Discharged Garbage
Prevent Breakdown in Compliance	Block Discharge of Garbage into Sea	Block Exposure to Discharged Garbage
Return monies from recycling to vessel crew for their discretionary use. Make shoreside disposal readily available.	Organize vessel support services to make compliance affordable. Review waste hauling schedules and contracts. Expand use of on-board equipment to reduce need for disposal at commercial ports.	

Protection Agency (EPA) under the Clean Water Act (P.L. 92-500), as amended, prohibit the discharge from platforms of floating solids and rubbish, trash, and other refuse. The transfer of garbage from platforms to supply vessels is regulated under both Annex V and the Clean Water Act.

Additional opportunities for government control are emerging in this sector. Supply vessels transporting and transferring platform garbage to port reception facilities are subject to the Shore Protection Act (SPA) of 1988[11], while the vessel's operational waste is covered by Annex V. Owners and operators of supply vessels must obtain SPA permits as commercial haulers of waste from the Coast Guard, which has been issuing conditional permits under an interim final rule (see 33 C.F.R. § 151) since 1989 and plans to finalize this rule. In the meantime, the EPA is drafting regulations to provide guidance for waste transfer and handling; supply vessels will have to comply with these requirements when finalized.

Another avenue for control may be record keeping. In addition to reporting to MMS items lost overboard, platform operators are required by the Coast Guard

---

[11]The SPA is Title IV of the Ocean Dumping Ban Act (P.L. 100-688), which prohibits the discharge of industrial waste and sewage sludge into the sea. This law is distinct from the Ocean Dumping Act (P.L. 95-535), which prohibits the transportation of any material for the purpose of dumping it into the ocean.

to maintain logs of garbage transactions. These logs can be reviewed during the Coast Guard's oversight inspections of 10 percent of offshore platforms annually.

Efforts to encourage Annex V compliance by the offshore industry are under way. Even before U.S. ratification of Annex V, MMS issued guidelines for reducing marine debris, recommending that offshore operators conduct worker training and awareness sessions, adopt waste reduction strategies, and implement control systems to account for the proper disposal of garbage, especially drums and hazardous items (Minerals Management Service, 1986). Federal officials also have spearheaded a number of other pollution prevention programs in the Gulf of Mexico, such as the Take Pride Gulf Wide campaign.

In addition, an industry organization, the Offshore Operators Committee (OOC), has established an Ad Hoc Task Group on Waste Handling and Recycling. As a result of OOC efforts, half of the platform operators have banned the use of foamed plastic offshore, to reduce the chances of this material being discharged into the gulf (Anderson, 1992).

Analysis of Interventions

Table 4-6 suggests possible interventions to improve Annex V implementation in the offshore oil and gas industry. Because this industry is so prevalent in the Gulf of Mexico and must anticipate operating in a special area, zero-discharge capability must be achieved or maintained.

Technological interventions may not be critical in this sector, because garbage is transported to shore regularly (thereby eliminating the need for treatment on platforms). Organizational interventions may be more useful. In establishing a corporate culture supportive of proper garbage management, platform operators could use only vendors committed to waste reduction and take measures to secure the plastic sheeting used to protect materials in transit. The OOC task group may develop other useful strategies, such as recycling programs.

Educational approaches include informing company managers about garbage handling mandates and strategies. These efforts could capitalize on the voluntary work of the OOC; the partial ban on use of foamed plastic could be held up as an example of how to eliminate or minimize waste—either voluntarily or by mandate.

Regulatory interventions include MMS examination of garbage and equipment handling practices during routine inspections and oversight, and Coast Guard review of garbage logs during occasional platform inspections. Increased surveillance of this sector seems justified, especially in view of concerns that the current level of environmental protection may decline as increasing numbers of independent operators enter the industry. On the other hand, the size of the offshore industry is shrinking.

Economic interventions include making offshore operators aware of the benefits of maintaining a positive public image through compliance with environmental regulations. Changes in provisioning practices, such as switching from small to large ketchup bottles, also can have economic benefits.

Navy Surface Combatant Vessels and Their Home Ports

Intelligence

The U.S. Navy collects a considerable amount and range of internal information, including data on ship garbage generation and management and the activities of the home ports that provide ships with supplies and services, including garbage disposal. In addition, the Navy has examined in depth its supply chain and the shipboard equipment options for treating garbage. The Navy's Annex V compliance plans also have been reviewed by the U.S. General Accounting Office (1994a, 1994b), which has criticized the Navy's planning and the large sums of money spent on technology projects that have not been deployed to the fleet. Thus, the level of detail available concerning the Navy's garbage generation and disposal practices exceeds that obtained for other maritime sectors. At the same time, it can be difficult to make generalizations about the Navy, because compliance strategies are not necessarily the same for every ship. For example, recycling practices vary by operating unit (U.S. General Accounting Office, 1994b).

Little is known about garbage management practices of the commercial or foreign ports sometimes used by naval vessels, or the commercial waste haulers that some home ports may be forced by local laws to employ. The Navy does not report inadequate port reception facilities using the IMO forms because public vessels are exempt from MARPOL requirements.

Control

The Navy's surface fleet is subject to direct federal control through both internal management practices and external congressional review. The Navy has an established command and control structure that has served as an effective mechanism for organizing fleetwide compliance with the MPPRCA. A range of interventions has been employed. Operational measures include the 3-day/20-day rule (described in Chapter 1) for holding plastics on board. The Navy supports its implementation efforts with a vigorous education program for ship and shoreside personnel (Koss et al., 1990; Koss, 1994) and an internal system of rewards and sanctions. (Violators have been punished [Ocean Science News, 1991].) Economic incentives include returning monies from recycling to ship crews.

Technical interventions are the key to full compliance in this sector. Federal control of progress in implementation currently is limited for very large ships, such as aircraft carriers, by both ship design and the need to undertake extended missions. Garbage cannot be treated adequately on these ships at present because

TABLE 4-6   Applying the Hazard Evolution and Intervention Model to Offshore Oil and Gas Industry Platforms, Rigs, Vessels, and Base Terminals

Hazard Evolution Model	Human Behavior that Encourages Generating Garbage	On-board Generation of Garbage
Intervention Model	Modify Behavior that Encourages Generating Garbage	Reduce Garbage Generation
Technological		Characterize garbage and conduct needs assessment.
Organizational and Operational		Voluntarily prohibit use of certain materials or items, such as foamed plastic and packing pellets. Sort garbage at site of generation. Use only vendors committed to packaging and storage techniques that minimize waste.
Educational (Target Population/ Content)	Establish housekeeping procedures for use during trips to and from shore. Establish garbage sorting systems at worksites.	
Government and Private Regulation and Enforcement		OOC is developing best practices guidelines for voluntary use by operators.
Economic (Market Forces)	Foster operators' awareness of economic benefit of good public image.	Revise provisioning practices.

Breakdown in Compliance	Discharge of Garbage into Sea (already prohibited by national law)	Exposure to Discharged Garbage
Prevent Breakdown in Compliance	Block Discharge of Garbage into Sea	Block Exposure to Discharged Garbage

Keep all equipment well maintained.		Install comminuters to reduce size of food particles discharged.
Assure that corporate culture discourages overboard disposal. Move all materials in sealed and covered containers to reduce chance of loss overboard. Keep records of garbage transactions.	Improve handling of large plastic sheeting (used to protect materials in transit) to reduce loss overboard. Develop strategies through Offshore Operators Committee (OOC) committee on waste management and OOC/API waste management practices project. Keep garbage confined during transit back to base terminals.	Retrieve large plastic sheeting found floating at sea.
Educate company managers and vessel operators about Annex V mandates and compliance strategies. Provide posters, placards, and worker training.		
Examine garbage handling practices and logs during routine inspections.	MMS regulations prohibit release of wastes into water.	

the equipment developed by Navy researchers has yet to be installed.[12] Part of the problem should be solved within several years. To comply with the 1988 MPPRCA deadline for halting overboard discharge of plastics, the Navy recently has focused its technical program on development of a shredder-heater-compactor system. The preproduction prototype was installed on an aircraft carrier in May 1994 and fleetwide installation is to be completed by late 1998.

But disposal of garbage other than plastics remains a problem. The Navy must operate in Annex V special areas, where no garbage except food waste may be discharged overboard. At present, this requirement is not causing major difficulties, because the Navy conducts few activities in the three special areas now in force (the Baltic and North seas and the Antarctic Ocean). However, the Navy must prepare for the entry into force of special area requirements in the Mediterranean, Gulf of Mexico, and elsewhere, where its operations are extensive. The Navy sought legislation that would have allowed its vessels to discharge pulped or shredded nonfood garbage in special areas, but the Congress did not authorize this change. As a result, the Navy has suspended plans to purchase and install pulpers and shredders[13] (the shredder technology now is used in the plastics processor).

While acknowledging that use of pulpers and shredders would be beneficial outside special areas (e.g., it would diminish evidence signaling vessel whereabouts to potential enemies and eliminate "aesthetically objectionable discharge of intact trash"), the Navy has determined it is not worth spending several hundred million dollars to retrofit ships with equipment that would not enable compliance with special area requirements (U.S. Navy, 1994). Instead, to meet these requirements, the Navy plans to solicit proposals from industry for technologies suitable for shipboard use. The Navy also is experimenting with several advanced garbage treatment technologies not included in its formal plan.[14]

---

[12]The Navy has been criticized for spending some 14 years and $52 million to research, develop, and produce on-board garbage treatment equipment without producing a plan for full compliance (Associated Press, 1994; U.S. General Accounting Office, 1994a, 1994b). Initially, the Navy planned to develop a vertical trash compactor, a solid waste pulper, and a plastics waste processor; in 1993, the compactor was abandoned in favor of the metal/glass shredder, since adapted to shred only plastics (U.S. General Accounting Office, 1994a, 1994b).

[13]Even if these technologies were installed, garbage management would remain a time-consuming duty on large ships. The Navy's pulper would have to operate approximately 19 hours per day to process the food waste, paper, and cardboard generated by an aircraft carrier with a crew of 5,600 (Swanson et al., 1994). The shredder would have to run for 4.8 hours per day to process all the glass and metal garbage, and the entire operation (including sorting, feeding, processing, and bagging) would take up to 11.5 hours (Swanson et al., 1994). The shredder might require repair about every two months (Swanson et al., 1994).

[14]These technologies include plasma arc, which uses an electronic arc as a heat source for converting materials to a gas or fused slag; molten salt destruction, which employs melted sodium in a closed container; and ram-jet incineration, a high-speed gas technology similar to rocket and jet engines (U.S. General Accounting Office, 1994a, 1994b).

Another option would be to revisit the decision to abandon use of on-board incinerators. The Navy's rejection of incineration has been attributed to concerns about crew safety with respect to use of older incinerators, shipboard space and weight constraints that may preclude installation of newer models, and possible air pollution. However, to the committee's knowledge the decision was not based on rigorous scientific and engineering evaluations. Such studies might be useful in view of the Navy's need for additional compliance strategies, the successful use of incinerators on large passenger cruise ships (described later in this chapter and in Chapter 5), and the availability of international standards for on-board incinerators (provided at the end of Appendix B).

Although the mission of protecting national security may appear to constrain the Navy's capability to attain full compliance with the MPPRCA, the same concerns are also an argument for accelerating compliance efforts. The Navy continues to discharge untreated garbage, including plastics, overboard, due to shortfalls in on-board storage space and treatment equipment. Such discharges create waste "signatures" of vessel activity, with undesirable consequences.[15] To the extent that ships can reduce generation of garbage and treat waste on board, overboard disposal and reliance on shore facilities can be minimized[16], with corresponding benefits to security.

Analysis of Interventions

Table 4-7 suggests possible interventions to improve Annex V implementation in the Navy fleet. Different interventions may be called for depending on the size and characteristics of a particular ship.

The Navy already is pursuing a number of technological and organizational strategies, such as modification of its supply system. One option not being pursued is on-board incineration, which could be reconsidered and evaluated through rigorous scientific and engineering tests. The Navy also could consider installing its pulpers and shredders for use where permitted, to make garbage discharges more benign. Compactors may be another option. Organizational interventions include reporting inadequate reception facilities encountered at commercial or foreign ports.

Education may provide means of leveraging the success achieved to date. For instance, Navy personnel could be encouraged to exchange information on

---

[15]As occurred during the 1991 Persian Gulf War, floating garbage—especially in the sea lanes—can be mistaken for floating mines, debris from damaged ships, or other, more sinister objects. The discharged materials also pose a security risk by leaving clues to the recent whereabouts of naval vessels.

[16]For example, use of shredders and pulpers outside special areas would reduce trash signatures, and use of compactors would reduce the need to return to port to off-load garbage.

TABLE 4-7   Applying the Hazard Evolution and Intervention Model to U.S.
Navy Combatant Surface Vessels and Their Home Ports

Hazard Evolution Model	Human Behavior that Encourages Generating Garbage	On-board Generation of Garbage
Intervention Model	Modify Behavior that Encourages Generating Garbage	Reduce Garbage Generation During Voyage
Technological	Use substitutes (where available) for plastic materials.	Continue converting supply system to limit plastics brought on board.
Organizational and Operational	Demonstrate management commitment to Annex V compliance. Establish shipboard regime for sorting garbage at point of generation.	Review shipboard activities to identify opportunities to reduce waste. Sort garbage at point of generation. Use only vendors committed to packaging and storage techniques that minimize waste.
Educational (Target Population/ Content)	Continue to educate shore personnel in how to modify the supply chain. The Navy has educated the Congress through fleet analyses; Congress has responded by showing serious commitment and establishing benchmarks.	Compliance by officers and crews is mandated; training is now needed in compliance strategies (both interim and permanent). Help shore support personnel develop implementation capabilities; monitor costs.

Breakdown in Compliance	Discharge of Garbage into Sea	Exposure to Discharged Garbage
Prevent Breakdown in Compliance	Block Discharge of Garbage into Sea	Block Exposure to Discharged Garbage
Keep shipboard systems well maintained.	Develop and install appropriate on-board garbage treatment equipment.	Install pulpers and shredders to block exposure to intact garbage.
Remind crew of Annex V regulations with posters and placards in prominent places on ships. Report inadequate reception facilities.	Follow interim plastic discharge restrictions, based on limits of shipboard hygiene and habitability (three days for food).	Use pulpers and shredders outside special areas, even where not required.
	Establish system for exchange of information on problems that encourage continued improper discharges. Establish recycling programs for items (cans) otherwise discharged overboard.	

TABLE 4-7   Continued

Hazard Evolution Model	Human Behavior that Encourages Generating Garbage	On-board Generation of Garbage
Intervention Model	Modify Behavior that Encourages Generating Garbage	Reduce Garbage Generation During Voyage
Regulatory		Restrict materials allowed on board (this may affect ship habitability).
Economic (Market Forces)	Solicit proposals for development of alternative packaging materials, such as edible packagings (now under study).	Off-load materials before departing home port (it may cost more to discard items later into a reception facility at another port). Require waste minimization in contracts and purchase orders and give preference to those with least waste.

implementation problems and solutions. In addition, vessel crews could be educated about the benefits of recycling even those items, such as cans, now legally discharged overboard.

Enforcement alternatives include the assessment of significant internal penalties against personnel who violate Annex V. Economic options, in addition to the present practice of giving crews any proceeds from recycling, include marketing the metal and glass wastes now collected and separated on board.

Breakdown in Compliance	Discharge of Garbage into Sea	Exposure to Discharged Garbage
Prevent Breakdown in Compliance	Block Discharge of Garbage into Sea	Block Exposure to Discharged Garbage
Discourage violations through peer pressure and peer enforcement of rules for on-board activities. The "3-day/20-day" rule for holding plastics reduces amount discharged overboard.	Establish and enforce internal penalties for violations.	Develop in-service guidelines and directives. Require internal records of legal discharges at sea. Keep records of garbage transactions.
Make shoreside disposal readily available at both military and commercial ports. Return monies from recycling to vessel crew for their discretionary use.	Explore marketing of metal wastes now collected and separated. Recycle plastics in commercial market (to strengthen market). Encourage on-board procedures to limit legal overboard discharges and improve centralized waste management.	

Passenger Cruise Ships and Their Itinerary Ports

Intelligence

Coast Guard and APHIS inspectors board passenger cruise ships, but, as is the case with cargo vessels, they do not collect data that would be useful in Annex V implementation. Such data might be collected if the Congress adopts proposed MPPRCA amendments that would allow requirements for garbage logs to be extended to foreign-flag vessels. In the meantime, general information about garbage handling practices is available from cruise line operators and printed sources.

Waste management practices on cruise ships vary, depending on company policies, geographical areas of operation, and the availability of adequate port reception facilities, but ships constructed recently are fitted with an array of garbage treatment equipment, and waste minimization and sorting procedures are elaborate (Whitten and Wade, 1994).

Control

The United States has direct enforcement authority over most of this fleet only in U.S. waters, as the majority of cruise ships fly a foreign flag. However, cruise ship operators typically are very image conscious and responsive to U.S. concerns. Recently, the threat of public embarrassment over citizen reports of Annex V violations has served as a control on Annex V compliance by cruise ships. Moreover, the cleanliness of the waters in which cruise ships sail is important to vessel operators, who are in the business of satisfying passenger expectations. Operators have employed a variety of strategies to comply with Annex V (see sidebar).

A major barrier to compliance in this sector lies in port reception facilities, which are rarely adequate to the task of serving a large passenger vessel. There are no reception facilities suitable for cruise ships, for example, at Mexican ports along the Caribbean or the Gulf of Mexico. No amount of control can assure compliance if reception facilities are inadequate, although vessel operators may overcome this problem, at least in part, by treating garbage on board.

Analysis of Interventions   .

Table 4-8 lists possible interventions to improve Annex V implementation among cruise ships. Most of the technological, organizational, and educational options listed in the matrix have been tried, apparently with success; use of these strategies could be expanded.

With so many crew members and passengers aboard cruise ships and the constant turnover, ongoing education and training is particularly important. To foster recycling and reduce legal overboard discharges, crews and passengers could be informed about the benefits of recycling items, such as cans, that otherwise may be discarded. Education is so critical in this sector that it might be mandated; proposed MPPRCA amendments would require Annex V posters, placards, and briefings on foreign-flag vessels while in U.S. waters.

A key regulatory intervention would be to work on resolving difficulties related to port reception facilities. Some efforts are under way in this regard (see Chapter 7). In addition, garbage handling logs could be maintained on foreign-flag cruise ships, either voluntarily or, if the proposed MPPRCA amendments are

---

GARBAGE DISPOSAL ON CRUISE SHIPS

Implementation of Annex V in this sector is facilitated by the partnership the cruise industry has cultivated with naval architects, shipyards, and equipment manufacturers to develop and improve shipboard garbage handling and treatment technology.

On vessels built before 1970, refrigerated storerooms are used to hold food-contaminated materials until their disposal in an APHIS-approved port reception facility. On vessels built between 1970 and the ratification of Annex V in 1987, incinerators are installed to destroy garbage that cannot be discharged overboard. On ships built recently, the garbage handling system is elaborate and often integrates different types of machinery, such as incinerators, pulpers, and grinders. (The Coast Guard intervened on behalf of cruise ships in Alaska, where use of incinerators is prohibited; the Coast Guard insisted that the state either waive the rule or provide suitable APHIS-approved disposal facilities.)

Cruise ship operators also are trying to reduce their reliance on plastic products, often substituting paper products. Plastic bags have been eliminated on older vessels. Almost all ships have discontinued use of foamed plastic materials and excess packaging such as individual bottles of shampoo. Some garbage, particularly food waste, is run through a pulper and then discharged overboard where permitted by Annex V. In addition, some recreational activities have been modified. For instance, the once-common practice of driving golf balls off the deck into the open ocean is now rare.

---

adopted, by mandate. Another possibility would be to pursue an international requirement that flag states issue waste management certificates to cruise ships, as suggested earlier with respect to cargo ships. Economic interventions include imposing internal (company) fines for violations of garbage handling rules, and improving garbage treatment equipment to reduce its costs.

Research Vessels and Their Ports of Call

Intelligence

Many research vessels are supported or owned and operated by the federal government, so information on vessel activities and personnel behavior can be obtained, even though vessels may be away from shore or in foreign waters for extended periods of time. The National Oceanic and Atmospheric Administration (NOAA) has conducted a fleetwide assessment of pollution prevention needs and a survey of available equipment; Annex V compliance strategies rely on food grinders and garbage compactors or incinerators (Art Anderson Associates, 1993). The EPA's research vessel stores garbage for entire two-week voyages, not a particularly sanitary solution. The EPA also has reviewed and revised its pur-

TABLE 4-8   Applying the Hazard Evolution and Intervention Model to
Passenger Cruise Ships and Their Itinerary Ports

Hazard Evolution Model	Human Behavior that Encourages Generating Garbage	On-board Generation of Garbage
Intervention Model	Modify Behavior that Encourages Generating Garbage	Reduce Garbage Generation during Voyage
Technological	Eliminate single-portion cosmetic amenities packaged in plastic. Discourage suppliers from delivering ship's stores packaged or bundled in plastic. Encourage suppliers to adopt alternate delivery packaging.	Eliminate disposable containers. Modify galley equipment or food and beverage equipment to reduce amounts of single-use items. Invest in reusable containers.
Organizational and Operational	Senior management commitment to Annex V compliance must be visible and converted into management directives.	Sort garbage at the site of generation. Use only vendors committed to packaging and storage techniques that minimize waste.
Educational (Target Population/ Contents)	Instill respect for clean ocean among passengers and crew. Impose expectation that each can comply with Annex V.	Train crew to sort garbage for recycling and proper disposal.

Breakdown in Compliance	Discharge of Garbage into Sea	Exposure to Discharged Garbage
Prevent Breakdown in Compliance	Block Discharge of Garbage into Sea	Block Exposure to Discharged Garbage
Keep shipboard systems well-maintained.	Develop and install appropriate garbage handling equipment or integrated waste management systems.	Pretreat legal discharges by grinding and shredding garbage, to minimize drifting in the water. Provide incinerator with ash storage space (to retain ash for off-loading into port reception facility or legal discharge in deep water).
Audit shipboard practices regularly. Provide many trash cans.		Prohibit discharges unless supervised by appropriate officer. Establish shipboard collection of recyclable materials for return to portside recycling networks.
Train crew in on-board garbage management. Provide Annex V placards, posters, and public address announcements.	Develop recycling programs for items (cans) that may be discharged overboard legally.	

TABLE 4-8   Continued

Hazard Evolution Model	Human Behavior that Encourages Generating Garbage	On-board Generation of Garbage
Intervention Model	Modify Behavior that Encourages Generating Garbage	Reduce Garbage Generation during Voyage
Government and Private Regulation and Enforcement		
Economic (Market Forces)		

chasing practices (e.g., no foamed plastic cups are used) and has reduced the amounts of packaging and plastics brought on board. Less is known about non-federal research vessels.

Garbage handling can be a problem due to vessel mode of operation. While at sea, sampling or monitoring tasks may require that a research vessel remain on station or restrict its motion and curtail overboard discharges; the vessel may be unable to return to port before garbage storage space is full. Oceanographic vessels are notoriously cramped, with every on-board space obligated to science missions or operational needs. Moreover, the duration of some expeditions—many over 10 days and some over 50 days—makes garbage storage difficult and untenable, and the mission profile of some oceanographic vessels leaves little space for garbage treatment equipment. Anecdotal reports and the NOAA survey (Art Anderson Associates, 1993) suggest that on-board equipment, such as incinerators, tends to be primitive.

The demands of Annex V are particularly taxing for research vessels operating in extreme situations. The NOAA ship *Surveyor*, homeported in Seattle, is obligated to conduct scientific missions in the Antarctic, designated as a special

Breakdown in Compliance	Discharge of Garbage into Sea	Exposure to Discharged Garbage
Prevent Breakdown in Compliance	Block Discharge of Garbage into Sea	Block Exposure to Discharged Garbage
Make illegal overboard discharge of garbage a firing offense. Require crew and passenger education for entry into U.S. waters or ports. Require flag states to issue waste management certificates.	Foster development of reliable and affordable port reception facilities. Keep records of garbage transactions.	
Impose internal fines for violations of garbage handling rules.	Improve on-board garbage treatment equipment, to reduce costs. Establish portside recycling networks.	

area (with zero-discharge restrictions). On the positive side, because research vessels are operated by small, cohesive communities, informal networks exist for the sharing of information on strategies for reducing waste and overboard discharges.

Some research vessels operate in the vicinity of home ports. Other vessels rarely visit their home ports and only infrequently call at any port. In instances where operations center around a home port, shoreside managers can address the unique challenges of complying with Annex V. For example, in Seattle, NOAA's waste reception requirements are met by a commercial contractor at NOAA's Pacific Marine Center, so it may be possible for managers to audit informally the materials discharged by their vessels.

In general, while sensitivity to environmental concerns has increased within the research fleet in recent years, there are anecdotal reports of continuing overboard disposal of items such as used, expendable scientific instruments. Annex V does not address disposal of research equipment but IMO implementation guidelines encourage the return of garbage to port reception facilities "whenever practicable."

Control

In recent years, the federal government has taken on increased responsibility for the environmental well-being of remote locations in which the United States conducts research. Antarctica, for example, is being cleaned up rapidly after decades of poor garbage disposal practices. Annex V offers a chance to effect a similar change within the government's oceanographic fleet.

In the United States, much of the active oceanographic fleet is federally funded. Through direct budget authority, NOAA's budget covers the waste management costs of the agency's fleets. National Science Foundation (NSF) sponsorship of research cruises pays for the waste management costs of the University National Oceanographic Laboratory Systems fleet. Thus, the federal government can exert budgetary control over shipboard practices and can include funding for Annex V compliance in the appropriations for research vessels. The government also can require that research proposals include information on how scientists plan to minimize and handle garbage and give priority to those with appropriate plans.

Direct on-site control is limited, however, because most research vessels are not subject to routine government inspections, boardings, or oversight. The principal exception is the NOAA fleet (see sidebar). On federally supported missions, the government can exert some control through selection of supplies and materials and requirements for MARPOL briefings. For instance, NSF has banned the use of foamed plastic "peanuts" as packaging materials for scientific gear aboard NSF-sponsored voyages. The EPA provides information about MARPOL to new ship personnel, researchers, and visitors along with the routine safety briefing. But control is limited when the vessel must rely on disposal facilities in civilian or foreign ports. In some instances, it may be difficult to obtain any garbage disposal services at all. The *Surveyor*, returning from an extended voyage in a zero-discharge zone, once arrived in a South American port and was refused permission to off-load any garbage. Jammed with about 10 cubic meters (13 cubic yards) of waste, the ship was dubbed "the garbage scow" by the local press.

Control also is limited by the characteristics of the current fleet. Engineering and space constraints make it awkward, at best, for owners and operators of oceanographic vessels to install expensive on-board treatment equipment. Expenses for routine maintenance and other repairs virtually preclude the possibility of finding sufficient equipment funds in a vessel's budget to cover refitting the vessel for Annex V compliance. If such upgrades are to be made without depletion of operating accounts, then special funds earmarked for Annex V equipment will need to be provided.

On the positive side, there may be minimal need for direct control of behavior in this sector, because marine researchers and oceanographic vessel crews tend to value environmental protection, and they have expressed willingness to comply with the legal mandates. They understand the importance of Annex V

CONTROL OF THE NOAA FLEET

The NOAA fleet is subject to complete government control, particularly when these vessels use their home ports. The complement aboard a NOAA vessel includes uniformed service officers, civilian merchant mariners, and visiting scientists; the officers, who answer directly to higher commands, have authority over the entire crew. As a matter of policy, each NOAA command develops its own solid waste management procedures, although they have begun to receive direct technical support from the central engineering staff, particularly with regard to selection of pollution prevention equipment.

All NOAA vessels are aging, and it will be difficult to retrofit them with either waste treatment equipment or on-board storage spaces to hold garbage for extended periods of time. Because it now appears that Navy R&D will not produce equipment appropriate for NOAA's missions, the research fleet will be compelled to use commercial equipment. However, NOAA has found available commercial incinerators to be unreliable, ineffective, and time consuming to operate.

NOAA plans to foster informal controls by introducing fleetwide Annex V awareness training for new officers and crews and well as visiting scientists. At present, no central MARPOL training is offered. Instead, each vessel's command is expected to provide a boarding briefing for all newly arriving personnel and visiting scientists. This briefing emphasizes emergency procedures but it also provides an opportunity to explain waste management practices and garbage disposal restrictions.

---

compliance and have helped present evidence of marine debris to other seafarers. Internal sanctions and peer pressure not only encourage compliance but also foster innovations and improvements in garbage handling practices.

Analysis of Interventions

Table 4-9 outlines possible interventions to improve Annex V implementation on research vessels. Among the technological options, it is obvious that improved on-board garbage treatment equipment and appropriate storage space are needed. These features could be designed into any new vessels and retrofitted where possible.

Promising organizational interventions include continued reduction in use of disposable supplies. In addition, where feasible or required, discharge of all garbage except food could be halted. For example, when adequate storage space and garbage treatment equipment is available (e.g., on short voyages or well-designed new vessels), the crew and guest scientists might be able to refrain from even legal overboard discharge of garbage, including used equipment. Federally supported research vessels could set an example in this regard.

Education is also important, particularly because of the turnover in guest scientists. Vessel operators also need to be educated about compliance strategies,

TABLE 4-9   Applying the Hazard Evolution and Intervention Model to
Research Vessels and Their Ports of Call

Hazard Evolution Model	Human Behavior that Encourages Generating Garbage	On-board Generation of Garbage
Intervention Model	Modify Behavior that Encourages Generating Garbage	Reduce Garbage Generation
Technological	Reduce or eliminate convenience packaging of supplies and foods.	Provide alternate packaging where possible (given packaging standards for electronic equipment).
Organizational and Operational	Modify comfort and convenience levels. Reduce number of daily meals (now set by union contract). Reduce crew sizes.	Use only vendors committed to packaging and storage techniques that minimize waste. Remove disposables from ship stores. Sort garbage at site of generation.
Educational (Target Population/ Content)	Inform crews of the need for and benefits of changes (in terms of health, nutrition, cost savings, environmental protection).	Inform managers of options for alternate packaging, provisioning, and deployment procedures. Inform crews and guest scientists of ways to minimize waste materials brought on board.
Government or Private Regulation and Enforcement	Renegotiate union agreement provisions that trigger waste generation. Amend voyage operating agreements to minimize equipment packaging scientists bring on board.	Prohibit use of disposable items.

Breakdown in Compliance	Discharge of Garbage into Sea	Exposure to Discharged Garbage
Prevent Breakdown in Compliance	Block Discharge of Garbage into Sea	Block Exposure to Discharged Garbage

Keep shipboard systems well maintained. Incorporate garbage handling equipment and storage spaces into new vessels.	Provide sufficient garbage storage space and efficient on-board garbage treatment equipment.	
Provide reminders with posters and placards.	Introduce efficient on-board garbage handling procedures. Assure that port reception facilities are adequate.	Implement a zero discharge standard where feasible or necessary.
Educate management about legal mandates, compliance strategies, and methods for educating and training personnel. Educate crews and scientists about mandates, compliance methods, environmental consequences of discharge, and penalties for violations.	Develop recycling programs for items (cans) otherwise discharged overboard legally. Promote recognition of marine debris problem at scientific conferences.	
Require garbage sorting and holding of certain materials for shoreside recycling.	Keep records of garbage transactions. Establish and enforce internal guidelines and penalties (fleet policies).	

TABLE 4-9   Continued

Hazard Evolution Model	Human Behavior that Encourages Generating Garbage	On-board Generation of Garbage
Intervention Model	Modify Behavior that Encourages Generating Garbage	Reduce Garbage Generation
Economic (Market Forces)	Require proposals for federal funding for ship time to describe garbage minimization and handling plans. Give priority to proposals with appropriate plans.	Demonstrate any cost benefits from switch to reusable packaging.

such as waste reduction. Researchers could help educate their peers by promoting recognition of the marine debris problem and Annex V compliance strategies at scientific conferences.

Regulatory interventions include limiting equipment packaging brought on board and requiring the holding of certain materials for recycling. In addition, logs of garbage transactions could be maintained, not only on research vessels covered by the present record-keeping regulations but also on voyages supported by the federal government. The utility of keeping logs on public vessels would have to be weighed, however.

Economic interventions are particularly important in this sector, to make it easier for researchers and vessel crews to comply. As suggested by the analysis of intelligence and control, funds need to be provided for on-board garbage handling equipment (where needed) and efforts need to be made to assure availability of port reception facilities. Absent such measures, the willingness of oceanographers to comply will be wasted. In addition, returning monies from recycling programs to vessel crews could foster voluntary compliance.

REFERENCES

Alverson, D. and J.A. June, eds. 1988. Proceedings of the North Pacific Rim Fishermen's Conference on Marine Debris, October 13–16, 1987, Kailua-Kona, Hawaii. Seattle, Wash.: Natural Resources Consultants.

Breakdown in Compliance	Discharge of Garbage into Sea	Exposure to Discharged Garbage
Prevent Breakdown in Compliance	Block Discharge of Garbage into Sea	Block Exposure to Discharged Garbage
Return monies from recycling to vessel crew for their discretionary use. Budget funds for on-board garbage handling equipment as means to meet legal mandate.	Make port reception facilities affordable and available. Provide affordable and reliable on-board treatment equipment. Encourage on-board procedures to limit legal overboard discharge.	

American Red Cross. 1991. American Red Cross National Boating Survey. Washington, D.C.: American Red Cross.

Anderson, C. 1992. Presentation by Carl Anderson, Minerals Management Service, to the Committee on Shipborne Wastes of the National Research Council, at the Governor Calvert House of the Historic Inns of Maryland, Annapolis, Md., May 7–8, 1992.

Art Anderson Associates. 1993. NOAA Fleetwide Shipboard Waste Management. Report prepared for the National Oceanic and Atmospheric Administration by Art Anderson Associates, Bremerton, Wash. Jan. 29.

Associated Press. 1994. Navy faulted for its garbage: Sailors may dump $26 million project overboard. The Washington Post. Aug. 24. A17.

Boat Owners Association of the United States (BOAT/U.S.). 1990. Water quality low, clean-up interest high. BOAT/U.S. Reports. 25:1. July.

Boudreaux, D. 1993. Presentation by Deyaun Boudreaux, environmental director of the Texas Shrimp Association, to the Committee on Shipborne Wastes of the National Research Council, at the University of Texas at Austin Marine Science Institute, Port Aransas, Tex., Feb. 16, 1993.

Bunch, P.A.. 1994. Vessel Environmental Compliance Program Plan. Internal planning document sent from Admiral Peter A. Bunch, chief, Office of Engineering, Logistics, and Development, to the commandant of the U.S. Coast Guard. Feb. 1. Photocopy.

Buxton, R. 1989. Plastic Debris and Lost and Abandoned Fishing Gear in the Aquatic Environment. Background paper prepared for the Canadian Department of Fisheries and Oceans (DFO), CANADA Working Group on Plastic Debris. Project DFO 085214. Available from DFO, 200 Kent Street, Ottawa, Ontario, K1A 0E6. April.

Center for Marine Conservation (CMC). 1989. Marine Debris Information Offices, Atlantic Coast/ Gulf of Mexico and Pacific Coast: Annual Report, October 1, 1988–September 30, 1989. Washington, D.C.: CMC.

Chang, T.J. 1990. Low technology (burn barrel) disposal of shipboard generated (MARPOL V) wastes. Pp. 915-920 in Proceedings of the Second International Conference on Marine Debris, 2–7 April 1989, Honolulu, Hawaii (Vol. II), R.S. Shomura and M.L. Godfrey, eds. NOAA-TM-NMFS-SWFSC-154. Available from the Marine Entanglement Research Program of the National Marine Fisheries Service (National Oceanic and Atmospheric Administration), Seattle, Wash. December.

DPA Group. 1989. Plastic Debris in the Aquatic Environment—Halifax Workshop Report, May 16–18, 1989, Halifax, Nova Scotia. Project DFO 085214. Available from the Canadian Department of Fisheries and Oceans, 200 Kent Street, Ottawa, Ontario, K1A 0E6. July.

Emshwiller, J.R. and M.J. McCarthy. 1993. Coke's soda fountain for offices fizzles, dashing high hopes. The New York Times. June 14. 1,9.

F.I.S.H. Habitat Education Program. 1994. Net Recycling Program Summary. Fact sheet prepared by the Fishermen Involved in Saving Habitat Education Program, Gladstone, Ore. September.

Green, E. 1993. Presentation by Ed Green, U.S. Coast Guard Marine Safety Office/Corpus Christi, to the Committee on Shipborne Wastes of the National Research Council, University of Texas at Austin Marine Science Institute, Port Aransas, Tex., Feb. 14–17, 1993.

Gulf of Mexico Program. 1991. Marine Debris Action Plan for the Gulf of Mexico. Dallas, Tex.: U.S. Environmental Protection Agency.

International Maritime Organization (IMO). 1993. International Management Code for the Safe Operation of Ships and For Pollution Prevention (International Safety Management [ISM] Code). Resolution A.741(18). Adopted November 4, 1993. Available from IMO, 4 Albert Embankment, London, SE1 7SR.

Kasperson, R.E. and K.D. Pijawka. 1985. Societal response to hazards and major hazard events: Comparing natural and technological hazards. Pub. Admin. Rev. 45:7-18. Special issue.

Kearney/Centaur, Inc., division of A.T. Kearney, Inc. 1994. Managing Oily Wastes and Garbage from Ships—A Guide to Waste Management Practices for Shipping Agents, Waste Haulers, Shipping Companies, and Port and Terminal Operators. Available from the U.S. Coast Guard, Marine Environmental Protection Division, Washington, D.C.

Koss, L., F. Chitty, and W.A. Bailey. 1990. U.S. Navy's Plastics Waste Educational Efforts. Pp. 1132–1139 in Proceedings of the Second International Conference on Marine Debris, 2–7 April 1989, Honolulu, Hawaii (Vol. II), R.S. Shomura and M.L. Godfrey, eds. NOAA-TM-NMFS-SWFSC-154. Available from the Marine Entanglement Research Program of the National Marine Fisheries Service (National Oceanic and Atmospheric Administration), Seattle, Wash. December.

Koss, L.J. 1994. Dealing With Ship-generated Plastics Waste on Navy Surface Ships. Paper prepared for the Third International Conference on Marine Debris, Miami, Fla., May 8–13, 1994. Office of the Chief of Naval Operations, Department of the Navy, Washington, D.C.

Minerals Management Service (MMS). 1986. Guidelines for Reducing or Eliminating Trash and Debris in the Gulf of Mexico. NTL No. 86-11. Notice to Lessees and Operators of Federal Oil and Gas Leases in the Outer Continental Shelf, Gulf of Mexico OCS Region. Available from MMS Gulf of Mexico OCS Region Office of Leasing and Environment, New Orleans, La. Nov. 17.

National Research Council (NRC). 1991. Fishing Vessel Safety: Blueprint for a National Program. Marine Board, NRC. Washington, D.C.: National Academy Press.

Ocean Science News. 1991. The U.S. Navy is responding to criticism of its vessels for dumping trash at sea. Ocean Science News 33(22):7. Aug. 10.

Pearce, J.B. 1992. Viewpoint: Marine vessel debris, a North American perspective. Marine Pollution Bulletin 24(12):586-592. December.

Recht, F. 1988. Report on a Port-Based Project to Reduce Marine Debris. NWAFC Processed Report 88-13. Available from the Marine Entanglement Research Program of the National Oceanic and Atmospheric Administration, Seattle, Wash. July.

Sutinen, J.G., A. Rieser, and J.R. Gauvin. 1990. Measuring and explaining noncompliance in federally managed fisheries. Ocean Development and International Law 21:335–372.

Swanson, R.L., R.R. Young, and S.S. Ross. 1994. An Analysis of Proposed Shipborne Waste Handling Practices Aboard United States Navy Vessels. Paper prepared for the Committee on Shipborne Wastes, Marine Board, National Research Council, Washington, D.C.

U.S. General Accounting Office (GAO). 1994a. Pollution Prevention: Chronology of Navy Ship Waste Processing Equipment Development. GAO/NSAID-94-221FS. Washington, D.C.: GAO National Security and International Affairs Division. August.

U.S. General Accounting Office (GAO). 1994b. Pollution Prevention: The Navy Needs Better Plans for Reducing Ship Waste Discharges. GAO/NSIAD-95-38. Washington, D.C.: GAO National Security and International Affairs Division. November.

U.S. Navy. 1993. Shipboard and Plastics Waste Management Program Plan (draft). Prepared by Naval Sea Systems Command 05V, Environmental Engineering Group, Washington, D.C. April.

U.S. Navy. 1994. Naval Message, Dept. of Navy: Shipboard Solid Waste Disposal. Message from CNO Washington DC/N4. R 031112Z May 94. UNCLAS/N05090. May 3. Photocopy.

Wallace, B. 1990. How much do commercial and recreational fishermen know about marine debris and entanglement? Phase 1. Pp. 1140–1148 in Proceedings of the Second International Conference on Marine Debris, 2–7 April 1989, Honolulu, Hawaii (Vol. II), R.S. Shomura and M.L. Godfrey, eds. NOAA-TM-NMFS-SWFSC-154. Available from the Marine Entanglement Research Program of the National Marine Fisheries Service (National Oceanic and Atmospheric Administration), Seattle, Wash. December.

Weikart, H. 1993. Presentation by Heather Weikart, National Marine Fisheries Service Observers Program, to the Committee on Shipborne Wastes of the National Research Council, Red Lion Inn, Seattle, Wash., July 15, 1993.

Whitten, D.H. and R.L. Wade. 1994. Environmental Challenges Faced by the International Cruise Industry. Paper prepared for the Annual Meeting of The Society of Naval Architects and Marine Engineers, New Orleans, La., Nov. 17–18, 1994.

5

Integrating Vessel and Shoreside Garbage Management

The preceding chapter addresses only part of a national Annex V implementation program—the part that applies to fleets. In addition to establishing performance standards for vessels, Annex V also mandates the provision of "adequate" garbage reception facilities at ports. Yet the crucial portside segment of the garbage management scheme conceived in Annex V is left undefined in the United States, with the result that compliance has been limited. If a comprehensive, effective Annex V implementation program is to be developed, then a systems perspective is needed that views vessels and their ports of call as part of the same system. Awareness of this need seems to be growing. Whereas the problem of marine debris once was viewed in isolation from broader waste management issues, there has been a trend over the past several years toward a more comprehensive systems-oriented perspective (Laska, 1994).

The vessel garbage management system has two elements: the vessel and the port, which is the transfer point to the landside solid waste management system. In general, vessels operate within and receive services from specific types of terminals. Just as vessels differ, so do terminals. Recreational boats use marinas, private docks, and launch ramps, while fishing vessels use fishing piers and terminals. General cargo vessels call at public ports (sometimes maintaining specified ports of call), while bulk vessels use private terminals and may operate only from selected home ports. All the materials delivered to and removed from the vessel must pass through the terminal's facilities.

As vessels become more specialized and diverse, so must terminals and the facilities they provide. Both vessels and terminals are costly to develop, build, and operate. Yet it is even more costly for a port to lose business to a competitor with better facilities, so port operators continually modify terminals, equipment, and services to reflect changes in vessels and shipping operations (Atkins, undated). Thus, there is a symbiotic relationship between vessels and their ports of call. Viewing the vessel and port as a system (henceforth referred to as the vessel garbage management system) significantly improves prospects for control and opens the door to solutions, fleet by fleet.

This chapter examines the vessel garbage management system, exploring each element and what is needed to integrate vessel garbage into the system for handling land-generated waste. The introduction describes the principles of integrated waste management and how they apply in the maritime setting. The core of the chapter has two parts: an assessment of on-board garbage handling practices and technologies, and an assessment of port reception facilities and practices. The challenge is to maximize the garbage handling capabilities of both the vessel and port and then establish a seamless interface. If this can be achieved, then the goal of full Annex V implementation can be achieved. The final section of the chapter examines four issues that pose barriers to the internal integration of the system: quarantine requirements for vessels arriving from foreign shores, implementation of the Coast Guard's Certificate of Adequacy (COA) program and other requirements for ports, port operators' liability for handling vessel garbage, and financing—both who should pay for garbage services and how they should pay.

PRINCIPLES OF INTEGRATED WASTE MANAGEMENT

The Environmental Protection Agency (EPA) defines an integrated solid waste management system (ISWMS) as "a practice of using several alternative waste management techniques to manage and dispose of specific components of the municipal solid waste stream. Waste management alternatives include source reduction, recycling, composting, energy recovery, and landfilling" (ICF, Inc., 1989). Managers of ISWMS for land-generated waste select and employ these technical alternatives based on analysis of their needs, careful planning, and technical and economic evaluations of options.

Implementation of Annex V to date has been guided—or misguided—by a perception that the effort to implement controls over vessel garbage should be separated from other initiatives to control land-generated solid waste. In fact, vessel garbage is simply a poorly controlled solid waste stream that, logic dictates, would best be managed using principles and systems similar to those developed for land-generated waste. Integration of the two systems, rather than development of redundant and parallel regimes for vessel garbage, could simplify implementation of Annex V and minimize the burdens on regulatory agencies and the regulated mariners and ports, in that all could pursue compliance with a consistent national standard, operating within a coordinated regulatory regime. This approach would require the establishment of professional standards for waste management throughout the vessel–port system, as well as oversight and enforcement comparable to that carried out for land-based systems.

It is clear that the general principles of integrated solid-waste management apply in the maritime setting. It is also clear that, with notable exceptions, these principles are not put to use consistently because there are important differences between land-based and maritime waste management. First, vessels may continue to discharge some garbage in the oceans legally, so long as they comply with Annex V. Second, waste treatment and storage capabilities are severely restricted on vessels, due to space and weight limits (this becomes an important factor in vessel design and retrofitting). Finally, vessels are mobile and may call at different ports, which has the effect of making garbage disposal demands more unpredictable and ad hoc than they are on land.

As conceived by the committee, the vessel garbage management system depends on the key players to carry out the following roles:

• The role of vessel operators is to minimize waste through source reduction and to dispose of garbage in compliance with the law through on-board techniques and, where permissible, disposal at sea, and by delivering all other garbage to a port reception facility.

• The role of terminal operators and the port reception facility is to receive the remaining garbage and provide a simple process to transfer it to the well-developed disposal system for land-generated waste.

• The role of the existing land-based systems and their operators is to integrate the needs of vessel garbage handling into the system and to transfer technologies and methods into the vessel garbage management system.

• The role of boat manufacturers and shipyards is to ensure that all new vessels are designed to incorporate convenient garbage storage spaces and, where appropriate, garbage treatment technologies.

• The role of state governments is to help port and terminal operators establish and maintain garbage reception facilities.

• The role of the federal government is to provide clear legislation, criteria, and guidelines to ensure that this intermodal transfer of waste is simple, cost-effective, and in compliance with the U.S. commitment to MARPOL Annex V.

The committee used this framework as a basis for identifying problems with existing procedures as well as potential solutions. The remainder of this chapter outlines how the disparate elements of the vessel–port transaction might be integrated into a process that meshes well with the prevailing national system for handling solid waste.

SHIPBOARD TECHNOLOGIES AND PRACTICES

To apply the principles of integrated solid waste management, vessel operators first conduct a needs assessment, which includes determining how much and what sort of garbage is generated and the disposal restrictions in the waters where the vessels operate. A waste management plan then is developed. More often than not, such plans have been developed on an ad hoc basis out of necessity rather than based on engineering expertise. To assure zero discharge of plastics, plans call for waste sorting. It appears that the requisite behavioral change is occurring and that sorting can become a universal practice. Where garbage sorting procedures have been implemented, training and educational efforts (such as posters and placards) and process simplification (such as color coding and labeling of receptacles) have been cited as factors determining success (Kauffman, 1992).

Many vessels have advanced and comprehensive waste management plans (National Oceanic and Atmospheric Administration Corps, 1993). In one instance, Navy personnel developed their own environmental compliance program—a "cookbook" on how to integrate garbage handling with other practices to meet environmental objectives (Gallop, undated). In addition, a number of fully integrated shipboard waste management systems have been designed. An example is the approach taken in constructing some of the newest passenger vessels, where the garbage handling and management system is designed concurrently with the vessel, to provide the best possible means of complying with Annex V. This approach elevates the mundane task of garbage handling to the same level of importance as all the other auxiliary systems considered during ship construction (Deerberg, 1990, 1993; Vie, 1990; Florida-Caribbean Cruise Association, 1993; Laitera, 1993; Whelpton, 1993).

Source Reduction

An important step in integrated waste management is the effort to reduce amounts of materials brought on board that will become garbage. As indicated in Chapter 3, this type of early intervention in the hazard evolution process has been largely overlooked in the past but is an important aspect of Annex V implementation. Source reduction demands the cooperation of vendors as well as vessel operators and crews.

A typical target in source reduction plans is plastic packaging. Each vessel operator tailors a source-control approach to fit the circumstances. Needs and supplies are examined, and excess packaging can be left on shore. The committee witnessed such source-control efforts at a cruise ship terminal. These procedures may create extra up-front work for the steward and staff but can reduce significantly the amount of garbage to be managed during the voyage. Another approach is to discontinue use of disposable plates, cups, and cutlery and equip the vessel with durable serving pieces. Waste minimization can be encouraged or required as a condition for bids, contracts, and purchase orders.

Consumables such as cleaning supplies and table condiments can be purchased in large receptacles for refilling smaller containers for daily use. Each operator devises bulk storage containers to store the necessary supplies without compromising health and safety.

As some fleet operators reported to the committee, it is important that the changes do not compromise shipboard comfort and living conditions so much that the crew and/or passengers begin to resent the source reduction effort. It is always important to sustain the morale of those who are confined together on a vessel at sea.

On-Board Storage

A vessel operator may satisfy the mandates of Annex V by holding any restricted wastes and all plastic until the vessel returns to its home port or reaches a port reception facility that provides affordable, prompt service. Some commercial maritime operators feel that no U.S. port they visit has done an adequate job of organizing reception facilities and services, in that each garbage transaction is awkward and difficult. The committee was told of two shipping lines that prefer to hold all garbage generated while in U.S. waters rather than deal with U.S. port reception facilities as they exist now.

The practice of storing wastes on board revives longstanding concerns over ensuring sanitation[1] on vessels at sea. When vessels were slower, crews were larger, and there was less reliance on shoreside food preparation than is currently the case, vessel operators and builders were attentive to details that might predispose a vessel to problems with vermin or communicable diseases. Today, smaller crews must cope with not only tight itineraries but also complicated requirements for handling garbage such as food-contaminated plastics, which must be stored on board for disposal ashore. This is an issue that affects all maritime sectors. Yet the only federal guidelines on this topic, developed by the Centers for Disease Control and Prevention (CDC) Vessel Sanitation Program, apply solely to passenger vessels with international itineraries (i.e., cruise ships).

Several incidents of serious contagious illness on passenger ships during the summer cruise season of 1994 (Dahl, 1994; Journal of Commerce, 1994) underscored the importance of safeguarding vessel sanitation. These incidents demonstrated that the government must retain a capability to monitor sanitation on all forms of domestic transportation and public accommodation. At present, such monitoring of vessels other than cruise ships is left to local and state health department personnel, who typically will respond to a request from a Coast

---

[1]For purposes of this report, sanitation refers specifically to the promotion of hygiene and prevention of disease through proper handling and storage of garbage (not sewage).

Guard boarding officer to examine a ship (J.M. Farley, U.S. Coast Guard, personal communication to Marine Board staff, September 1993). One way to help ensure sanitation would be to strengthen the federal program of vessel inspections in ports, through either the CDC or the Food and Drug Administration's existing Program on Interstate Travel Sanitation. The operations manual used to check sanitation on cruise ships (Centers for Disease Control, 1989) is an example of an approach that could be integrated into vessel inspection programs. The provision of standard guidelines for maritime sectors other than cruise ships could help assure that sanitation is not compromised in the pursuit of Annex V compliance. Some fleets also may need technical assistance in developing safe and efficient on-board storage procedures.

On-board garbage storage facilities can be designed to provide for quick and easy off-loading at ports while preventing unintended loss overboard. Facilities range from secured plastic bags for day trips to large dumpsters requiring mechanical off-loading (Mike Prince, marine superintendent, Moss Landing Oceanographic Laboratory, personal communication to Marine Board staff, February 4, 1994). Waste storage areas on vessels can be designed or modified to isolate certain types of wastes, minimize odors, and prevent vermin infestation. The Navy is experimenting with odor-barrier bags for storing food-contaminated plastics on board (Koss, 1994).

Shoreside Recycling

Assuming adequate on-board storage space is available, port waste disposal volumes can be reduced if recyclable materials are separated. Easily recycled materials include aluminum and steel cans, glass bottles, plastic bottles, newspapers, and cardboard packaging. Other materials that may be recycled include metal parts, fishing nets, ropes, and other gear.[2]

As noted earlier, sorting is best accomplished with standard, color-coded containers and simple, appropriate training programs (Princess Cruises, 1993). Each vessel operator tailors training to fit the circumstances. Short videotapes, followed by practice and demonstrations, greatly assist in crew training. Similar educational programs may be developed for passengers, emphasizing the need for their cooperation in improving the vessel's waste disposal practices. Obviously, recycling only makes sense if the port reception facility and the land-based ISWMS can accept the specific, separated recyclable materials.

A few pilot programs have demonstrated the feasibility of recycling vessel garbage, but few permanent arrangements are in place (Middleton et al., 1991; Kauffman, 1992).

---

[2]A recycling infrastructure has evolved for many land-generated waste materials as the popularity of recycling has grown (Grove, 1994).

Some of the most advanced vessel garbage handling procedures and equipment can be found on cruise ships. The top photo shows a garbage sorting area. Sorting is essential both to ensure that plastics are held on board and to separate recyclable materials from other garbage. The bottom photo shows a commercially available compactor that reduces aluminum to 1/30th of its former volume and tin to 1/10th of its former volume. These materials then are baled and brought ashore for recycling. Credit: Princess Cruises

Treatment/Destruction

If there is room to install appropriate equipment and organize on-board storage, a range of technologies for treating or destroying garbage is available and in use on vessels. The amount of garbage generated, as well as Annex V operating restrictions, may dictate which methods and technologies are employed. Many types of commercial equipment can be purchased for shipboard use, although little testing and evaluation has been carried out to determine whether the size and ruggedness meet shipboard needs. The International Maritime Organization (IMO) *Guidelines for Implementation of Annex V* encourage the further development of shipboard technologies, acknowledging that the present state of the art is wanting.

Vessel operators in some sectors, such as the passenger cruise industry, work with equipment vendors and engineers to meet individual needs, but the potential markets for many of the needed technologies, such as those for commercial fisheries, are too small to attract commercial developers. The cruise ship industry has invested heavily in state-of-the-art equipment, including shredders, pulpers, compactors, and incinerators. These technologies have been retrofitted on existing ships and incorporated into the design and construction of new ships. In addition, the industry works closely with naval architects, shipyards, and equipment suppliers to improve the technology.

The Navy is the only federal agency that has been able to develop, test, and evaluate shipboard garbage handling technologies. The results have not been widely applicable to either the civil maritime sector or other public vessels, so there remains a need for product and systems development to support Annex V implementation. The maximum benefits could be derived from garbage treatment technologies, both existing and new, if information about them were exchanged promptly among the various maritime sectors.

Compactors

A compactor is a powered device used to reduce the volume of garbage, to facilitate storage during a voyage. Many such units are available commercially and most are sized to fit the needs of vessels; the committee observed successful shipboard use of compactors purchased at retail outlets. The Navy began developing compactors in 1979 but cancelled this research in 1993, deciding it was no longer necessary (U.S. General Accounting Office, 1994a, 1994b). Compactors are the backbone of the Coast Guard's Annex V compliance plans.

High-volume, low-density materials, such as plastic bottles, containers, and sheeting can be compacted easily to as little as 10 percent of their former volume. Other recyclable materials, such as metal cans and even paper products, can be reduced to 25 percent of their original volume. Volume reduction of glass (e.g., bottles, containers, bulbs, plate glass) is best achieved with glass crushers, which shatter rather than compress the materials.

A plastics processor, such as the one being developed by the Navy, is a hybrid of a shredder, a compactor, and a thermal treatment device. Plastic materials are shredded and then compressed and heated (not combusted) to form fused bricks of plastic. Developers claim the process produces sterile blocks that meet the federal quarantine standards for food-contaminated plastics. If such claims can be substantiated, then the device may be attractive to maritime operators struggling to satisfy both Annex V and Animal and Plant Health Inspection Service (APHIS) mandates.

Compactors may reduce the cost of waste disposal by reducing the volume of materials to be handled. Wastes destined for quarantine may be suitable for compacting, but APHIS treatment standards are calibrated on normal-density, uncompacted wastes. Those standards fail to meet the complete needs of the Annex V regime. To support Annex V implementation and expanded use of compactors aboard ships on international voyages to the United States, APHIS could arrange a series of calibration tests to establish appropriate quarantine treatment of compacted wastes. (A range of calibrations might be needed to allow for differences in compaction among units.)

Compactors are considered safe and efficient and are suitable for vessels that remain at sea for up to two or three days. These units effectively compact most shipboard garbage into paper or plastic containers, which can be sealed and stored safely on board for short periods until disposal ashore. Builders of small vessels could consider offering compactors as part of an integrated on-board garbage management system. Such technology could be incorporated readily into the design and construction of new vessels, and retrofitting may be a viable option on some existing vessels.

Comminuters, Pulpers, and Shredders

A comminuter is an oversized garbage disposal that reduces food scraps to a finely chopped residual, which is rinsed out of the unit with a steady stream of water. The effluent is a slurry of water and food bits. Commercial devices are made specifically for marine use. Annex V permits discharge via a comminuter, which is the single piece of shipboard equipment for which the Annex establishes a performance standard (see Appendix B, Annex V, Regulation 3). Therefore, disposal of food wastes is not a problem, because they can be ground up and discharged into the ocean[3]; the organic detritus can be assimilated into the envi-

---

[3]A greater problem is disposal of food-contaminated cellulosic material, such as paper, waxed paper, paperboard, cartons, and cellophane. Cellulosic materials also can be ground up, but it is difficult to separate out plastic coatings and film to prevent their discharge.

ronment. Even so, food discharges are prohibited within 3 miles of the coast (12 miles in all special areas except the Wider Caribbean).

A pulper is a powered device that reduces paper, cardboard, and other readily pulped materials into a mush that resembles papier-mâché. This pulp is rinsed out of the unit with a heavy, continuous stream of water, and the effluent is a slurry of pulp and water. A commercial unit has been manufactured for years. The Navy improved this device for its own shipboard use, developing both a small unit and a large unit designed for continuous heavy use. The small pulper can process up to 64 kilograms (kg) (140 pounds [lbs.]) per hour of mixed wastes, including paper, cardboard, and food wastes. The large pulper can process up to 308 kg (680 lbs.) of mixed wastes per hour. Both units capture plastics and metal and prevent their discharge. This equipment is designed to occupy the least amount of space possible and can be maintained in place; even so, the units resemble large industrial washing machines set on angled foundations.

Use of pulpers can reduce the aesthetic problems caused by intact garbage and permit discharges closer to shore than otherwise would be allowed. Some Navy personnel even see pulpers as an acceptable means for discharging wastes other than food (or plastics) into special areas. They assert that the biodegradable, pulverized, cellulosic effluent poses no harm, even in highly sensitive environments. The Navy is conducting research on this issue. At present, however, Annex V and the Marine Plastics Pollution and Control Act (MPPRCA) prohibit discharge of nonfood wastes into special areas.

As noted in Chapter 2, little is known about the behavior or effects of pulped garbage, paper, or cardboard in the marine environment. Larger, denser particles such as bone and seeds settle rapidly, while small particles could become widely dispersed in the surface water layers. Some fraction of pulped waste may float and eventually be found on beaches, while some accumulation of pulped waste could be expected on the sea floor of shallow special areas, such as the Persian Gulf and Baltic and North seas (Swanson et al., 1994).

Another way to treat paper on board vessels is with shredders, machines with rotating blades that also can be designed to shred bones, metal, glass, and plastics. One cruise line employs four types of shredders: a bone shredder and crusher; a paper shredder used upstream of an incinerator to improve combustion; a glass shredder and crusher; and a plastics shredder used prior to storage of this material (Richard Wade, Princess Cruises, personal communication to Marine Board staff, August 29, 1994). The Navy's shredder originally was designed to process 272 kg (600 lbs.) of glass and metal per hour (Swanson et al., 1994). The pieces were to be placed in burlap bags and thrown overboard. This plan has been abandoned and the shredder technology is now part of the Navy's plastics processor.

Once thrown overboard, bags of metal and glass tend to settle to the ocean floor. Swanson et al. (1994) estimated that, to ensure sinking, the ratio of metals to glass in a bag should be at least 1 to 2. Some bags may be recovered by fishermen using trawls or scallop or clam dredges. Other bags will deteriorate eventually and the contents will become part of the sedimentary record.

Incinerators

Incineration devices range from primitive "burn barrels" to complex dual-chamber systems with sophisticated emission controls. True incineration uses controlled combustion to achieve near-total destruction of waste with minimal emissions, so the more rudimentary burn barrels and older "fireboxes" seen on some ships are not representative of the technologies now available (Chang, 1990). The latest models have multiple chambers in order to maximize combustion of the waste and consume the resulting gases, and some of the exhaust heat can be reclaimed for other uses (Whitten and Wade, 1994). The IMO recently adopted standards for shipboard incinerators in order to document the technologies acceptable under Annex V and establish combustion performance standards in line with modern capabilities (the standards may be found at the end of Appendix B). The government of Bermuda also has established standards and licensed two ships to use incinerators (T. Sleeter, senior surveyor, Bermuda Ministry of the Environment, personal communication to Marine Board staff, June 2, 1994; Bermuda Ministry of the Environment, undated).

Properly designed and operated incinerators can burn successfully most types of garbage, including paper, cardboard, and, under certain conditions, plastics (metal and glass cannot be burned). A number of acceptable, purpose-built marine designs are manufactured and sold for commercial use. Several units with tailor-made sorting and ash-handling systems are now in service (see Figure 5-1). These integrated systems have enabled passenger vessels to comply with Annex V in situations where compliance would have been unmanageable otherwise. Many recently constructed cruise ships have one or two high-capacity incinerators (Whitten and Wade, 1994), and some government vessels are equipped with incinerators as well. The National Oceanic and Atmospheric Administration, despite some poor experiences with units installed on its fleet, has been advised to equip its vessels with either trash compactors or appropriately designed incinerators (Art Anderson Associates, 1993). The Coast Guard has purchased and installed a prototype unit meeting IMO standards on one of its cutters (Sara Ju, Coast Guard, personal communication to Marine Board staff, August 18, 1994).

It is important that the equipment selected be appropriate, that seafarers learn to use it proficiently, and that the units not be misused. Controlled combustion must be sustained in order to get good waste destruction; an incinerator is not appropriate for a vessel that generates very little or erratic amounts of waste. If fed too little waste, or surges of waste, an incinerator can perform poorly; either destruction may be inadequate or operating problems may arise within the unit. Thus, there are many instances where a vessel operator would do well to avoid relying on an incinerator. On the other hand, incinerators may be appropriate for

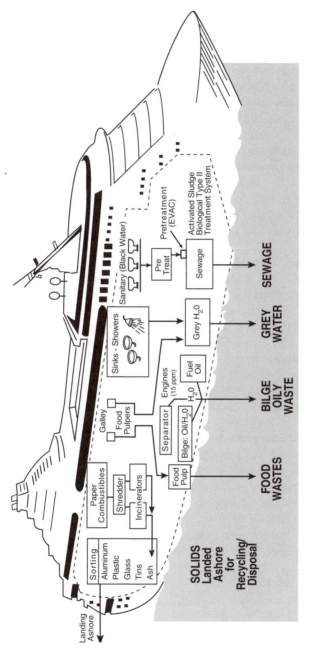

Figure 5-1 Cruise ship waste management systems. Source: Princess Cruises.

use on ships with large populations or that generate waste streams sufficient to sustain uniform combustion. Vessels in this category include ships carrying more than 100 persons, large fish processing or factory ships, and research ships on long voyages.

In the United States, there has been some public uncertainty about incineration of wastes due to concerns that emissions and residues resulting from use of this equipment cause more problems than they solve. Surveys by industry (an example is Kiser et al., 1994) suggest there is increasing public acceptance of and satisfaction with incineration as part of an integrated waste management strategy. The committee did not find any studies on this issue conducted independent of industry or groups opposed to incineration. While old incinerators could not meet current standards, the state of the art has progressed, and, in addition to the guidelines issued recently by IMO for on-board incinerators, stringent new operating standards have been established for units employed on land.

The primary concerns are whether the stack emissions or the ash resulting from fuel combustion pose hazards. Because its focus was on solid waste, the committee did not examine the air emissions issue in depth, but it deserves attention. The IMO guidelines for Annex V implementation recognize the potential for air pollution and therefore discourage use of incinerators in ports in or near urban areas. The Coast Guard plans to conduct emissions tests on its prototype unit. There is some legitimate concern among American ship operators that future restrictions on air emissions (International Maritime Organization, 1994a) could make existing shipboard incinerators obsolete. An additional concern is that the performance of incinerator technology installed to enable Annex V compliance is difficult to monitor; however, reliable instrumentation and recording units are available that document whether emissions are within regulatory standards.

The committee did examine the ash issue and found little cause for concern. Combustion of solid waste produces bottom ash (pieces of glass and metal, vitrified clays, and "clinkers") and fly ash (fine, lightweight particles). Tests conducted on the ash from a cruise ship incinerator showed that key contaminant levels not only were non-hazardous according to EPA standards but also were lower than those for ash produced by a municipal waste-to-energy plant (see Table 5-1).[4] (The materials burned on the ship did not include plastics.) Thus, at least for this particular batch of materials burned in this specific incinerator, the ash did not appear to pose a hazard, whether discharged overboard or in a landfill. Incinerator ash, including clinkers, is considered operational waste under Annex

---

[4]Shipboard incineration becomes even more attractive considering that it generates only one-third as much ash as is produced by a municipal waste-to-energy plant. The difference is due to the lighter weight of ship-generated garbage (e.g., light plastics, cardboard) as compared to shore-generated waste, which may include heavy materials such as wood and leather.

TABLE 5-1 Comparison of Contaminant Levels in
Ash from a Municipal Waste-to-Energy (WTE) Plant
and a Cruise Ship Incinerator (in milligrams per liter)[a,b]

Chemical	WTE Plant Ash	Ship Ash
Arsenic	0.093	not detected
Cadmium	0.012	not detected
Chromium	0.009	0.62
Copper	0.157	not measured
Lead	0.121	not detected
Mercury	0.0009	not detected

[a]The ash tested was obtained from the Delaware County Resource
Recovery Facility and the ship *Fascination* of the Princess Cruises fleet.

[b]Tests were conducted using the EPA's Toxic Characteristics Leaching Procedure, which discriminates between non-hazardous ash (which
can be landfilled with other solid waste) and ash classified as hazardous waste. In the leaching process, any metallic compounds in the ash
are dissolved in water.

V and therefore may be discharged at least 12 nautical miles from shore (except
in special areas). However, the IMO implementation guidelines recommend holding on board the ash from combustion of some plastic products that may contain
toxic residues.[5]

The committee's findings with respect to the ash are echoed by Bermudian
research, which was motivated by that island's dwindling land space for garbage
disposal. Those studies assessed the effects of dumping incinerator ash into landfills by focusing on the leachate expected to seep into the nearby marine environment. Results indicated no demonstrable increase in the concentration of metals
in the water column (Knap et al., 1992; Hjelmar, 1993). The government has
issued permits to two cruise ships for the use of incinerators while in Bermudian
ports and while underway in their waters. Discharge of ash is not permitted
within Bermuda's Exclusive Economic Zone.

If conducted in accordance with the IMO standards, incineration offers an
opportunity for ships generating large amounts of garbage to achieve self-con-

[5]The landing of incinerator ash in the United States could be affected by a recent U.S. Supreme
Court ruling (92-1889, issued May 2, 1994) that all ash from municipal incinerators is assumed to be
hazardous unless proven otherwise (Whitten and Wade, 1994).

tained waste management, and use of this technology may reduce the need for disposal at port reception facilities. But ship operators need to recognize that incinerators must be operated within design limits by trained personnel, and that the unit's performance must be audited. In addition, there is a need for U.S. regulators to establish performance standards for on-board incinerators used to enable Annex V compliance. One option would be for the EPA to accept officially the IMO incinerator standards, which were based in part on specifications submitted by the U.S. delegation.

Enhancing Shipboard Technology Development and Use

As the preceding summary shows, many on-board garbage handling and treatment options are available. But present technologies are not designed for every type of vessel, because the fleets are too diverse to provide for a commercial equipment market that meets every need. What is required is a mechanism for adapting available technology to the full spectrum of vessel types and, where necessary, developing new technology.

The federal government could provide such assistance by establishing a program to develop, test, and evaluate shipboard technologies for wide application. Research on maintenance and operating practices would need to be part of this effort, because problems in these areas have been identified as leading to breakdowns in Annex V compliance (Burby and Patterson, 1993). A possible lead agency is the Maritime Administration (MARAD), which already conducts a broad-based research and development (R&D) effort through its Office of Technology Assessment. Garbage treatment technology would seem to be a pertinent topic for the MARAD program, which, among other things, identifies and stimulates the transfer of advanced technologies from other areas into the maritime environment, and serves as a focal point to bring advanced technical expertise to bear on issues of concern. In addition, the Maritime Administration has five large cargo ships that are supplied to state maritime academies for training purposes. Each ship carries 200–500 persons on two-month voyages. These ships might be used as research platforms for on-board garbage treatment technologies.

Another option would be to expand the Navy R&D program to develop on-board garbage handling and treatment technologies for both military and commercial use. Although military technology development has shifted to a dual-use focus, the Navy has yet to develop on-board equipment for commercial fleets. Doing so would correspond to the current administration's emphasis on defense conversion, but it might interfere with the Navy's effort to develop its own environmentally sound fleet.

The government also could provide financial assistance for research on and installation of garbage treatment technology. The National Marine Fisheries Service (NMFS) already offers similar assistance to the fisheries fleets through its Capital Construction Fund Program. Extending such assistance to cover garbage treatment equipment might be advisable, in that some operators, particularly in smaller fisheries, may not be able to afford the requisite improvements on their own. The capital construction program requires that (1) at least 80 percent of reconstruction expenditures be classifiable as "capital" expenses for tax purposes, and (2) costs be either at least $100,000 or, if less, equal to at least 20 percent of the original acquisition cost of the vessel involved. The NMFS has waived the second requirement for vessel improvements to conserve energy; a similar waiver is being considered for improvements to increase vessel safety. Given this philosophy of using the program to support other federal mandates, waivers for pollution abatement also may be appropriate.

GARBAGE MANAGEMENT IN PORTS

The port reception facility is the link between the international Annex V regime and the U.S. integrated solid waste management system. That interface needs to be as seamless and transparent to users as possible. The committee developed and sent a short questionnaire to a variety of port authorities, port users, and other waterfront facility operators. The responses indicated that the ship/shore interface in the United States is clumsy, inadequate, and at times non-existent. Each individual port or terminal has to devise its own means to comply, and each has to pay for any related expansion. Rarely has a port had either the funding or the technical preparation to execute the task alone.

The problem is not simply that port reception facilities are lacking, although this is sometimes the case. In fact, a recent Coast Guard survey found that reception facilities are readily available on the East and Gulf coasts (North, 1993). The poor interface can be attributed to a variety of factors, including whether a port will allow the vessel to off-load garbage, whether the vessel operator knows that reception facilities exist and where they are located, and whether the facilities are convenient and affordable.

The committee found it impossible to gauge the overall level of activity in all the U.S. ports and terminals that must comply with Annex V.[6] Port information for the nine maritime sectors examined by the committee is mostly anecdotal. Available public information has been collected mainly for U.S. Customs purposes; only commercial vessels (both passenger and cargo) are monitored closely.

There is little evidence of strategic planning to support the provision of adequate reception facilities, other than IMO's recent efforts to begin to provide

---

[6]The states of Texas and Louisiana have sponsored an extensive survey of garbage reception facilities in the Gulf of Mexico through two Sea Grant studies (Hollin and Liffman, 1991, 1993). The scarcity of data on waste management spurred the National Solid Wastes Management Association (NSWMA) to initiate a program in 1993 to develop improved estimates (Gene Wingartner and Allan Blakey, NSWMA, personal communication to Marine Board staff, September 25, 1992).

much-needed guidance.[7] In the United States, port governance is highly decentralized. Indeed, ports are far more likely to compete than to cooperate. Public ports, through the American Association of Port Authorities (AAPA), exchange information about MARPOL compliance; however, AAPA's membership is largely public and often does not include owners of private terminals. It has not been industry practice to organize internally to coordinate implementation of Annex V or any other international agreement. As a result, terminal operators employ a variety of strategies to handle vessel garbage.

The range of current port practices may be best illustrated by specific examples. The norm for the United States is the Port of Charleston, South Carolina. About 200 commercial cargo ships call at the port per month, and each ship's agent makes separate arrangements for vessel services, including garbage disposal; the cost is not included in the port fees. Very little vessel garbage is separated for recycling in the community system. Two companies handle quarantined garbage, which is bagged, boxed, labeled, taped, and hauled 60 miles to an APHIS-certified incinerator.

The decentralization of port governance in the United States is quite different from the approach taken in most other countries, where centralized port systems allow for more effective intervention at the national level. Internationally, the state of the art is the highly effective garbage service provided by the Port of Rotterdam in the Netherlands, the largest port in the world (Port of Rotterdam, 1992). When a vessel arrives in the port, the port office collects from the ship's agent a deposit to cover port fees and waste collection and disposal charges. Any fine levied against the vessel also is charged against the deposit. When the vessel leaves the port, any unspent portion of the deposit is returned. To handle garbage, the port issues licenses to four firms for the provision of reception facilities. Each licensee charges a tariff for services rendered, with charges based on tonnage and material type. Garbage usually is transported by barge; garbage containers are lowered into the barge, and the average weight of all the containers is estimated for billing purposes. The port imposes a separate environmental fee. The port's participation in the vessel garbage management system ensures that services are carried out in a predictable manner.

Several years ago, a regional regime was created by port states to manage marine pollution in Europe. This effort has helped coordinate inspections of merchant vessels among 14 European nations. The inspections focus on struc-

---

[7] The IMO Marine Environment Protection Committee approved the text of a *Manual on Reception Facilities* at its March 1994 meeting. The manual provides advice on developing a waste-management strategy; planning of reception facilities and choice of location; equipment for garbage collection, storage, and treatment; recycling and disposal; financing and cost recovery; and the needs of small vessels (International Maritime Organization, 1994b). This guidance should assist in U.S. implementation of Annex V.

tural integrity and the condition of on-board pollution control equipment. In mid-1993, the Port of Rotterdam went even further, initiating a partnership with several other port authorities for the exchange of information concerning amounts of garbage on board a ship when it leaves a port. This program, known as Port Promotion MARPOL, initially will include port authorities in Bremen, Germany; Felixstowe, United Kingdom; and Barcelona, Spain.

Correspondence received by the committee indicates that some local and national governments in other countries have integrated waste management at ports into planning for municipal and regional waste systems. Both Bremen, Germany and Copenhagen, Denmark established Annex V port reception facilities as part of overall municipal waste management (Federal Republic of Germany, 1990; Larsen and Borrild, 1991). Even the Port of Manila, in the Philippines, where vessel operators historically discharged garbage overboard without punishment, now mandates that operators off-load ship's garbage into port reception facilities (Fairplay International Shipping Weekly, 1993).

An example of well-managed garbage management at U.S. marinas may be found at the Port of Oakland on San Francisco Bay. Recreational boat owners have cooperated with the 10 port-owned marinas in curtailing trash discharges at sea. The port operator, based on polls of boater tenants, determined that it would be sufficient to supply ample dumpsters at each marina and arrange for regular pickups of the garbage (Irvin-Jones, 1992).

The special needs of fishing ports merit some attention, for two reasons. First, fishing seasons are shortened artificially by the federal management regime (Pacific Associates, 1988), so fishing ports must plan for fluctuating demand for garbage reception facilities. The surge loads on landside facilities may overwhelm local capabilities on occasion. Second, fishing fleets sometimes operate from terminals that are managed privately or by the local government, rather than by a public port authority. As a result, development of reception facilities in fishing ports has been uneven, even though notable pilot projects have been undertaken with federal funding (Recht, 1988; Recht and Lasseigne, 1990) and state and local government funding (Bayliss and Cowles, 1989).

In the Gulf of Maine, a regional campaign is under way to encourage fishermen to bring debris back to shore and deposit it into reception facilities at piers (Pearce, 1992). In the Gulf of Mexico, the shrimp fleet in Aransas Pass, Texas operates out of a harbor owned by the municipality, and the city manager oversees the fishing port. The town installed additional dumpsters at the docks to enable Annex V compliance by shrimpers. By contrast, some sport fishing piers in the United States have no dumpsters or other waste receptacles anywhere in sight.

Garbage Management Strategies

Although it is difficult to generalize about garbage handling practices in

TABLE 5-2   Providing Port Reception Facilities

Facility	Disposal	APHIS Recycling	Key Waste	Needs
Residential dock	Household waste receptacle	Curbside or drop-off	n.a.	Coordinate with shoreside ISWMS
Boat ramp	Litter containers[a]	Off-site drop-off	n.a.	Coordinate with shoreside ISWMS
Marina	Dumpster[a]	On- or off-site drop-off	n.a.	Coordinate with shoreside ISWMS
Captive pier/ terminal	Dumpster[a] or storage facility	Off-site drop-off	Yes (with interim storage)	Coordinate recycling with local requirements; coordinate APHIS with ISWMS
Commercial pier or port	Commercial (for fee) pickup[b]	On-site drop-off	Yes (on demand)	Build cost of pickup into disposal fee; coordinate APHIS with ISWMS
Large harbor complex (Naval base)	Commercial pickup (for fee or contract)[b]	On-site drop off or commercial pickup	No	Pickup using truck or barge

[a]Affix placard identifying location of and materials accepted by nearest drop-off center(s)
[b]Provide fact sheet (multilingual) to ship captain or agent identifying available services, fee structures, requirements (including APHIS requirements), and locations of and materials accepted at drop-off recycling centers.

ports, the basic strategies can be characterized according to port type, as indicated in Table 5-2. All nine categories of vessels examined in this report use one or more of the port facilities listed. It is important to remember that ports vary widely in terms of size, types of vessels served, and management organization—all factors that affect choice of garbage management strategies.

Disposal approaches (the second column) can be as simple as putting a trash can on a dock. On the other hand, disposal can become complicated for commercial vessels calling at many different ports, due to variations in garbage-handling practices, restrictions, and fee structures. Costs vary, as they reflect local disposal costs for land-generated garbage and may be based on tons, cubic meters, truckloads, or other measures, depending on local practice. The cost in 1992 was $400 for 36 cubic meters in Honolulu; $1,300 per ton in San Diego; free of charge for the first truckload in Goa, India; and $1,288 for a 60-meter barge full in Hong Kong. Docking fees typically cover docking costs and do not include garbage disposal services or costs; some managers believe this is a source of inefficiency (Robert N. Shepard, Fennell Container Company, personal communication to Marine Board staff, July 1, 1992).

Recycling (the third column) may be easy to arrange in a community that offers curbside recycling to residents and businesses. On the other hand, a regional or national infrastructure is required to direct the materials from widely dispersed terminals into the recycling network. Several local fishing communities have demonstrated that fishing nets can be collected and recycled, but it is difficult to sustain such efforts without access to either a regional network of recyclers or waste exchanges that can help locate markets for the recycled materials. When the Coastal Resources Center (CRC) conducted a recycling pilot project at a recreational marina in California (Kauffman, 1992), the effort proved more difficult than had been anticipated due to swings in the markets for reselling the collected materials and the amount of work required to deliver the materials to the recyclers.[8] Such problems may tend to limit the popularity of recycling as an option for handling vessel garbage. Strengthening this infrastructure would be a way to promote recycling.

Some garbage materials may be exactly what a manufacturer needs or can use as feedstock. Waste exchanges have evolved with EPA support to encourage those with waste materials to locate others who can use the materials (U.S. Environmental Protection Agency, 1994). Such efforts can help identify potential markets for recyclable materials that otherwise would be returned as garbage to port reception facilities. Mariners and port operators could participate in these waste exchanges. Some efforts are under way in the United States to improve the prospects for recycling: Researchers at New Jersey Institute of Technology are exploring the use of old plastic fishing nets (and maybe nylon and plastic fishing line) to form a matrix in asphalt.

Key needs (the fifth column) include coordination within the port and between the port and the local or regional ISWMS.

ENHANCING THE VESSEL GARBAGE MANAGEMENT SYSTEM

Some factors contributing to successful garbage management by ports have been identified. For instance, while flexibility is necessary to accommodate local needs and resources and the wide variance in vessel practices, it is essential that one entity take charge of identifying and planning the port's waste management activities and work toward their implementation (Recht and Lasseigne, 1990; Kearney/Centaur and Martinez, 1991). However, in most U.S. ports, no one person affiliated with the port or local government is responsible for port-wide garbage management planning. As a result, vessel garbage is handled on an ad hoc basis.

---

[8]The CRC overcame these difficulties, and the project has become a permanent fixture at the harbor. Indeed, comprehensive recycling has had a number of benefits, including reductions in garbage hauling costs and the amount of staff time spent on waste management.

This situation reflects a larger issue related to Annex V implementation. The U.S. port system is decentralized at both the local and federal level, and there is great diversity among ports. Even in a single port, all facilities are not managed centrally; terminals in Boston harbor, for example, are run by the Massachusetts Port Authority and myriad other public and commercial groups. The lack of a national port governance system—which most countries have—impedes U.S. implementation of Annex V, because MARPOL assumes a direct link between "the government of each Party to the Convention" and the local port reception facility. No such link exists in the United States; the federal government has indicated repeatedly that it will rely on the free market to provide port reception facilities. Control is limited further by the lack of any requirement that ships off-load garbage upon either arrival at or departure from U.S. ports.[9] Such a requirement, potentially a straightforward way for the government to exert additional control, would be particularly useful in the case of large commercial vessels, such as cargo and cruise ships, that generate sizeable amounts of garbage.

Apart from requiring off-loading of garbage by commercial ships, the federal government could take additional steps to improve vessel garbage management. There are four primary barriers to the internal integration of the system. These barriers need to be overcome if the system is to function effectively. The four salient issues are quarantine requirements for vessels arriving from foreign waters; implementation of the Coast Guard COA program for ports; port operators' liability for handling vessel garbage; and financing—both who should pay for garbage services and how they should pay.

Quarantine Requirements

The Animal and Plant Health Inspection Service, a unit of the U.S. Department of Agriculture (USDA), prohibits the off-loading of any garbage that has come in contact with either animal or plant products originating in or transported through a foreign country, unless the garbage is handled under strict procedures to ensure quarantine (9 C.F.R. §94.5; 7 C.F.R. §330.400). The objective of these controls is to prevent contaminated animal or plant material from bringing new diseases into the country. A vessel operator may comply by retaining suspect materials on board, in tight waste containers "inside the rail" of the weather deck, so nothing can be dropped accidentally on the pier. The regulated garbage may be off-loaded only in tight receptacles and must be incinerated or sterilized prior to disposal in an approved landfill. In addition, the process must take place under the supervision of either a USDA inspector or a contractor who has signed a compliance agreement (in the latter case, APHIS monitors the process occasion-

---

[9]An informal Coast Guard survey of port reception facilities on the East and Gulf coasts indicated that fewer than 20 percent of vessels off-load garbage (59 Fed. Reg. 18,700 [1994]).

ally). Before Annex V came into force, mariners complied with APHIS rules by discharging garbage overboard before entering U.S. waters; obviously, for many, this is no longer a legal option. Therefore, the quantity of foreign garbage to be held has increased.

Garbage subject to APHIS inspection is generated mostly in the galley area[10]; the vessel must maintain an additional and separate sorting, packaging, storage, and disposal system for this material. Under Annex V, plastic food-packaging materials may be disposed of only on shore, a requirement that effectively adds to the amount of APHIS waste that either must be retained on board or must be discharged to a certified port reception facility. Due to its specialized nature, APHIS waste handling is much more expensive and difficult to obtain than is ordinary garbage disposal. Cargo vessel operators often strive to avoid use of U.S. ports for APHIS waste disposal, due in part to the high cost, which may range from $250 to more than $1,000 per pickup.[11] There is also considerable confusion among vessel operators concerning what types of waste must be quarantined and the basis for the disposal charges.[12] An additional concern with respect to Annex V implementation is the need to make separate arrangements for shoreside disposal of APHIS waste and Annex V garbage; the lack of full integration of the two garbage management regimes adds to the burden on vessel operators and may be a deterrent to compliance. Yet, at the same time, APHIS has contributed to implementation of Annex V. There is a standing agreement for APHIS boarding officers to assist the Coast Guard in monitoring arriving vessels for compliance with Annex V. Inspectors ask four questions related to Annex V, and this assistance has resulted in numerous Annex V violation reports (U.S. Coast Guard, 1993).

The APHIS regulations, which specify methods for packaging, transporting, and disposing of the waste, were developed separately from the national ISWMS. But APHIS has modernized and partially integrated its program with other garbage management systems. Since the advent of Annex V, for example, APHIS handling and transportation requirements have been altered to comply with procedures for handling hospital waste.[13]

---

[10]The other main source is spoiled cargoes of animal products.

[11]The cost is high for two reasons: The APHIS waste stream is small in comparison to amounts of Annex V garbage, and the required handling techniques are relatively expensive.

[12]This confusion can increase both actual and perceived disposal costs for Annex V garbage. Many vessels operators do not realize that they must separate Annex V garbage from APHIS waste and as a result must pay the higher APHIS disposal fees for mixed waste (U.S. Coast Guard, 1993). In addition, some shipping company operators have misinterpreted an APHIS inspection fee (instituted in 1991) as related to Annex V (Coe, 1992).

[13]The medical waste management system, created since the late 1980s (partly in response to syringes and other medical waste washing up on beaches), requires strict chain of custody and controlled destruction of materials capable of transmitting pathogens dangerous to humans.

Managers of APHIS programs also have attempted to respond to the sudden increase in the need for their services resulting from Annex V and the COA program. Before Annex V, APHIS waste haulers typically were not allowed to transport quarantined wastes through rural areas. This policy was changed in 1988 to allow certified waste haulers to transport containers of garbage through rural areas and for long distances. Also in 1988, following passage of the MPPRCA, the Coast Guard was directed by law to require ports to prove they could provide reception facilities for quarantined garbage; without such a capability, a port was unlikely to receive a COA verifying its compliance with Annex V. In 1987, only 32 ports had facilities that were approved to handle garbage under USDA regulation; by 1992, most U.S. ports had USDA-approved garbage handling procedures and appropriate equipment (Caffey, 1993).[14] Access to proper equipment may be limited in certain cases, however. Some hospitals and international airports have the technology, for example, but their operators, fearing damage to the equipment, generally decline to make it regularly available for vessel garbage (Carangelo and Buch, 1993).

The same compliance agreements and supervision provisions are used to handle both vessels and passenger aircraft, and the airport side of the quarantine program seems to function well.[15] However, there is at least one major difference between APHIS operations at airports and those at seaports: Quarantined garbage is removed from aircraft at the end of each flight due to the lack of on-board storage space, while ships do not necessarily off-load any waste in port. Furthermore, compliance agreements at airports are with caterers, who personally board aircraft and remove regulated garbage, whereas compliance agreements at seaports are with waste haulers, who do not board vessels and therefore have no control over what is off-loaded. All of this means that waste haulers at ports, because their APHIS services are in less demand than are those of airline caterers, have less market incentive to comply with regulations and thereby maintain and attract business. The airport practices may have the effect of increasing control over garbage management, in addition to freeing up on-board space for storage of additional garbage. Both of these effects are desirable.

In summary, the committee identified four basic problems related to the

---

[14]Because APHIS allows waste haulers to transport waste for long distances, numerous ports deemed capable of providing APHIS waste reception facilities actually have no such facilities. Instead, the garbage is transported to facilities elsewhere. That extra shipment certainly increases the disposal cost to ship operators using those ports.

[15]Airports enter into agreements that make compliance easy and routine. The quarantine practices do not cause delays in flight operations. The garbage removal is performed largely by airline caterers in well-monitored kitchens on the airport premises (Carol Heaver, Ogden Aviation Services, personal communication to Marine Board staff, October 17, 1991). In addition, APHIS prepares training materials in multiple languages for new catering employees, to reinforce the need to adhere to quarantine practices (Caffey, 1991).

APHIS regime. One, high disposal costs, is outside the committee's scope. The second problem, the confusion over what types of garbage are subject to quarantine, is relevant to the committee's task, in that Annex V compliance depends in part on widespread understanding of proper garbage handling practices. The third problem, the lack of full integration of the APHIS and Annex V regimes, is directly relevant to the present study because it is further evidence of the need for a systems approach to vessel garbage management. The overlay of Annex V on APHIS regulations may compound confusion and compliance problems among vessel operators.

The fourth problem, the lack of a requirement for off-loading of APHIS waste at U.S. port calls, is related to the need for integration of the APHIS and Annex V regimes. As noted earlier, vessels are not required to off-load Annex V garbage either, although other nations have adopted such mandates and there may be good arguments for doing so. If vessel operators were required to off-load Annex V garbage, then the adoption of parallel requirements for APHIS waste would have the multiple benefits of fostering integration of the two regimes, freeing up much-needed space on board, and bringing the seaport side of the quarantine program into line with the airport side. This concept is applicable primarily to cargo and passenger cruise ships, which may generate large amounts of garbage, including APHIS waste, and routinely call at commercial ports.

Port Accountability

Also of concern are the significant gaps in port controls. The COA program and the related requirements covering smaller terminals are meant to assure the existence of a complete garbage management plan that covers, among other things, the handling of APHIS waste. But the certification process only shows that the structure for compliance exists within a port serving large tankers or fishing vessels[16]; there is little verification that the structure actually functions as described. Similarly, while reception facilities also are required at small fishing piers, recreational marinas serving 10 or more boats, and terminals serving offshore oil and gas operations, the Coast Guard neither inspects the facilities nor requires that COAs be obtained. Still another problem is that regulations do not identify clearly the parties responsible for implementing APHIS requirements in a terminal or port. Because the current regulations are not comprehensive (the many small, unattended piers and launch ramps are not covered), do not assign responsibility for port improvements, and do not require record keeping or inspections, the system of controls is primarily an exercise in paperwork.

---

[16]As noted in Chapter 1, the COA program applies to ports and terminals serving vessels of 400 gross tons or more carrying oil or noxious liquid substances, or those that serve fishing vessels that cumulatively off-load more than 500,000 pounds of commercial fishery products during a calendar year.

Nonetheless, if the overall garbage management system is to be strengthened, then the COA program is a logical starting point. Several MPPRCA amendments have been proposed that would require inspections of COA facilities when the owner or operator changes, make COAs valid for a five-year time period, require inspections before issuance of new certificates, and mandate examinations of all non-COA holding facilities. If adopted, these provisions may be helpful. Even so, designing and administering the COA program is a heavy burden on the Coast Guard, which has no expertise in waste management and might be overwhelmed by the attempt to ensure that the more than 10,000 U.S. ports provide garbage reception facilities that are truly adequate. The more logical authorities for overseeing the land side of the vessel garbage management system are the EPA, which has extensive expertise in handling waste of all types, and the states, which develop solid waste management plans authorized by the Resource Conservation and Recovery Act (RCRA) (P.L. 94-580), as amended.[17] (States must submit these plans, which detail regulations and strategies, in order to avoid having EPA take over their programs.) Unlike the Coast Guard, the EPA and the states employ waste professionals who are engaged full-time in managing regimes for solid and industrial waste.

At present, the Coast Guard is the primary government authority with official responsibility for overseeing port reception facilities. But the committee has obtained a legal opinion stating that RCRA and the regulations are sufficiently broad that they arguably could allow a state's solid waste management plan to cover a vessel docked at a port in the state (Dana J. Schaefer, Parkowski, Noble and Guerke [Dover, Delaware], personal communication to a member of the Committee on Shipborne Wastes, March 23, 1994). The EPA could establish technical standards for determining whether port reception facilities are "adequate," and states could assure that the standards were met as part of their waste management planning process. The EPA has supported similar technical assistance when other waste streams have been brought under federal control (U.S. Environmental Protection Agency, 1990a and 1990b; Council of State Governments, 1992). Certainly in Texas and New Jersey, where reducing waterborne and beach debris is a top public priority, full integration of port reception facilities into the state ISWMS would be a logical approach. Either legislation or a regulatory directive might be required to bring the EPA into this process.[18]

To supplement the COA program, other government agencies that regulate ports could help assure the adequacy of port reception facilities. State govern-

---

[17]These provisions are codified at *United States Code*, Title 42, Sections 6941-6949.

[18]The EPA interprets current requirements as addressing *permanent* disposal structures (40 C.F.R. §258 establishes minimum criteria for landfills, which are considered permanent structures with lasting impact on the environment). Dumpsters and other temporary facilities are considered disposal *practices*, which the EPA has chosen not to regulate.

ments issue permits related to matters such as waterfront construction and environmental regulations and could review or require port reception facilities as a condition of granting permits to ports. In addition, the U.S. Army Corps of Engineers routinely surveys and approves new docks and other port structures and could review the adequacy of port reception facilities as a part of this process. Review of port reception facilities as part of existing regulatory processes could help foster Annex V compliance without overburdening government agencies.

Liability

Since the late 1970s, the U.S. Congress has enacted several laws (e.g., RCRA and the Comprehensive Environmental Response, Compensation, and Liability Act [CERCLA] [P.L. 96-510], known as the "Superfund" law) that changed the legal responsibilities of those who create wastes and those who handle, transport, or treat wastes. The intent has been to remedy problems caused by old practices and to halt the use of ineffective practices. In fact, many practices have been abandoned under the new laws, and the government has supported substantial technical research to help develop new, more reliable techniques for handling wastes of all kinds.

The RCRA and CERCLA regimes also have gained public recognition because of their emphasis on punishing offenders and allocating liability for damages resulting from poor waste disposal practices. Legal precedents have been established in this arena that expand the range of entities that can be held accountable for a polluting event, well beyond the obvious candidates. In particular, the "cradle-to-grave" model that forms the basis of these regimes establishes legal liability for everyone who comes into contact with a waste material. Many businesses, concerned that they might become entangled unwittingly in the legal consequences of poor waste handling, have imposed strict audits and controls on their own waste generation and on the haulers who service their facilities.

It should be no great surprise, then, that fear of being saddled with liability for vessel waste handling is impeding implementation of Annex V (Pisani, 1989). Both public and private port operators are concerned that a more active role by public authorities in developing and overseeing a vessel garbage management system would expose ports to liability, particularly with regard to APHIS and hazardous wastes. As a result, vessel owners have been on their own in identifying and implementing waste disposal alternatives, at least in the United States. Those who drafted Annex V did not anticipate placing ports in legal jeopardy; nevertheless, this issue requires attention if ports are to become active players in the development of an effective vessel garbage management system.

In the judgment of the committee, concerns over port liability are not well justified at this stage. If port reception facilities were integrated into the national ISWMS, then much of the uncertainty over liability would be eliminated. Management systems for other forms of waste seem to work well and address liability concerns. For example, in the medical and hazardous waste programs, records are kept that can link the generator to the disposal process. The chain of custody is established, and liability is shared by all those involved; if problems arise, then the waste can be tracked and the culprit identified.

Who Should Pay?

There is considerable debate over who should pay for vessel garbage services. On one level, the question is whether these services are a public responsibility, to be funded by government regardless of the amount of garbage or level of service use, or a private responsibility, to be paid for only by those who use it. On another level, the question is, what funding mechanism should be used? These questions need to be answered if the vessel garbage management system is to be effective and efficient.

Fundamentally, there are three options. One is for each vessel or agent to arrange for garbage services individually, with no port involvement beyond the provision of adequate reception facilities. This is the current approach. While in keeping with the government's free-market policy, this method has allowed for wide variations in disposal fees and, in some cases, inadequate facilities. The inconsistency among disposal fees and the perception that costs are too high have discouraged some vessel operators from off-loading garbage,[19] which then may end up in the ocean or at ports with less-expensive fees but inferior disposal practices. With respect to the adequacy of facilities, the U.S. debate on this issue has not addressed the true costs (including debt service) of providing additional garbage services. It may be that these costs deter port operators from upgrading facilities. This situation needs to be examined in detail, to determine whether the free market can provide for adequate facilities.[20] In other pollution-control arenas, the federal government has offered a variety of incentives and financing

---

[19] A Coast Guard survey revealed that vessel operators may avoid using U.S. port reception facilities for several reasons, including a perception that disposal costs are exorbitant, confusion over the distinction between Annex V garbage and APHIS waste, cost differences among states, and variations in the types of containers used (North, 1993; U.S. Coast Guard, 1993). The Coast Guard has suggested that one way to improve Annex V compliance would be to reduce the cost of garbage disposal options (Eastern Research Group, 1992).

[20] The experience in Corpus Christi suggests that a port making a large investment in garbage services is unlikely to see corresponding returns. The Port of Corpus Christi Authority constructed a modest steam boiler, which beginning in mid-1989 was operated full-time by a port employee as an APHIS-certified facility. Costs were high: In addition to the initial $100,000 capital investment, the port had to assume liability for waste treatment (Carangelo and Buch, 1993). The facility was shut down in early 1994 because, ironically, the waste hauler found it cheaper to truck quarantined materials to Houston.

vehicles to make funding available to compensate for market hesitation. Such an approach might be warranted as part of Annex V implementation.

The second option is for garbage services to be covered by the tariff that, in some ports, is paid by all vessel operators, regardless of whether garbage is deposited. This approach is the simplest option to administer but distributes the costs among parties who do not benefit directly. Moreover, ports are moving away from tariffs. The third option is to impose a standard fee for garbage services actually used. This approach conforms to the pattern for other services, which vessel operators pay for on a fee-for-service basis. It has been suggested that ports should participate in setting or capping garbage-disposal rates.

Both the second and third options would require that a port arrange for garbage services for all vessels calling there and assure that the costs of those services were covered in some manner by tariffs, fees, or some other revenue source. Then the question becomes whether private terminals should be subject to the same rules as public terminals.

In the absence of a cohesive national port system, the federal government may need to initiate discussions of these issues as part of its effort to assure that adequate port reception facilities are provided. Regardless of which option is pursued, it may be prudent and indeed necessary for port authorities to work cooperatively on a wide scale (either regional or national) to establish a common fee for garbage hauling, independent of the port receiving the ship. Because the crucial parameter is the prevailing cost in the land-based disposal system, there is little basis for neighboring ports to diverge much from those fees. Long-term arrangements might offer an economic benefit; a ship operator who purchased waste-hauling services only occasionally would pay a higher fee than would a customer who negotiated a long-term contract.

SUMMARY

The preceding analysis of vessel garbage management as a system identifies numerous opportunities for improving the system and thereby the implementation of Annex V. On the vessel side of the system, the government could provide three general types of assistance designed to foster Annex V compliance:

• *Technology Assistance.* A range of on-board garbage handling and treatment technologies is available. In some cases, commercial equipment can be purchased, but available technology may be inappropriate, due to its size or operating features. Some vessel operators may require assistance in locating available equipment or adapting or developing improved or more appropriate units. The federal government could facilitate technology transfer, so that all maritime sectors could make maximum use of information about Navy and cruise industry R&D, as well as equipment designed for land use. The government also could establish a program to develop, test, and evaluate shipboard technologies for wide application. The Maritime Administration could lead the effort, or the Navy R&D program could be expanded to develop on-board garbage handling and treatment technologies for commercial use.

• *Guidance on Key Issues.* Federal agencies could take steps to resolve issues that may be impeding safe garbage storage and expanded use of garbage treatment equipment. Guidelines on shipboard sanitation could be developed and technical assistance provided for all fleets to ensure that on-board storage procedures were safe. To support expanded use of trash compactors, APHIS could develop standards based on compacted waste. And, to foster proper use of incinerators, the EPA could adopt the IMO standards for shipboard incinerators.

• *Financial Assistance.* Some fleets, notably the fisheries sector, may require financial assistance in order to achieve compliance. The NMFS could offer financial assistance for research on and installation of garbage handling and treatment technology to fisheries fleets. To expand access to this assistance, the NMFS could consider waiving the $100,000 minimum expenditure requirement for the Capital Construction Fund Program.

Turning to the port side of the system, there is little evidence of strategic planning to support the provision and use of adequate garbage reception facilities. Steps could be taken in five areas to improve the vessel–port interface:

• *Require Cargo and Cruise Ships to Off-Load Garbage at U.S. Port Calls.* Such a requirement would help ensure that large commercial ships use port reception facilities and thereby increase government control over the vessel garbage management system.

• *Strengthen the Recycling Infrastructure.* The vessel garbage management system could benefit from an improved infrastructure for recycling, to take advantage of this now-standard mechanism for reducing waste streams.

• *Transfer Oversight of Port Reception Facilities to EPA and the States.* Responsibility for port reception facilities could be assumed by waste management experts within EPA and state governments; EPA could set technical standards, and states could assure that the standards were met as part of the waste management planning process under RCRA. State governments also could review or require port reception facilities as a condition of granting permits to ports.

• *Improve Integration of the Annex V and APHIS Regimes.* The federal government could make it easier for vessel operators to comply with all applicable Annex V and APHIS regulations. Such an effort would involve educating mariners about both Annex V and quarantine requirements, ensuring that any off-loading requirements were parallel, and working toward a system that would require vessel operators to make only one arrangement for handling of both types of garbage in a port.

• *Address Payment Issues.* Attention to the question of who should pay for garbage services, and how, could help ensure that port reception facilities are adequate and that vessels use them, instead of the oceans, for disposal of garbage. As part of this process, ports may need to cooperate in setting fees. The federal government may need to initiate discussions of these issues.

REFERENCES

Art Anderson Associates. 1993. NOAA Fleetwide Shipboard Waste Management. Report prepared for the National Oceanic and Atmospheric Administration by Art Anderson Associates, Bremerton, Wash. Jan. 29.

Atkins, W.H. Undated. Modern Marine Terminal Operations and Management. Oakland, Calif.: Port of Oakland.

Bayliss, R. and C.D. Cowles. 1989. Final Report on the Impact of MARPOL Annex V Upon Solid Waste Disposal Facilities of Coastal Alaskan Communities. NWAFC Processed Report 89-20, prepared for the Southwest Alaska Municipal Conference. Available from the Marine Entanglement Research Program of the National Marine Fisheries Service (National Oceanic and Atmospheric Administration), Seattle, Wash. October.

Bermuda Ministry of the Environment. Undated. Standard Operating Conditions for Cruise Ship Incinerators. Available from the Ministry of the Environment, Government Administration Building, 30 Parliament Street, Hamilton HM 12, Bermuda.

Burby, R. and R.G. Patterson. 1993. Improving Compliance with state environmental regulations. Journal of Policy Analysis and Management 12(4):753-772.

Caffey, R.B. 1991. Presentation by Ronald B. Caffey, Animal and Plant Health Inspection Service, to the Advisory Panel for the Shipping Industry Marine Debris Education Project of the Marine Entanglement Research Program (National Oceanic and Atmospheric Administration), at the Kearney/Centaur office, Alexandria, Va., Feb. 11, 1991.

Caffey, R.B. 1993. Testimony of Dr. Ronald B. Caffey, assistant to the deputy administrator, Veterinary Medical Office, Plant Protection and Quarantine, Animal and Plant Health Inspection Service, before the Subcommittee on Superfund, Ocean, and Water Protection of the Committee on Environment and Public Works, U.S. Senate, 102nd Congress, Second Session, Washington, D.C., Sept. 17, 1992. Pp. 28-30 in Implementation of the Marine Plastic Pollution Research and Control Act. S. Hrg. 102–984. Washington, D.C.: U.S. Government Printing Office.

Carangelo, P. and T. Buch. 1993. Presentation by Paul Carangelo and Tex Buch, Port of Corpus Christi Authority, to the Committee on Shipborne Wastes of the National Research Council, at the Port of Corpus Christi Authority, Corpus Christi, Tex., February 17, 1993.

Centers for Disease Control. 1989. Vessel Sanitation Program Operations Manual. Available from the U.S. Public Health Service, Vessel Sanitation Program, 1015 N. America Way, Miami, Fla. 33132.

Chang, T. 1990. Low technology (burn barrel) disposal of shipboard-generated (MARPOL V) wastes. Pp. 915-920 in Proceedings of the Second International Conference on Marine Debris, 2–7 April, 1989, Honolulu, Hawaii (Vol. II), R.S. Shomura and M.L. Godfrey, eds. NOAA-TM-NMFS-SWFSC-154. Available from the Marine Entanglement Research Program of the National Marine Fisheries Service (National Oceanic and Atmospheric Administration), Seattle, Wash. December.

Coe, J. 1992. Presentation by James Coe, National Marine Fisheries Service, to the Committee on Shipborne Wastes of the National Research Council, at the Governor Calvert House of the Historic Inns of Maryland, Annapolis, Md., May 7–8, 1992.

Council of State Governments. 1992. Model Guidelines for State Medical Waste Management. Lexington, Ky.: Center for Environment of the Council of State Governments.

Dahl, J. 1994. Tracking travel: A failed inspection won't dry-dock a cruise. Wall Street Journal. Aug. 2. B1.

Deerberg, G.J. 1990. Waste handling on board of ships. Pp. 183-188 in Proceedings of IMAS 90: Marine Technology and the Environment, May 23–25, 1990, London. London: Institute of Marine Engineers.

Deerberg, G.J. 1993. Multipurpose Waste Management on Board Passenger Ships for the Year 2000. Paper presented at the Cruise and Ferry Conference, May 11–13, London, organized by BML Business Meetings Limited, 2 Station Road, Rickmansworth, Hertfordshire WD3 1QP, England. Cited in Fulford, C. 1993. Environmental pleas for better waste handling. The Naval Architect. E404–E407. September.

Eastern Research Group, Inc. 1992. Report to Congress on Compliance with the Marine Plastic Pollution Research and Control Act of 1987. Report prepared for the U.S. Coast Guard by ERG, Arlington, Mass. (now Lexington, Mass.) June 24.

Fairplay International Shipping Weekly. 1993. Raw prawn. (5721):319. July 8.

Federal Republic of Germany. 1990. Collection of Garbage in the ports of Bremen. Information paper 11 for the 30th negotiating session (MEPC 30/INF. 11), International Maritime Organization (IMO), Marine Environment Protection Committee. Available from IMO, 4 Albert Embankment, London SE1 7SR.

Florida-Caribbean Cruise Association. 1993. The Cruise Industry's Role in Waste Management. Position paper available from the association, Miami, Fla. April.

Gallop, M. Undated. USS Theodore Roosevelt Environmental Compliance Program Cookbook. Available from commanding officer, U.S. Theodore Roosevelt, homeport in Norfolk, Va.

Grove, N. 1994. Recycling. National Geographic 186(1):92-115.

Hjelmar, O. 1993. Assessment of Environmental Impact of Incinerator Ash Disposal in Bermuda, Executive Summary. Report prepared for the Bermuda Ministry of Works and Engineering by the Water Quality Institute (VKI) and the Bermuda Biological Station for Research, Inc. Available from Ministry of the Environment, Government Administration Building, 30 Parliament Street, Hamilton HM 12, Bermuda. February.

Hollin, D. and M. Liffman. 1991. Use of MARPOL Annex V Reception Facilities and Disposal Systems at Selected Gulf of Mexico Ports, Private Terminals and Recreational Boating Facilities. Report to the Texas General Land Office by Dewayne Hollin, Texas A&M University Sea Grant College Program, and Michael Liffman, Louisiana State University Sea Grant College Program. September.

Hollin, D. and M. Liffman. 1993. Survey of Gulf of Mexico Marine Operations and Recreational Interests: Monitoring of MARPOL Annex V Compliance Trends. Report to the U.S. Environmental Protection Agency, Region 6, Gulf of Mexico Program by Dewayne Hollin, Texas A&M University Sea Grant College Program, and Michael Liffman, Louisiana State University Sea Grant College Program.

ICF, Inc. 1989. Decision-Makers Guide to Solid Waste Management, 2nd ed. Report prepared for the U.S. Environmental Protection Agency, Office of Solid Waste. EPA/530-SW-89-072. Washington, D.C.: U.S. Government Printing Office. November.

International Maritime Organization (IMO), 1994a. Draft Annex VI MARPOL 73/78. Attachment to Working Paper #10, Draft Report of the Bulk Chemical (BCH) Subcommittee. Forwarded to the Marine Environment Protection Committee at the 24th session of the BCH, London, Sept. 19–23. Available from IMO, 4 Albert Embankment, London, SE1 7SR.

International Maritime Organization (IMO). 1994b. Reception facilities manual approved. IMO News (U.K.) 2:5.

Irvin-Jones, L. 1992. Presentation by Louise Irvin-Jones, harbormaster, Port of Oakland, to the Committee on Shipborne Wastes of the National Research Council, at a public marina, Oakland, Calif., Oct. 15, 1992.

Journal of Commerce. 1994. Passenger dies, 400 stricken on cruise ship. Sept. 6. 1B.

Kauffman, M. 1992. Launching a Recycling Program at Your Marina. San Francisco: Coastal Resources Center. February.

Kearney/Centaur, Division of A.T. Kearney, Inc., and L. Martinez. 1991. Revision of the Port Reception Facility Section of the IMO Guidelines for the Implementation of MARPOL Annex V. Report prepared for the Marine Entanglement Research Program of the National Oceanic and Atmospheric Administration, Seattle, Wash. March.

Kiser, J.V.L., M.A. Charles, J.P. Bridges, G.L. Carr, and C.O. Velzey. 1994. Waste-to-energy: Citizens respond to plants in their neighborhood. Solid Waste Technologies 8(3):18-22. May/June.

Knap, A.H., C.B. Cook, S.B. Cook, J.A.K. Simmons, R.J. Jones, and A.E. Murray. 1992. Marine Environmental Studies to Determine the Impact of the Mass Burn Incinerator Proposed for Tynes Bay Bermuda. Prepared for the Bermuda Ministry of Works and Engineering. Available from the Ministry of the Environment, Government Administration Building, 30 Parliament Street, Hamilton HM 12, Bermuda. March.

Koss, L.J. 1994. Dealing With Ship-generated Plastics Waste on Navy Surface Ships. Paper presented at the Third International Conference on Marine Debris, Miami, Fla., May 8–13, 1994. Office of the Chief of Naval Operations, Department of the Navy, Washington, D.C.

Laitera, J. 1993. The non-discharge ship—what is involved? Paper presented at the Cruise and Ferry Conference, May 11–13, London, organized by BML Business Meetings Limited, 2 Station Road, Rickmansworth, Hertfordshire WD3 1QP, England. Cited in Fulford, C. 1993. Environmental pleas for better waste handling. The Naval Architect. E404–E407. September.

Larsen, I. and K. Borrild. 1991. A Coherent Regulatory System for Commercial and Industrial Wastes in the City of Copenhagen. Paper prepared for the Conference on An Integrated Approach to Solid Waste Management, Toronto, Canada, Oct. 29–31, 1991. Reprints available from the Agency of Environmental Protection, Copenhagen, Denmark.

Laska, S. 1994. Exploring a Wide Range of Interventions for Recreational Users by Applying the Hazards Evolution Model. Paper prepared for the Third International Conference on Marine Debris, Miami, Fla., May 8–13, 1994. University of New Orleans.

Middleton, L., J. Huntley and J.J. Burgiel. 1991. U.S. Navy Shipboard-Generated Plastic Waste Pilot Recycling Program. Report available from the Council for Solid Waste Solutions, a program of the Society of the Plastics Industry, Inc., Washington, D.C. March.

National Oceanic and Atmospheric Administration (NOAA) Corps. 1993. Shipboard Waste Management Plan for the NOAA Research Ship *Miller Freeman*. Available from the commanding officer, *Miller Freeman*, homeport in Seattle, Wash. Jan. 25.

North, R.C. 1993. Prepared statement of Capt. Robert C. North, chief, Office of the Marine Safety, Security, and Environmental Protection, U.S. Coast Guard, before the Subcommittee on Superfund, Ocean, and Water Protection of the Committee on Environment and Public Works, U.S. Senate, 102nd Congress, Second Session, Washington, D.C., Sept. 17, 1992. Pp. 146-149 in Implementation of the Marine Plastic Pollution Research and Control Act. S. Hrg. 102–984. Washington, D.C.: U.S. Government Printing Office.

Pacific Associates. 1988. A Report to the Alaska Department of Environmental Conservation on The Effects of MARPOL, Annex V, on the Ports of Kodiak and Unalaska. NWAFC Processed Report 88-26. Available from the Marine Entanglement Research Program of the National Marine Fisheries Service (National Oceanic and Atmospheric Administration), Seattle, Wash.

Pearce, J.B. 1992. Viewpoint: Marine vessel debris, a North American perspective. Marine Pollution Bulletin 24(12):586-592. December.

Pisani, J. 1989. Port Development in the United States (Status, Issues, and Outlook). Presentation by John Pisani, U.S. Maritime Administration, to the 16th World Ports Conference of the International Association of Ports and Harbors, Miami Beach, Fla., April 22–28, 1989.

Port of Rotterdam. 1992. Safety and Environment Protection: Essential Quality Factors in Port Management. White paper prepared by the Shipping Directorate of the Port of Rotterdam, Netherlands.

Princess Cruises. 1993. Waste Management. Instructions to shipboard employees. Internal guidelines prepared by Princess Cruises, Los Angeles.

Recht, F. 1988. Report on A Port-Based Project to Reduce Marine Debris. NWAFC Processed Report 88-13. Available from the Marine Entanglement Research Program of the National Marine Fisheries Service (National Oceanic and Atmospheric Administration), Seattle, Wash. July.

Recht, F. and S. Lasseigne. 1990. Providing Refuse reception facilities and more: The port's role in the marine debris solution. Pp. 921-934 in Proc. of the Second International Conference on Marine Debris, 2–7 April, 1989, Honolulu, Hawaii (Vol. II), R.S. Shomura and M.L. Godfrey, eds. NOAA-TM-NMFS-SWFSC-154. Available from the Marine Entanglement Research Program of the National Marine Fisheries Service (National Oceanic and Atmospheric Administration), Seattle, Wash. December.

Swanson, R.L., R.R. Young, and S.S. Ross. 1994. An Analysis of Proposed Shipborne Waste Handling Practices Aboard United States Navy Vessels. Paper prepared for the Committee on Shipborne Wastes, Marine Board, National Research Council, Washington, D.C.

U.S. Coast Guard. 1993. Written answers to questions posed by the Subcommittee on Superfund, Ocean, and Water Protection of the Committee on Environment and Public Works, U.S. Senate, Dec. 3, 1992. Pp. 151-153 in Implementation of the Marine Plastic Pollution Research and Control Act. S. Hrg. 102–984. Washington, D.C.: U.S. Government Printing Office.

U.S. Environmental Protection Agency (EPA). 1990a. Medical Waste Management in the United States: First Interim Report to Congress, Executive Summary. EPA/530-SW-90-051B. Washington, D.C.: EPA Office of Solid Waste, Waste Minimization Branch. May.

U.S. Environmental Protection Agency (EPA). 1990b. Medical Waste Management in the United States: Second Interim Report to Congress, Executive Summary. EPA/530-SW-90-087B. Washington, D.C.: EPA Office of Solid Waste. December.

U.S. Environmental Protection Agency (EPA). 1994. Review of Industrial Waste Exchanges. EPA-530-K-94-003. Washington, D.C.: EPA Office of Solid Waste. September.

U.S. General Accounting Office (GAO). 1994a. Pollution Prevention: Chronology of Navy Ship Waste Processing Equipment Development. GAO/NSAID-94-221FS. Washington, D.C.: GAO National Security and International Affairs Division. August.

U.S. General Accounting Office (GAO). 1994b. Pollution Prevention: The Navy Needs Better Plans for Reducing Ship Waste Discharges. GAO/NSIAD-95-38. Washington, D.C.: GAO National Security and International Affairs Division. November.

Vie, R.H. 1990. Shipboard generated waste—treatment and disposal—operational experience on passenger cruise vessels. Pp. 175-182 in IMAS 90: Marine Technology and the Environment, London, May 23–25, 1990. London: Institute of Marine Engineers.

Whelpton, P.G. 1993. Preventive waste management—A state of mind. Paper presented at the Cruise and Ferry Conference, May 11–13, London, organized by BML Business Meetings Limited, 2 Station Road, Rickmansworth, Hertfordshire WD3 1QP, England. Cited in Fulford, C. 1993. Environmental pleas for better waste handling. The Naval Architect. E404–E407. September.

Whitten, D.H. and R.L. Wade. 1994. Environmental Challenges Faced by the International Cruise Industry. Paper prepared for the Annual Meeting of The Society of Naval Architects and Marine Engineers, New Orleans, La., Nov. 17–18, 1994.

6

Education and Training

Education and training have important strategic roles to play in the implementation of MARPOL Annex V, as previous chapters have demonstrated. Most significantly, many opportunities for intervening in the hazard evolution model (Chapter 4) involve these approaches. Moreover, it is clear that, given the vast expanse of the sea, violations of Annex V are and will continue to be difficult to detect and prosecute; accordingly, implementation must rely heavily on motivation and education of seafarers, to persuade them to comply voluntarily and give them, through training, the requisite skills and tools. Furthermore, regulatory authorities alone cannot control land-based sources of marine debris. What is needed is behavioral and ethical change.

This chapter outlines strategies for initiating and sustaining the various types of education and training needed to promote successful implementation of Annex V. As defined in this report, education refers to informal, formal, and professional communications for all types of audiences, as well as information exchange programs aimed at disseminating experiences with Annex V implementation strategies and technologies. Training is a specific type of education focused on development of skills in repetitive tasks and practices.

The first section of the chapter outlines opportunities for education and training in implementing Annex V, briefly highlighting where these efforts are needed in various maritime sectors. The second section assesses past experience with education and training to support Annex V implementation and outlines a model program. The last section describes the key elements needed from government if the full potential of education and training is to be exploited.

OVERVIEW OF OPPORTUNITIES FOR
EDUCATION AND TRAINING

Education and training play important roles in Annex V implementation throughout all fleets. The two approaches are mutually reinforcing. Early environmental education motivates young sailors to comply with the Marine Plastic Pollution Research and Control Act (MPPRCA), for example, while Navy training enables them to carry out the mandates. This example also illustrates that, to effect behavioral change in groups, education and training programs must be long term. It is important to remember that such programs, as valuable as they are, cannot easily overcome failure of the port side (or any other element) of the vessel garbage management system.

Education is a key tool for influencing recreational fishermen and boaters and is also critical for commercial fisheries, due to limited enforcement capabilities and the difficulty of reaching these sectors in any other way. Information exchange programs need to reach all sectors, to maximize the benefits of knowledge gained about Annex V implementation strategies and technologies. Training of crews on large commercial and military ships is essential if proper garbage-handling procedures are to be followed consistently.

Beyond the practical arguments for conducting Annex V education and training programs, there are political reasons as well. Education is one of the most accepted interventions for dealing with environmental hazards (Laska, 1994). Even so, direct government appropriations for support of educational programs are rare.

Types of Education and Training

There are three basic audiences for Annex V education and training: the public; employees and/or visitors on vessels, in ports, and in the supply chain; and managers of vessel, port, and supply operations. Different types of programs must be developed for each audience. The goal of all three types of programs is implementation of Annex V, but the objectives vary depending on audience characteristics. The three types of programs are described briefly here.

Public Awareness Campaigns

Public awareness campaigns are directed at informing the general public about Annex V and fostering support for compliance. The ultimate goal of such campaigns is social and cultural change. An example would be a multimedia campaign in coastal areas explaining the ecological harm caused by marine debris.

Education and Training for Employees and Visitors

Education and training for employees and visitors on fleets and in ports can help ensure that proper waste reduction, sorting, and disposal procedures are followed. This form of instruction is designed to control both vessel and land-based sources of marine debris. Examples include education of fishermen concerning the harmful effects of discarded or lost nets and traps (including reductions in commercial fish and shellfish stocks) and the economic losses incurred when debris is caught in trawls; education of cargo and cruise ship personnel concerning the types of garbage subject to Annex V as contrasted with U.S. quarantine regulations; boating safety courses that include Annex V information; education of cruise passengers to convince them to forego certain amenities for the sake of the environment; and education of waste haulers who otherwise might dispose of ship garbage by dumping it illegally.

Employee education and training also can target product and service suppliers for ports and fleets. For example, vendors and packaging designers can be educated about environmentally conscious design techniques. Experience shows that market pressures alone are not enough to stimulate production of environmentally conscious products; suppliers need to understand the nature of a problem before they will respond. Education played a role in the redesign of bait boxes used by commercial fishermen to eliminate plastic strapping bands.

Management Education and Training

Annex V education and training programs must target management, including owners and operators of vessels and shore-based garbage management systems as well as government managers. These are the agents of change—professionals who oversee and influence others and establish organizational culture. Because they select organizational practices and materials, managers must be the key audience for information exchange programs.

This category of programs includes education to introduce vessel operators to Total Quality Management principles; meetings to improve coordination and share information among the federal agencies responsible for Annex V and quarantine inspections, and among individuals involved in on-board and shore-based garbage management; training for employers focusing on the benefits to a company's image accruing from environmental initiatives; efforts to disseminate information about the shipboard garbage treatment technologies developed by the Navy or the passenger cruise ship industry; and education of port operators and local government concerning the garbage disposal facilities they need to provide.

EXPERIENCE BASE RELATED TO ANNEX V

During the past decade, numerous educational and some training programs have been carried out to combat the problem of marine debris, and a variety of Annex V materials have been developed for these purposes. These efforts, while limited in scale, have been critical in the success to date of Annex V implementation. Because marine debris comes from a variety of land-based sources as well as mariners at sea, the educational campaign, by necessity, has been waged on many fronts.

The early educational programs were developed as a result of the 1984 International Workshop on the Fate and Impact of Marine Debris (Shomura and Yoshida, 1985), the first comprehensive effort to examine the impacts of marine debris on living marine resources. Among other things, workshop participants identified an urgent need to educate vessel operators and others about the marine debris problem. Shortly thereafter, the U.S. Congress directed the National Marine Fisheries Service (NMFS), part of the National Oceanic and Atmospheric Administration (NOAA), to help define and resolve the problem, and, in consultation with the Marine Mammal Commission (MMC), to develop a plan of activities defining priority research and management needs (Herkelrath, 1991).

Over the ensuing years, the NMFS provided funds through the Marine Entanglement Research Program (MERP) to carry out the action plan, which includes mariner education and public awareness efforts (Herkelrath, 1991). A number of non-profit organizations have been awarded funds to conduct public education projects, and the state Marine Advisory Services, funded in part by NOAA's National Sea Grant College Program, have maintained public awareness efforts. State and local governments also have participated, through sponsorship of beach cleanups and public education.

After MERP was established, the MPPRCA recognized the importance of education in remedying the marine debris problem. The MPPRCA directs the Coast Guard, along with NOAA and the Environmental Protection Agency (EPA), to develop public awareness programs and citizen monitoring groups. However, no funds have been appropriated under the Act for public education, so federal agencies have been constrained in carrying out their mandate. Therefore, MERP officials have continued to spearhead efforts to educate and persuade mariners and the general public to safeguard the marine environment (Coe, 1992). Without question, MERP has led the way in federal Annex V education efforts (while also laying a strong foundation in other areas). A number of other agencies also have contributed. The Coast Guard, for example, distributes Annex V information through several existing channels, such as contacts with vessel crews during routine boardings and inspections as well as interactions with boaters during boating safety campaigns.

The committee reviewed past and ongoing marine debris education and training programs. In general, successful programs have targeted defined populations and involved cooperation among federal agencies and between the public and private sectors. The role of government has been to provide leadership and limited funding. The private sector, including non-profit organizations, scientists, teachers, and community activists, has provided the experience and expertise needed to design and implement the programs.

The federal government has made a start in fulfilling its role, but it is acknowledged widely that much more is needed. The objectives of the various agencies with respect to Annex V have not been defined clearly, and perceptions of these goals certainly are not uniform. No agency has established meaningful objectives that would enable measurements of progress. Similarly, no agency has articulated an appropriate organizational theme for rallying the available work force and attracting adequate resources to implement a long-term education and training program that would assure full Annex V compliance at reasonable costs.

Following is a summary of the major education and training activities that have been initiated, highlighting lessons learned that may be useful in developing future programs. The summary is not exhaustive but it offers a sense of the characteristics, accomplishments, and diversity of the efforts. These efforts constitute perhaps the richest reservoir of Annex V implementation experience available, as pilot projects have touched every maritime sector.

Marine Debris Information Offices

The NMFS determined early on that education would have to be a major component of Annex V implementation efforts, because few fishermen or recreational boaters recognized the adverse effects of discharging garbage overboard. The greatest impact of the early educational efforts was believed to be in helping fishermen and others recognize that improved handling of vessel garbage was in their self interest. Several successful MERP programs were aimed at commercial mariners, both in the United States and internationally (Kearney/Centaur, 1989; Recht and Lasseigne, 1990; Wallace, 1990). To reach the wider community of seafarers, MERP sought the assistance of other federal agencies and the private sector.

The keystone of the MERP educational program is a pair of Marine Debris Information Offices (MDIOs), established in 1988 and run by the non-profit Center for Marine Conservation (CMC) (Center for Marine Conservation, 1989). The EPA also provides funding for the MDIOs, and the Coast Guard and NOAA cooperate on some of the individual projects.

The MDIOs have evolved into international clearinghouses for information and print materials developed by MERP and other organizations. The MDIOs develop and disseminate print material to approximately 11,000 educators, government and industry personnel, and media organizations annually; create, prepare, and distribute information packets aimed at 18 specific maritime groups; and distribute thousands of brochures to recreational boating, fishing, and shipping industry audiences. The establishment of the MDIOs has accomplished one fundamental government objective, by making information available to not only the general public but also other educators and specific maritime communities.

An example of a cooperatively produced project distributed through the MDIOs is a grade-school curriculum, U.S.S. My School, developed by the CMC with Navy funding. Students pretend they are on board an aircraft carrier for a week and must develop plans to manage garbage. Techniques learned through this curriculum can be applied in the students' homes.

Sea Grant Activities

The MERP effort spawned related work within the National Sea Grant College Program, which funds some 29 programs involving approximately 300 colleges, universities, and marine research institutions. Most programs address marine debris issues through research, outreach, and education (e.g., Liffman, 1987; Louisiana State University Sea Grant Program, 1987, 1989). Annex V activities include beach cleanup efforts, marina recycling programs, and MARPOL information programs. Sea Grant offices produce their own brochures, radio spots, videos, and bibliographies; they also disseminate materials produced by others.

Sea Grant's Marine Advisory Service supports several hundred coastal marine extension agents,[1] who transfer information and technologies to marine users, especially fishermen and boaters, in most U.S. ports. Among other marine pollution education activities, agents have worked to persuade fishermen in Oregon, New Jersey, and other states to return plastic garbage to port. But there are clearly opportunities for this program to include more Annex V-related activities, particularly in work with commercial and recreational fishermen. Agents also could work with local governments and agencies to reduce littering by beach goers.

Efforts Targeting Boaters

Through the annual National Safe Boating Week campaign (see sidebar), the Coast Guard taps into 25,000 volunteer coordinators who run local boat safety campaigns, staff booths at boat shows and fairs, develop relationships with the press and local employers, and conduct other types of education and outreach. Since 1989, these activities have involved Annex V materials developed by the CMC. In 1991, the federal government funded the reprinting of existing Annex V educational materials for distribution through the Coast Guard Auxiliary, a group of boat owners who voluntarily help foster boating safety. These packets were

---

[1]There were 284 local agents and university-based specialists in 1992 (National Oceanic and Atmospheric Administration, 1993).

A Model for Educational Leadership

The National Safe Boating Council is a group of private citizens and boating organization representatives who advise the Coast Guard in a variety of matters. The council's most visible public activity is National Safe Boating Week, an annual public awareness campaign aimed at recreational boaters. The council is an example of a private, independent group that collaborates with federal authorities to achieve common educational objectives.

The format of National Safe Boating Week, which has taken years to evolve, deserves some examination as a model for other broad-based maritime educational efforts. The program has three simple strengths. The first is coordinated preparation. The private groups make all the decisions and handle all the preparations; both the producers of the educational materials and the boating organizations that distribute them work together in a consistent manner year after year. This coordination helps the large-scale effort succeed despite any personnel changes.

A second merit of this program is that costs are shared in predictable ways. The government provides consistent funding. Each year, funds are appropriated through the Coast Guard to pay for administration and the coordinated distribution of all 25,000 packets of information to groups across the country. However, those funds do not cover all production expenses or the mailing of additional packets requested. Therefore, private organizations that want to participate know they must pay the full costs for any materials they want to include in the packets.

A third strength of this program is that the educational materials originate within the communities of boaters. The recipients recognize that the information has been prepared by groups knowledgeable about boating, not a regulatory agency or some other "outside" group. Readers' identification with the authors makes the message more palatable.

---

distributed widely during the 1991 boating season. The Coast Guard also gives out MARPOL fact sheets to callers requesting such information from the Boating Safety Hotline (1-800-368-5647).

The Coast Guard also is developing its own educational materials for use by the auxiliary. Although they do not enforce the law, members of the auxiliary supplement Coast Guard safety patrols, conduct public education courses in boating safety, and carry out other tasks that augment federal boating safety resources. With appropriate guidance and support, the auxiliary could become more active in educating recreational boaters about Annex V.

Port Projects

From January 1987 through March 1988, MERP supported a pilot project (described by Recht, 1988) to demonstrate port reception facilities for commercial vessels in Newport, Oregon. The fishing port management and fishing vessel owners collaborated to encourage vessel operators to return garbage, obsolete netting, and other gear to the port. Receptacles for the nets were placed at dockside, and the collected materials were handled in a variety of ways. Some were taken by homeowners for use as decorations and for protecting fruit trees from pests, while other materials were shipped for recycling and remanufacture into other products. Other port recycling programs have been organized by the Pacific States Marine Fisheries Commission and the New Jersey Sea Grant Program, which also has conducted research to identify ways to increase use of the recycled materials. For a recycling project in California, the Coastal Resources Center developed a comprehensive set of educational materials, including information for port tenants that was translated into Vietnamese for the local Asian fishing community.

In 1990, MERP launched an initiative to design and implement a model port/marina Annex V implementation project in Fajardo, Puerto Rico. This project, supported by Sea Grant personnel in Puerto Rico and New Jersey, was of strategic importance in that it extended MERP's focus from the North American coast to the Wider Caribbean region. Spanish-language educational materials, debris education programs, and an adequate port reception facility have been developed (Wypyszinski and Hernandez-Ariba, 1994).

In another type of outreach effort, the Coast Guard recently initiated the SEA-KEEPERS Campaign, a six-month pilot program in which some 270 reservists assigned to 47 port communities throughout the nation educated both civilian and military marine users about marine environmental protection laws, regulations, and strategies. Target groups included port operators, shipping agents, waste haulers, commercial fishing vessel operators, and recreational boaters. The Department of Defense funded this pilot program as part of the federal government's defense conversion strategy.

Efforts Involving Industry

In 1988, MERP launched the Shipping Industry Marine Debris Education Plan. This program had two major components. The first involved development of a binder of informative materials to assist vessel operators in complying with Annex V. The binder, distributed to shipping organizations, included MARPOL placards, the plastic control and minimization plan, sample waste management plans, examples of port reception facilities, international guidelines for implementation of Annex V, various regulation and policies regarding MARPOL implementation, and commercial telephone numbers for Coast Guard Captains of the Port (Wallace, 1990).

The second component of the plan, included in the activities of the MDIOs, was liaison with cruise line operators and owners. Activities included writing Annex V articles for cruise trade journals, presenting Annex V information at cruise trade meetings, producing and distributing brochures on the problem of vessel garbage, and development of a workshop presentation about Annex V compliance for members of cruise industries.

The Society of the Plastics Industry, Inc., has worked cooperatively with CMC, EPA, and NOAA to plan and fund the Clean Ocean Campaign. This public service effort targeted five separate audiences (commercial fishermen, recreational fishermen, the maritime industry, recreational boaters, and the plastics industry). The campaign included full-page advertisements, brochures, posters, buttons, television announcements, and development of a citizen's guide to plastic marine debris (O'Hara et al., 1988).

Programs Involving Government Fleets and Personnel

The Navy has carried out a number of educational efforts (which have not proven adaptable to applications outside the military). Indeed, the Navy has been relying on education to change shipboard practices until on-board garbage treatment technologies can be upgraded. Practices for handling plastics (see Chapter 1) were instituted with a fleetwide education program. The plastics education package sent to all Navy ships includes posters, a videotape, and a guide that addresses the problems caused by plastics in the marine environment, pertinent Navy requirements, essential elements of a successful shipboard program, and a list of plastic and substitute nonplastic items (Koss et al., 1990; Koss, 1994). The sailors have understood rapidly both the new controls and the reasons for them.

Several training programs have been developed by public agencies. For example, the EPA developed a training session on enforcement and implementation of marine protection laws, including MARPOL, for participants from various federal agencies. The Coast Guard is pursuing a "train the trainers" strategy to show members of its auxiliary how to reach out into their communities to train others in maritime debris management approaches.

Public Awareness Programs

The Center for Marine Conservation has become the dominant national environmental advocacy organization working to reduce marine debris and implement Annex V.[2] The CMC takes a broad educational perspective in addressing marine debris and Annex V, but the most visible efforts are beach cleanups. The CMC initiated and maintains the annual International Beach Cleanup Program (now supported by several federal agencies) and works to improve statistics on debris. The CMC concentrates on cooperative educational approaches, such as its citizen pollution patrols (efforts to inform boaters about regulations and how to

---

[2]Established in 1972, the CMC is a national, non-profit organization funded by foundations, corporations, members, and government grants. It runs five large programs addressing species recovery, marine protected areas, biodiversity, fisheries management, and pollution prevention. Programs involve research, policy, and education projects.

detect and report violations). The CMC also works through the media to persuade mariners to change their behavior (Weisskopf, 1988; Stoller, 1992).

Beach cleanups also are sponsored by a number of other groups, such as the Texas General Land Office, the American Littoral Society, and myriad environmental advocacy organizations across the country. Other types of public awareness efforts have been mounted as well. For example, the National Aquarium in Baltimore recently launched a marine debris education project that includes a documentary about the rescue, treatment, and return to the wild of a pygmy sperm whale that had ingested plastics (Craig Vogt, EPA, personal communication to Marine Board staff, August 5, 1994). The EPA contributed funds for the video.

International Efforts

Among its other Annex V implementation efforts, the EPA participates in the Gulf of Mexico Program (GOMP), which developed a Boater's Pledge Program to educate boaters about MARPOL and initiated a Take Pride Gulf Wide educational campaign that includes fact sheets and brochures. The GOMP also conducts a public awareness program aimed at marinas in region IV, in the Eastern Gulf of Mexico (U.S. Coast Guard, 1994). The EPA also has produced a marine debris curriculum, available in Spanish, with a chapter on MARPOL.

Information Exchange

The exchange of information of all varieties has been crucial to the development and implementation of Annex V from the start. Indeed, the scale and scope of marine debris as an environmental pollutant first became clear to government authorities after scientists met to exchange disparate observations and data sets, which yielded a composite picture of harm involving many bodies of water, many ecosystems, and many sources of debris. And where Annex V implementation initiatives have succeeded, considerable credit must go to persistent, aggressive, and largely informal efforts to exchange information.

The principal forums for formal information exchange have been three international conferences on marine debris, held in 1984, 1989, and 1994. Sponsors of these conferences have included federal agencies, universities, industry, international organizations, agencies of foreign governments, and research and development institutes. The papers presented and reports of workshops held at these conferences constitute much of the literature base supporting Annex V implementation efforts.

Among U.S. government information-exchange efforts, the Marine Debris Roundtable persevered from 1987 to 1990 after a task force failed to produce a formal interagency arrangement to implement Annex V. Through this informal roundtable, mid-level federal managers assembled with representatives from environmental advocacy organizations and the newly regulated maritime sectors.

Because little formal communication took place among the agencies, the roundtable served as a clearinghouse for ideas, pilot projects, data analysis, and coordination. At the suggestion of the Marine Mammal Commission and in consultation with the Coast Guard and NOAA, the EPA plans to build on this concept in establishing a marine debris coordinating committee involving all appropriate federal agencies. The committee will address MARPOL-related issues and study all sources of marine debris.

Another effective but temporary forum for information exchange was the Navy's Ad Hoc Plastics Advisory Group (described in Chapter 1), through which congressional staff and environmental organizations were able to share concerns about the military's garbage disposal practices and discuss alternatives.

Notwithstanding the benefits of the international conferences and short-term government efforts, the lack of formal, ongoing information exchange reaching all maritime sectors clearly is holding back Annex V implementation. Although a variety of technologies and methods is available for managing marine debris, the committee found that knowledge about them is not widespread. For example, the Navy's experiences in developing shipboard garbage treatment equipment appear to be largely unknown within other government agencies and the private sector.[3] And organizers of fishing net recycling efforts could benefit from knowledge of the EPA's waste exchanges, which could help locate markets for used nets. Information exchange could foster the development of a national infrastructure for recycling fishing gear.

A MODEL ANNEX V EDUCATION AND TRAINING PROGRAM

Based on its assessment of opportunities for intervention in each maritime sector (Chapter 4) and its review of past and ongoing education and training programs, the Committee on Shipborne Wastes developed a model strategy to support Annex V implementation. Many elements of this model may be found in the MERP program; the key element missing from that effort is adequate long-term resources to carry out comprehensive, nationwide education. What is needed is an aggressive, coordinated education program that modifies the ethics and behaviors of all who use and profit from the marine environment.

In developing this model, the committee relied heavily on the professional expertise of several of its members as well as findings from the Second International Conference on Marine Debris, at which education was addressed in a workshop. The findings of that workshop (O'Hara, 1990) underscored the point that education is not a last resort to be employed when all else fails but rather a

---

[3]Some information can be located, but only with effort and only if one knows where to look. For example, the Navy's 1993 report to Congress outlines the on-board technology development strategy and status and the Navy has participated in international conferences on marine debris.

strategic tool for fostering voluntary Annex V compliance and, as such, is deserving of adequate, long-term funding. The 1989 Education Working Group also found that dissemination of marine debris educational materials could be enhanced through existing organizations, such as the Coast Guard Auxiliary and licensing and registration procedures for fishing and boating. The need for program evaluation also was emphasized, because it may be necessary to prove that education is a productive investment of time and money. Indirect evaluation includes long-term monitoring of beach debris and use of port reception facilities; direct evaluation includes surveys conducted through re-licensing programs to assess changes in attitude and behavior.

The 1989 working group stressed five criteria that must be satisfied to create strong educational programs: (1) involve members of the target audience when developing materials and organizing distribution; (2) identify specific, discrete behavior for individuals; (3) set realistic goals; (4) make educational experiences positive and enjoyable; and (5) involve individuals familiar with the target audience (each target group must identify its educator as well-known and reliable, expert, and sympathetic to the group's needs and concerns).

In the judgment of the Committee on Shipborne Wastes, an Annex V education and training program of an effective scope could be created and implemented only under the aegis of a *single entity* that would have the ultimate responsibility for directing, coordinating, and funding the program. This lead organization would coordinate the efforts of all other government agencies and private organizations. The program could include the following elements:

• **Targeted, Coordinated Efforts to Reach Multiple Audiences.** Education and training programs need to be well-defined; "shotgun" efforts are more expensive and less effective. The lead organization could organize, coordinate, and encourage the participation of a group of educators qualified to represent and identify with target groups. There is a particular need to educate managers and to expand the types of groups targeted beyond those that generate marine debris. To address the marine debris problem fully, innovations are needed in packaging design, garbage treatment equipment, and approaches employed in operations and enforcement. Therefore, future educational efforts need to include groups such as the packaging industry, government officials, and fishing tackle manufacturers.[4] The lead office could transform empirical data and information into a series of selected educational campaigns, which each educator could deliver to

---

[4]Groups that have been or currently are targeted for Annex V implementation education include plastics manufacturers and processors, offshore oil and gas workers, commercial fishermen and processors, military personnel, solid waste managers, port and terminal operators, commercial shipping companies, recreational fishermen, recreational boaters, charter vessel operators, and cruise ship operators and passengers.

Additional groups that have been identified for future targeted education programs include the

his or her respective group. This approach would avoid duplication of effort and expand the benefits derived from available resources. Still, to be effective, this work will require more money than currently is appropriated.

• **Appropriate Messages, Media, and Settings.** Education and training are most effective when the message clearly defines the problem in terms relevant to the target group, identifies with and responds to the specific needs of the target group, and offers viable solutions to the problem. In addition, the educational vehicle (e.g., electronic media, person-to-person communication, special event) must suit the audience; experience with a target audience is critical in determining which educational tool is most appropriate. Timing and delivery also are critical. The educator must use the appropriate setting (e.g., formal academic classes, informal youth or adult groups), and timing is important because many maritime activities are seasonal.

• **Train the Trainers.** The lead office could explore ways to enable the newly educated members of target groups, particularly unorganized groups such as recreational boaters, to become agents of change. These individuals could be taught how to conduct training for others and be given access to educational and other materials provided by the lead agency. Outreach agents, perhaps Sea Grant marine extension agents, could provide support and training as necessary.

• **Evaluation.** The program must include an evaluation process that emphasizes the strategic impact of different activities. To date, the best effort to monitor marine debris education activity comes from the MDIOs, but they only monitor what has been disseminated. A tracking obligation could spur more people to collect the data and provide national accountability on the effectiveness of the educational materials. However, routine tracking by the federal government would require specific approvals that would be difficult to obtain.

In addition to these elements, the model program would include a formal information exchange network reaching all maritime sectors, to assure that decisionmakers have access to knowledge about the latest Annex V education and training strategies, garbage treatment equipment, and data.

THE FEDERAL ROLE IN ANNEX V EDUCATION AND TRAINING

Leadership

To build on past success and exploit the potential of Annex V education and training programs, improved leadership appears to be essential. Leadership is

---

packaging industry; government officials and enforcement agencies; coastal tourism industries; tackle manufacturers; operators of small ports, docks, marinas, and yacht clubs; suppliers of stores for vessels; boat manufacturers; employees of retail stores (including fast-food and convenience stores, and fishing and boating stores); environmental and conservation organizations; employees of shipyards; longshoremen; and coastal hunters.

**A NATIONAL GOVERNMENT LEADERSHIP MODEL:
SMOKEY BEAR**

In the early 1940s, with over 30 million acres of forest land burned every year due to carelessness and Japanese wartime shelling of the Pacific Coast, the U.S. Forest Service recognized the need for a program to help prevent person-caused forest fires. The agency obtained support from the Advertising Council, a coalition of advertising executives working on public interest projects, and the National Association of State Foresters.

Since then, the three partners have worked together on the Cooperative Forest Fire Prevention Program, the symbol of which is Smokey Bear. The program is managed by the Forest Service and funded by federal appropriations (roughly $1.5 million a year), but decisions are made cooperatively, and the Advertising Council donates expertise in the development of a media campaign.

The success of the program is reflected in Smokey's high profile: 94 percent of adults and 77 percent of children recognized the bear in a 1988 survey. (Data also indicate that the acreage burned has declined, to less than 5.4 million acres in 1990.) Officials attribute this accomplishment to the clear, concise message; the effectiveness of the Smokey Bear symbol; and the longevity and non-controversial nature of the program (Elsie Cunningham, program manager, Cooperative Forest Fire Prevention Program, personal communication to Marine Board staff, February 4, 1994).

---

needed to ensure that the relevant government agencies, companies, and individuals are informed fully about Annex V requirements, given technical and operational information routinely, and provided with educational and other materials designed to improve compliance and reduce enforcement costs. Leadership also is needed to coordinate regional, national, and international information exchange.

What is lacking is a central office providing long-term leadership, focus, coordination, and stimulus for collaboration. An example of the type of program needed is the Smokey Bear campaign (see sidebar). The MPPRCA gives the Coast Guard the major responsibility for enforcing Annex V requirements yet provides little guidance on how to handle other aspects of implementation. The result is that no single agency "owns" the issue. This problem is especially visible with respect to education, training, and information exchange, where so much needs to be accomplished but only assorted small efforts have been carried out. It is difficult to envision the present collection of education programs, which are largely informal and short term, evolving into the broad, long-term education and training program needed to support an Annex V implementation strategy. Marine debris is more than a litter problem, so education needs to accomplish more than teach mariners how to be tidy.

There are three ways to execute an Annex V education and training program.

One is to maintain the status quo, and thereby continue the strategy of piecemeal, short-term projects that are not necessarily informed by or coordinated with similar efforts conducted elsewhere. Experience has shown that this course of action does not lead to the coherent, long-term effort needed to implement Annex V fully.

The other two options involve the establishment of a central manager—the preferred way to create and sustain a coherent, long-term program. Apart from having access to the necessary expertise, a coordinating authority for Annex V education and training would have to be able to reach all maritime sectors, either directly or indirectly. The most obvious option would be to make official the leadership role now played by NOAA. The MERP and Sea Grant efforts have led the way in educating mariners and the public about marine debris and have proven that NOAA has vision. This agency is particularly effective in dealing with debris generated by fishing activities. However, NOAA would require assistance in dealing with the needs of some other groups (e.g., recreational boaters, cargo and cruise ships, the packaging industry) as well as port and technology issues. Furthermore, NOAA could not be expected to expand and enhance its current education and training efforts without additional resources designated for this purpose over the long term—something that has not been available in the past and is unlikely to materialize in the near future.

A third option would be to seek congressional action to establish a quasi-governmental private foundation chartered to focus on education, training, and information exchange related to Annex V implementation. There is precedent for this approach to coordinating national programs. The National Safe Boating Council and the Fish and Wildlife Foundation are examples. A foundation could bring together all the requisite expertise and would be less likely to be distracted from education and training than would overburdened federal agencies. A foundation also would have more flexibility than would a government agency in dealing with the private sector and pursuing national and international efforts.

Carefully drafted, the charter for the foundation could articulate clearly defined goals and objectives supporting Annex V implementation. The foundation could develop a coherent program to be executed through appropriate channels, making the best use of past experiences in the field. Grants could be awarded to private industry and associations, academic institutions, public agencies, and non-profit organizations to develop and carry out programs.

Secure Funding

Funding for education and training is a significant problem. Perhaps as a result of the leadership vacuum, government agencies appear to have limited their investments in education and training at a time when such efforts could be particularly effective. The social ferment and the growth of environmental awareness and activism over the past two decades has created a climate that may be conducive to behavioral change. In addition, education is one of the most economical ways of encouraging compliance (Wypyszinski, 1993) and thus become even more attractive given the need to leverage federal spending and obtain maximum impact from every effort.

In the current budget climate, dedicated long-term funding for comprehensive Annex V education and training is unlikely to be obtained through one or even several federal agencies. The foundation concept may offer the best hope of establishing secure funding in that the existence of such an entity would serve to emphasize the importance of Annex V education and training and, of equal or greater significance, support could be obtained from the private sector. The Congress could provide a one-time endowment and/or modest annual appropriations, perhaps using a portion of existing maritime fees (e.g., fuel taxes or tariffs on imported fishing equipment). A nominal federal investment in this area could yield significant dividends.

Innovation

While many Annex V education and training programs have been developed, there are needs for new concepts that might succeed with marine users who are difficult to reach or persuade, and needs to target audiences who can help develop innovative technological, organizational, operational, regulatory, and economic strategies. Innovation requires not only knowledge of past education and training efforts and gaps in Annex V implementation in each maritime sector, but also the time, money, and mandate to go beyond the ordinary and foster development and testing of promising new concepts. Again, this is unlikely to be accomplished by an existing federal agency or group of agencies, simply because they must contend with many routine demands and distractions. A foundation that supports education and training may be the most effective means of fostering innovation and, through dissemination of the results, bringing overall Annex V implementation to a higher level.

Ideally, education and training programs would extend beyond groups that cause the marine debris problem to those whose can help solve it. This approach would encourage the development of innovative strategies, with particular emphasis on "upstream" interventions in the hazard evolution model (described in Chapter 3). The possibility of achieving integrated innovation by providing national leadership for all Annex V activities is addressed in Chapter 7.

SUMMARY

Two basic findings can be drawn from the preceding discussion. First, education and training have important strategic roles to play in Annex V implementation, and a permanent capability is needed to develop and implement such programs at all levels in all maritime sectors. As environmental protection has become a responsibility of every industry, it is especially important that these efforts target all senior managers in order to foster organizational change.

Second, numerous Annex V education and training programs have been carried out, but these efforts clearly need to be elevated to a higher level in order to meet the challenges involved in implementing the international treaty and the U.S. law. Strong national leadership, secure funding, and innovation will be required to coordinate and enhance education and training. Given the current budget climate and the many distractions faced by federal agencies, the most promising alternative may be for the Congress to charter and endow a foundation to coordinate a sustained, long-term, national program devoted to Annex V education and training.

REFERENCES

Center for Marine Conservation (CMC). 1989. Marine Debris Information Offices, Atlantic Coast/ Gulf of Mexico and Pacific Coast: Annual Report, October 1, 1988–September 30, 1989. Washington, D.C.: CMC.

Coe, J. 1992. Presentation by James Coe, National Marine Fisheries Service, to the Committee on Shipborne Wastes of the National Research Council, Annapolis, Md., May 7, 1992.

Herkelrath, J. 1991. Description and Status of Tasks in the National Oceanic and Atmospheric Administration's Marine Entanglement Research Program for Fiscal Years 1985–1991. AFSC Processed Report 91-12. Available from the Marine Entanglement Research Program of the National Marine Fisheries Service (National Oceanic and Atmospheric Administration), Seattle, Wash. April.

Kearney/Centaur Division of A.T. Kearney, Inc. 1989. Model Plastics Refuse Control and Minimization Plan for Ships. Report prepared for the Marine Entanglement Research Program, Northwest and Alaska Fisheries Center, Seattle, Wash. December.

Koss, L., F. Chitty, and W.A. Bailey. 1990. U.S. Navy's Plastics Waste Educational Efforts. Pp. 1132-1139 in Proceedings of the Second International Conference on Marine Debris, 2–7 April 1989, Honolulu, Hawaii (Vol. II), R.S. Shomura and M.L. Godfrey, eds. NOAA-TM-NMFS-SWFSC-154. Available from the Marine Entanglement Research Program of the National Marine Fisheries Service (National Oceanic and Atmospheric Administration), Seattle, Wash. December.

Koss, L.J. 1994. Dealing With Ship-generated Plastics Waste on Navy Surface Ships. Paper prepared for the Third International Conference on Marine Debris, Miami, Fla., May 8–13, 1994. Office of the Chief of Naval Operations, Department of the Navy, Washington, D.C.

Laska, S. 1994. Exploring a Wide Range of Interventions for Recreational Users by Applying the Hazards Evolution Model. Paper prepared for the Third International Conference on Marine Debris, Miami, Fla., May 8–13, 1994. University of New Orleans, New Orleans, La.

Liffman, M.M. 1987. Prepared statement of Michael Liffman, Louisiana State University Sea Grant Program, for the National Ocean Policy Study and the U.S. Senate Committee on Commerce, Science, and Transportation. Photocopy. July 29.

Louisiana State University Sea Grant Program. 1987. Marine litter: More than an eyesore. Aquanotes 16(2):1-5. June.

Louisiana State University Sea Grant Program. 1989. Saltwater anglers did research. Aquanotes 18(1):1-4. June.

National Oceanic and Atmospheric Administration (NOAA). 1993. Sea Grant Review: 1990 through 1992. Silver Spring, Md.: NOAA National Sea Grant College Program.

O'Hara, K.J. (chair). 1990. Report of the working group on marine debris education. Pp. 1256-1260 in Proceedings of the Second International Conference on Marine Debris, 2-7 April 1989, Honolulu, Hawaii, R.S. Shomura and M.L. Godfrey, eds. NOAA-TM-NMFS-SWFSC-154. Available from the Marine Entanglement Research Program of the National Marine Fisheries Service (National Oceanic and Atmospheric Administration), Seattle, Wash. December.

O'Hara, K.J., S. Iudicello, and R. Bierce. 1988. A Citizens Guide to Plastics in the Ocean: More Than a Litter Problem. Washington, D.C.: Center for Environmental Education (now the Center for Marine Conservation).

Recht, F. and S. Lasseigne. 1990. Providing refuse reception facilities and more: The port's role in the marine debris solution. Pp. 921-934 in Proc. of the Second International Conference on Marine Debris, 2–7 April, 1989, Honolulu, Hawaii (Vol. II), R.S. Shomura and M.L. Godfrey, eds. NOAA-TM-NMFS-SWFSC-154. Available from the Marine Entanglement Research Program of the National Marine Fisheries Service (National Oceanic and Atmospheric Administration), Seattle, Wash. December.

Recht, F. 1988. Report on a Port-Based Project to Reduce Marine Debris. NWAFC Processed Report 88-13. Available from the Marine Entanglement Research Program of the National Marine Fisheries Service (National Oceanic and Atmospheric Administration), Seattle, Wash.

Shomura and Yoshida, eds. 1985. Proceedings of a Workshop on the Fate and Impact of Marine Debris, 27–29 November, 1984, Honolulu, Hawaii. NOAA-TM-NMFS-SWFC-54. Available from the Marine Entanglement Research Program of the National Marine Fisheries Service (National Oceanic and Atmospheric Administration), Seattle, Wash.

Stoller, G. 1992. Garbage overboard. Conde Nast Traveler (June):17-18.

U.S. Coast Guard (USCG). 1994. Managing Waste at Recreational Boating Facilities: A Guide to the Elimination of Garbage Disposal at Sea. Washington, D.C.: USCG Marine Environmental Protection Division, Environmental Coordination Branch.

Wallace, B. 1990. Shipping industry marine debris education plan. Pp. 1115-1122 in Proceedings of the Second International Conference on Marine Debris, 2–7 April, 1989, Honolulu, Hawaii (Vol. II), R.S. Shomura and M.L. Godfrey, eds. NOAA-TM-NMFS-SWFSC-154. Available from the Marine Entanglement Research Program of the National Marine Fisheries Service (National Oceanic and Atmospheric Administration), Seattle, Wash. December.

Weisskopf, M. 1988. In the sea, slow death by plastic. Smithsonian 18(12):58-67.

Wypyszinski, A.W. 1993. Prepared Statement of Alex. W. Wypyszinski, director, Sea Grant Marine Advisory Service, Rutgers University, for the Subcommittee on Superfund, Ocean, and Water Protection of the Committee on Environment and Public Works, U.S. Senate, 102nd Congress, Second Session, Washington, D.C., Sept. 17, 1992. Pp. 60-67 in Implementation of the Marine Plastic Pollution Research and Control Act. S. Hrg. 102–984. Washington, D.C.: U.S. Government Printing Office.

Wypyszinski, A.W. and M.L. Hernandez-Ariba. 1994. Latin American Marine Debris Public Awareness Project—Final Report. PRU-T-94-001. Report by the University of Puerto Rico Sea Grant Program, Mayaguez, Puerto Rico.

7

Overarching Issues Affecting
Annex V Implementation

The preceding chapter established that leadership is critical to successful Annex V education and training. In fact, strong national leadership is essential for the entire Annex V implementation program, whether it involves developing and deploying on-board technology, assuring the adequacy and use of port reception facilities, informing vessel crews and passengers about compliance methods, or enforcing the law. This chapter examines why leadership is so important and suggests how it might be provided.

Following an analysis of the need for leadership, the chapter explores two broad and significant compliance challenges that demand leadership—U.S. enforcement of Annex V at sea, and issues related to the Wider Caribbean special area. These problems are considered overarching because they are relevant to all fleets, strong national leadership will be required to resolve them, and international considerations are involved. They are also interrelated, in that coordination of enforcement is particularly problematic in the Wider Caribbean special area. Options for addressing these problems are outlined.

THE NEED FOR LEADERSHIP

As noted throughout this report, many federal agencies have become involved in implementation of Annex V, yet there is no lead agency for the overall effort. Furthermore, many steps that could be taken to improve implementation would require the cooperation of two or more agencies. For example, development of nationwide standards, regulations, rules, or information networks might be of great benefit, but no agency has broad enough capabilities to tackle such projects across all maritime sectors. What is needed, rather than simply regulation and enforcement of existing rules, is a leader that can view Annex V implementation from a broad systems perspective and implement comprehensive and, where necessary, innovative measures to effect change in all relevant areas.

Strategically, there are four possible ways to organize Annex V implementation. One is to maintain the status quo, which essentially means each agency will conduct Annex V activities on its own, as budget and mission priorities allow, and there may be some incremental improvements in how vessel garbage is handled. The problems with this approach are documented throughout this report, in terms of missed opportunities to improve Annex V implementation.

The second option is for the Congress or the Administration to assign to one agency the formal task of coordinating the entire program. But in the committee's judgment, no single agency has the requisite breadth of expertise, jurisdiction, and resources to assume this responsibility in full. This situation is reflected in Table 7-1, which brings together and summarizes information provided at various points in this report concerning federal activities related to Annex V implementation. As the table shows, no single agency is active in all key areas. Even agencies that are active in many or most areas lack important capabilities and expertise, not to mention the resources to assume additional duties.

For example, the Coast Guard clearly has broad capabilities, including the legal authority to enforce Annex V and oversee all other fleets (see chapters 1, 4, and the forthcoming section on enforcement in this chapter), as well as experience with education and training, both for its own fleet and others, including the public (see Chapter 6). However, the Coast Guard's core mission is policing and enforcement, meaning it has neither the funds or the expertise to carry out technology research and development, scientific monitoring of pollution, or comprehensive (i.e., for all levels of all maritime sectors and the public) Annex V educational program development and information and technology exchange. Moreover, the Coast Guard's mission and proficiency concern activities that take place on the water. Thus, even though the agency has the authority to oversee the disposal of vessel garbage in ports, it lacks the knowledge base and resources to replace the Environmental Protection Agency (EPA) in addressing land-based waste management (see Chapter 5). Because the port side of the vessel garbage management system is a key problem area inhibiting full Annex V implementation, it seems advisable to have experts in land-based waste management—EPA officials—take charge of finding a solution to that aspect of the problem.

At the same time, it is clear that EPA cannot assume full leadership in Annex V implementation because its relevant expertise and authority is limited to waste management, environmental monitoring (see Chapter 1), a research fleet consisting of one vessel (see Chapter 2), and some aspects of education and training (see Chapter 6). The EPA has limited contact with mariners, no Annex V enforcement authority, and, while it has expertise in pollution-control equipment, the focus has been on land-based rather than maritime applications. Both the Navy and the

TABLE 7-1 Federal Agency Areas of Authority and/or Expertise Related to Annex V Implementation

Key areas Agency[b]	Enforces Annex V	Has own fleet[a]	Oversees other fleets in some way	Conducts on-board tech. R&D	Oversees natural resource or waste stream	Conducts Annex V educ. & training	Collects marine debris info.
Coast Guard (1,2,4,5,6,7)	x	Milit.	All fleets		Vessel garbage	x	
DOS (1,7)	x		Foreign-flag vessels				
EPA (1,4,5,6)		Res.			All land waste	x	x
MARAD (1,5)		Trng.	Cargo ships	x			
MMS (1,4)			Offshore industry		Resource (OCS)[c]		
Navy (1,2,4,5,6)		Milit.		x		x	
NOAA/NMFS/ME RP (1,2,4,6,8)		Res.	Fishing vessels		Resource (fish stocks)	x	x
NPS (1,2,8)					Resource (parks)		x
USDA (1,5,7)			Cargo, cruise ships		Quarantine waste		

[a]Abbreviations stand for military (milit.), research (res.), and training (trng.) fleets.
[b]The numbers in parentheses indicate the chapters (including forthcoming sections of Chapter 7) in which the agency's relevant activities are described.
[c]Outer Continental Shelf.

cruise ship industry have far more experience with on-board garbage treatment technology than does EPA (see Chapter 5).

Another possible lead agency might be the National Oceanic and Atmospheric Administration (NOAA), which has broad expertise and experience in marine debris education, research, and information exchange (see Chapter 6) and environmental monitoring (see chapters 2 and 8). However, the only fleets over which NOAA can exert control are its own research vessels and, through the National Marine Fisheries Service (NMFS), commercial fisheries (see Chapter 4). Most importantly, NOAA lacks authority to enforce Annex V or manage garbage generated by other fleets.

None of the other agencies has sufficient breadth of involvement in Annex V-related activities to be a serious candidate for providing comprehensive leadership. The Department of State (DOS) focuses on Annex V enforcement as it relates to foreign-flag vessels (as discussed later in this chapter), on special area designations, and on other international and intergovernmental issues. The Maritime Administration (MARAD), the Minerals Management Service (MMS), and the Navy each are engaged primarily in oversight of single maritime sectors (cargo ships, the offshore industry, and the Navy, respectively), although MARAD and the Navy also have programs dedicated to technology development (see Chapter 5). The National Park Service's sole activity related to Annex V is environmental monitoring (see chapters 2 and 8), while the U.S. Department of Agriculture (USDA) manages the handling of quarantined garbage in the cruise and cargo ship sectors (see Chapter 5).

A third option would be to establish an interagency task force, such as the marine debris coordinating committee being formed by EPA. (That committee will address land-based sources of marine debris as well as MARPOL-related issues.) A clear legislative mandate would be required to establish the overview authority of the task force and outline its responsibilities. This concept is attractive in that it would combine all the requisite expertise in a single panel, which could serve as a forum for government-wide information exchange and decision making related to Annex V. But an interagency task force, while it could accomplish much of value, would neither go far enough in assigning leadership (in terms of human and fiscal resources) nor go very far in garnering support from the private sector for Annex V implementation. There would still be divided federal leadership, with no clear line of authority and responsibility, and most likely no resources to accomplish much beyond maintenance or reshuffling of existing programs.

The fourth option is to establish a permanent national commission to coordinate all aspects of Annex V implementation. Such a commission would symbolize a commitment to Annex V implementation and demand attention to the problem. The U.S. Congress has established numerous permanent commissions to address other major problems. Examples in marine affairs include the Marine

Mammal Commission (MMC),[1] state and regional marine fisheries commissions, and river basin commissions. Some commissions seek to increase public awareness, advocate resource management, and develop educational materials designed to achieve a specific goal, such as pollution control and protection of living resources. Others provide assistance to states and federal agencies on environmental, natural resource, and conservation issues. Some provide recommendations on policies, public complaints, and directions on various issues of interest to a specific agency.

A commission guiding implementation of a single international agreement would be unusual, but federal agencies responsible for Annex V implementation could provide the necessary support. (Indeed, the work of the commission would be assisted by the formation of an interagency task force, described earlier as the third option.) Such an unusual mechanism may be the only way to concentrate on and meet fully the challenges inherent in gaining the cooperation of so many individuals in such diverse maritime sectors. An independent commission would have greater flexibility than would federal agencies or task forces in working with the private sector. A commission not only could marshal the efforts of federal agencies with different missions as well as private organizations, but also could serve as a high-level focal point for U.S. leadership internationally, overseeing the nation's efforts to guide the global community toward increased standards of performance. A commission would be well-positioned to address international issues such as U.S. enforcement of Annex V as it applies to foreign violators, dissemination of Annex V-related information and technology to other nations, and development of innovative programs with neighboring nations. (These issues are examined later in this chapter.)

In the committee's judgment, cost probably would not be a barrier to pursuing this option. In fact, establishing a commission likely would cost *less* than assigning all the tasks it might pursue to individual agencies. This assumption is based on the committee's knowledge of the operating budgets of other commissions, such as the MMC, rather than on a formal cost analysis.

A national commission addressing Annex V implementation would require a clear legislative mandate establishing its overview authority and outlining its responsibilities, which could include (1) reviewing information on the sources, amounts, effects, and control of shipborne garbage; (2) providing leadership for federal agencies to assure that they carry out their roles and responsibilities and share relevant information; (3) making recommendations to agencies on actions or policies related to identification and control of sources of shipborne garbage;

---

[1]The MMC was established under the Marine Mammal Protection Act of 1972 (P.L. 92–522). The commission is an independent agency of the Executive Branch charged with developing, reviewing, and making recommendations on the actions and policies of all federal agencies with respect to marine mammal protection and conservation, and with carrying out a research program. Annual appropriations are approximately $1 million.

(4) conducting research, regulatory, and policy analyses; (5) periodically providing the Congress with a report on the state of the problem, progress in research and management measures, and factors limiting the effectiveness of response; (6) overseeing a long-term program of Annex V education, training, and information exchange; and (7) overseeing international aspects of Annex V implementation. Legislation also would need to authorize and appropriate funding sufficient for the commission to carry out its duties.

If a foundation were created to coordinate Annex V education and training (as suggested in Chapter 6), then its relationship to the commission would need to be defined clearly. The commission would oversee the foundation and seek to integrate its activities with those of other agencies and organizations. For example, an innovative educational program funded by the foundation could be combined and tested with new organizational strategies or other types of interventions outlined in the hazard evolution model (described in Chapter 3). The commission would have to be responsive to the need for integrated and innovative implementation strategies of all kinds.

It might be possible to achieve some economies by, for example, combining the administrative staffs of the commission and the foundation into a single office. But it would be important to make clear distinctions between the roles and activities of the commission and those of the foundation to assure that the mandated functions of each were carried out.

U.S. ENFORCEMENT OF ANNEX V

As mandated in the Marine Plastic Pollution Research and Control Act (MPPRCA), the U.S. Coast Guard submitted an annual report to Congress in 1992, summarizing the status of enforcement of Annex V and identifying obstacles to compliance (Eastern Research Group, 1992). The report concluded that the two principal weaknesses in enforcement capabilities were "the difficulty of obtaining eyewitness accounts" and "the limitations imposed on prosecution of foreign vessels."

Some difficulties described in the Coast Guard report as "inherent" in enforcement are applicable to all vessels. As with implementation of any agreement aimed at curbing pollution from vessels, comprehensive surveillance is impossible because ocean space is far too large to monitor. Direct observation of violations by enforcement agents is unlikely on the high seas. On several occasions, a basis for proceedings has been provided by self-incriminating statements from vessel masters or crew members and complaints filed by other mariners, port operators, or foreign officials. But incriminating statements cannot be relied upon as a basis for enforcement because there are obvious psychological and economic disincentives for confessing or informing on other professional mariners.

The Coast Guard report also mentions a number of more subtle difficulties applicable to all vessels. Among the problems cited were the low national priority given to the problem of shipborne wastes; the complexity of administrative procedures for proceeding against violators; and the shortcomings in training of Coast Guard personnel, especially with respect to international shipping. The economics of compliance also present complex challenges.[2] The Coast Guard report notes that "penalties are sufficiently large to be considered significant but the likelihood of getting caught is considered low."

Concerning remedies for the enforcement problems, the Coast Guard report suggested that Annex V compliance rates depend on factors other than government efforts—specifically, the levels of environmental consciousness in the industry and among the public. Still, two avenues for improving compliance were proposed: lowering the cost of shoreside garbage disposal and pursuing international cooperation. The committee addressed the disposal cost issue in Chapter 5; international enforcement issues are discussed here.

Depending on the location of an alleged violation, the Coast Guard either takes direct action against a foreign-flag vessel or refers the case to the DOS for transmittal to the flag state. It is the handling of these latter cases that has been a key weakness in U.S. implementation of Annex V. As shown in Table 7-2, flag states seldom comply with the MARPOL requirement to notify the United States of the outcome of forwarded cases, much less impose penalties. Provoked by this poor response, in 1992 the United States expanded its exercise of port state enforcement authorities recognized under international law, thereby reducing the proportion of cases forwarded to flag states. Prior to that time, the Coast Guard took action against foreign vessels only when they discharged garbage within 3 miles of the coast; all other cases were transmitted to flag states. Under the new policy, the United States pursues direct civil or criminal action in all cases where jurisdiction can be established. That is, if there is evidence the violation took place within the Exclusive Economic Zone (EEZ), territorial sea, or internal waters, action is taken. In addition, penalties for Annex V violations have been increased, with the criminal offense upgraded to a felony. Port officials are authorized to withhold clearance for departure. Violations detected outside 200 nautical miles, as well as cases where the location of discharge cannot be established, continue to be referred to flag states; Table 7-2 suggests that the responsiveness of flag states may have improved slightly since the new policy was implemented.

Notwithstanding these changes, the inherent need to rely on circumstantial evidence makes it extremely difficult for the United States to proceed directly against foreign-flag vessels. If circumstantial evidence is the only indicator of a

---

[2]Cost is a major barrier to Annex V compliance and enforcement in developing countries, which view as onerous and unfair the expense of upgrading their ships, installing port reception facilities, and establishing the requisite administrative bureaucracy (Schrinner, 1992).

TABLE 7-2   Flag State Responses to U.S. Reports of Alleged Annex V
Violations by Foreign-flag Vessels (since December 31, 1988)[a]

	As of 6/92	As of 6/94
Reports transmitted	111	365
No acknowledgement	76 (68.5%)	203 (55.6%)
Acknowledged but no other information given	23 (20.7%)	84 (23.0%)
Fines levied by flag state	2 (1.8%)	20 (5.5%)
Other[b]	10 (9.0%)	58 (15.9%)

[a]The 1994 figures include cases referred to flag states under both the old and the new (post-October 1992) U.S. enforcement policies. Because the referral rules changed, the 1992 and the 1994 data do not reflect exactly the same types of cases. However, the two data sets are comparable in that both include only referrals to flag states and exclude direct enforcement actions taken by the United States.

[b]Includes all other cases, including those that were investigated and dropped, those in which warnings or reprimands were issued, and those in which the flag state was not a party to MARPOL.

Sources: 1992 data obtained from a report submitted to IMO (United States, 1992); 1994 data provided by the Department of State, Office of Ocean Affairs.

violation, then often the location of illegal disposal cannot be established adequately for direct U.S. enforcement action.

Options for Improving Annex V Enforcement

Annex V establishes simple performance standards, but the sheer number of garbage transactions taking place overwhelms any capability for direct surveillance, by either the Coast Guard or any other authority. Therefore, other alternatives need to be employed where possible, or compliance falls short. A new balance is needed that fosters robust compliance capabilities among vessels, ports, and governments and enhances the effectiveness of existing enforcement mechanisms. Over the long term, this approach would lay the foundation for a strengthened enforcement capability. Most of the options discussed here were mentioned in previous chapters.

Clarify Extent of Port State Authorities

The Coast Guard informed a Senate subcommittee in 1992 that notice of its new enforcement policy was submitted to the Marine Environmental Protection

Committee (MEPC) of the International Maritime Organization (IMO) (North, 1993). The new policy was seen as consistent with authorities recognized under international law and as not requiring consent by the international community. However, the Coast Guard also stated that the United States did not intend to undertake enforcement actions with respect to violations taking place in international waters. Such actions would "disrupt the institutional arrangements and multinational agreements that led to the existing level of international cooperation" (Eastern Research Group, 1992).

Principles of jurisdiction articulated in Article 218 of the 1982 Third United Nations Convention on the Law of the Sea (UNCLOS III) establish that a port state may take action against foreign-flag vessels that violate applicable international rules and standards—even when the violation takes place beyond the port state's EEZ. Thus, to date, the United States has chosen not to exercise the full complement of enforcement authorities recognized under international law. The broad context, including the impact on legal precedents for U.S. enforcement with respect to violations taking place on the high seas, would need to be examined. The DOS could review this issue as part of the federal effort to enhance the effectiveness of U.S. enforcement of Annex V.

In addition, the United States could examine with other nations the rights and responsibilities of port states to initiate actions against foreign-flag vessels. Through the appropriate diplomatic channels, clear rules could be developed so that port states can exercise fully and efficiently their jurisdictional authorities in a manner that assures compliance with the international standards set forth in Annex V and general principles of international law.

Simplify Handling of Civil Cases

The Coast Guard is experimenting with a simplified enforcement procedure for civil cases involving oil discharges that violate the Federal Water Pollution Control Act (FWPCA) (P.L. 80-845), as amended.[3] The committee believes this procedure might prove useful in enforcing Annex V. In the pilot project, conducted in three ports[4], alleged violators of the FWPCA were issued a Notice of Violation, similar to a traffic ticket, containing a proposed penalty. A penalty was assessed if an investigation established the elements of a civil case. The violator then had the option of paying the fine within 30 days, requesting a determination by a hearing officer, or not responding at all, in which case the file was sent to the

---

[3]The Interim Final Rule (59 Fed. Reg. 16,558 [1994]) took effect April 7, 1994. The procedures, which are related to FWPCA Section 309(g)(2)(A), apply only to discharges of less than 100 gallons of oil.

[4]The six-month pilot program was initiated in the spring of 1994 in the ports of Charleston, South Carolina; Galveston, Texas; and Los Angeles, California.

district commander for review before processing by the hearing officer according to current procedures (33 C.F.R. 1).

By allowing notices to be issued in the field and cases to be settled quickly, the procedures were expected to save time and money, improve the deterrent effect of the sanction, and expedite corrective actions. If the pilot projects are successful, then the final rule will be implemented nationwide. If this occurs, then the Coast Guard could explore using these procedures to handle civil cases[5] involving Annex V violations. This approach could be especially effective in the fisheries and recreational boating sectors, which pose special Annex V implementation problems. The ticketing strategy could free the Coast Guard from some extended paperwork duties and make the point among mariners that violators will be prosecuted.

A similar method is used by APHIS, which authorizes boarding officers to issue "spot fines" to vessel operators found to violate the standards of the quarantine program. Violators have 72 hours to pay. This authority allows APHIS to enforce its requirements, collect fines, and then release violators quickly so the vessel is not detained for extended periods. Simultaneously, the vessel and the operator are identified throughout the entire APHIS organization, so the vessel can be reinspected at every port if necessary. The spot fine policy is considered an "excellent deterrent" and is credited with improving the attitudes of ship personnel and reducing the number of violations[6] (Ronald B. Caffey, assistant to the deputy administrator, APHIS Plant Protection and Quarantine, personal communication to Marine Board staff, July 26, 1994).

Track and Punish Repeat Violators

To maximize the utility of its past successes in identifying and prosecuting Annex V violators, the Coast Guard could input the names of offending vessel operators and shipping companies into a centralized database. (This approach could be either combined with the ticketing strategy proposed earlier or employed with the current enforcement program.) The database could be used to identify and assess special penalties against those who repeatedly disobey the law. Such a system would be similar to that used by police departments to keep track of motor vehicle operators.

---

[5]The difference between a civil and a criminal case is largely a matter of intent; inadvertent violations are handled under civil law, while willful violations prosecuted under criminal law and could not be handled through the ticketing process.

[6]The annual number of APHIS violations by vessels ranged from 323 to 404 in the fiscal years 1990 to 1993; APHIS collects 98 percent of the fines owed (Ronald B. Caffey, assistant to the deputy administrator, APHIS Plant Protection and Quarantine, personal communication to Marine Board staff, July 26, 1994).

For example, a point system could be established; the number of points assessed would vary based on the degree of seriousness of the infraction (with overboard disposal of plastics carrying the highest penalty). Each time a new violation is detected, the names of the vessel operator and shipping company could be checked against the database. Any vessel operator or shipping company accumulating a threshold number of points could be required to pay a heavy fine upon entering the U.S. EEZ. If they were observed within the EEZ but had not paid the fine, then the vessel operator could be arrested and the vessel detained within U.S. jurisdiction pending resolution of the case in court.

The APHIS system already tracks high-risk vessels and assesses extra penalties against repeat violators. The "blacklist" primarily includes vessels that have violated the quarantine standards in the past 12 months.[7] These vessels are boarded by APHIS inspectors upon all arrivals at U.S. ports for one year after the most recent violation. Initial fines are in the $100 to $200 range and may increase as much as fivefold for repeat violations in a 12-month period (Ronald B. Caffey, assistant to the deputy administrator, APHIS Plant Protection and Quarantine, personal communication to Marine Board staff, July 26, 1994).

Monitor Garbage Handling Practices

Until recently, there was no way to verify where vessel garbage was discharged. Coast Guard officials could not confirm the claims of vessel operators who said, for example, they had off-loaded garbage in the last port. The Coast Guard has addressed this problem in part by requiring garbage logs on ocean-going, U.S.-flag commercial vessels over 12.2 meters (about 40 feet) in length, as well as fixed and floating platforms. Legislation has been proposed that would allow this requirement to be extended to foreign-flag vessels, thereby filling major gaps in accountability in the cargo and cruise ship sectors. Still, it could be difficult and time-consuming to verify the accuracy of the logs. This problem could be remedied if ports were required to provide receipts for garbage off-loaded into their reception facilities, and if the Coast Guard examined these receipts when reviewing vessel logs.

Northern European countries have taken even more direct action to monitor potential violators of Annex V. Before departure from Rotterdam, the Netherlands (which is in the North Sea special area), all vessel operators are obliged to off-load garbage or declare their intentions for disposal in a later port of call. That information is recorded in a regional database and can be used to ensure that vessel operators conform with their plans. To further support U.S. monitoring of

---

[7]The violation list is not shared with the Coast Guard. The APHIS program also maintains a separate list of vessels calling at certain Russian ports where Asian gypsy moths may be found. The Coast Guard assists in identifying and tracking those vessels.

disposal practices by commercial ships, it might be advisable to require cargo and cruise ships to off-load Annex V and APHIS garbage at every U.S. port call, as suggested in Chapter 5, or to declare their intentions, as in Rotterdam.

Surveillance by Government Authorities

Although comprehensive surveillance is impossible, various government authorities already monitor the actions of certain fleets and might be able to include Annex V on their lists of concerns. The NMFS stations observers on vessels in some fishing fleets to monitor compliance with fisheries management plans. The MMS routinely inspects offshore oil and gas operations. And state marine police encounter recreational and fishing boats during the course of ordinary duties.

Additional Annex V enforcement capabilities would be useful with respect to fisheries fleets, offshore operations, and recreational boaters. If authorities monitoring these fleets were informed about Annex V and methods for reporting violations, perhaps they could provide these extra capabilities.

Surveillance by Ship Operators

Ship operators have every reason to want to assure the adequacy of garbage reception facilities, and they could be encouraged to help the government monitor ports. The IMO has a form[8] ship operators may use to report inadequate port reception facilities, but these forms seldom are filled out. There may be a way to encourage use of these forms, collect the data, and pursue violators on national and international levels—similar to the way the United States keeps track of how its Annex V violation reports are handled by other flag states. This type of voluntary monitoring is condoned by MARPOL and would assist primarily with enforcement focusing on vessels from signatory states.

In the United States, Coast Guard or APHIS officers boarding a vessel could hand the crew a report form. Operators of cruise ships and military vessels could obtain forms upon every departure from the United States. In smaller ports and marinas, availability of the IMO forms could be publicized through the Coast Guard and its auxiliary; boating and fishing groups; and education programs, such as the NOAA's Marine Debris Information Offices. In the offshore oil and gas industry, the Offshore Operators Committee could publicize the availability of the forms and circulate them to operators of platforms and service vessels.

---

[8]The form is provided in Appendix B, last page of the implementation guidelines.

Surveillance by Citizens

Given the vastness of the oceans, it is clear that the U.S. Coast Guard cannot singlehandedly enforce the requirements of Annex V at sea. The MPPRCA envisioned that additional "eyes" for witnessing and reporting violations might be provided by seafarers, beach goers, and vessel passengers. The MPPRCA includes an unprecedented provision that empowers *anyone* to report a violation. The Act further rewards citizen reporting by authorizing the courts to give some of the fines collected to those reporting the violation.

Citizen reporting has proven to be worthwhile. Beginning in 1990, EPA funded a pilot program conducted by the Center for Marine Conservation to develop, test, and evaluate a Citizen Pollution Patrol Program (Podlich, 1992). In addition to educating the maritime community about marine debris and related federal and state regulations, the program was designed to involve citizens in reporting Annex V and MPPRCA violations. A standard form was developed to assist eyewitnesses in documenting suspected violations. In the most highly publicized incident of this type to date, citizen reports led to the criminal conviction of a cruise line operator and the maximum fine allowed—$500,000—for illegal discharge of garbage from a ship (U.S. Department of Justice, 1993). In that case, cruise passengers witnessed and videotaped more than 20 plastic bags of garbage being discharged into the sea near the Florida Keys (U.S. Department of Justice, 1993).

If more citizens were educated in how to recognize violations of Annex V and report them, their tips could assist in enforcement. In fact, as mere awareness of the provision for citizen reporting increases, would-be violators may be deterred from carrying out illegal discharges (Weikart, 1993).

All mariners should know that they are encouraged to report Annex V violations by any vessel, just as if they had witnessed any other illegal act. The Coast Guard recently added Annex V violations to the types of reports handled by the National Response Center.[9] Through the Coast Guard Auxiliary, a campaign recently was initiated in several states to foster public awareness of how to recognize violations and report them to the center's toll-free telephone number (1-800-424-8802). Plans to expand the campaign nationally should be encouraged.

ISSUES RELATED TO SPECIAL AREAS

As noted in Chapter 1, MARPOL permits the designation of special areas where overboard discharge of garbage other than food waste is prohibited. The

---

[9]The EPA also has recognized the value of citizens as "watch dogs" for ensuring implementation and enforcement of environmental regulations (U.S. Environmental Protection Agency, 1988).

convention does not spell out in detail the criteria and characteristics to be considered in designating special areas. Such direction is provided, however, under guidelines recently adopted by the IMO (International Maritime Organization, 1991).[10] Significantly, even when the MEPC adopts a proposal for a special area, the requirements become binding only when IMO determines that sufficient numbers of adequate port reception facilities are provided in the region.

Eight special areas have been designated under Annex V, although the rules have entered into force in only three.[11] Of particular interest to the United States, for reasons of proximity, is the Wider Caribbean special area, which includes the Gulf of Mexico (see Figure 7-1). The United States pushed for the designation of that area[12] and has a distinct interest in minimizing pollution there. The U.S. Navy also is concerned with other special areas, such as the Mediterranean Sea, where its missions may demand frequent transits or extended stays.

The existence of special areas means that vessels using those waters must achieve zero-discharge capability. An operator can treat garbage on board the vessel, bring the garbage to reception facilities in ports surrounding the special area, hold this garbage for legal discharge at sea or in ports outside the special area, or some combination of these options. All vessels, including U.S.-flag research vessels and cruise ships, have to contend with this mandate when they sail in internationally recognized special areas. Fixed platforms in the Gulf of Mexico already are operating at zero discharge. Eventually, as on-board garbage handling technologies and procedures evolve, awareness of Annex V grows, more special areas come into force[13], and adequate port reception facilities become more widely available, zero-discharge capability may become the operating norm.

It will be important, therefore, that the U.S. Annex V implementation strat-

---

[10]Among factors to be considered are oceanography, ecological characteristics, social and economic value, scientific and cultural significance, environmental pressures (including those of ship-generated pollution), and measures already in place to protect the local environment.

[11]The rules are in force in the Antarctic Ocean, the Baltic Sea, and the North Sea. The other five special areas are the Black Sea, Mediterranean Sea, the Persian Gulf, the Red Sea, and the Wider Caribbean.

[12]The United States initially proposed special area status for the Gulf of Mexico in response to public outcry over debris washing up on Texas beaches. Studies indicated that much of the debris was of foreign origin. In reaction to the U.S. proposal, a regional workshop was held in Venezuela, and participants called for extending the special area proposal to include the entire Wider Caribbean, to assure that vessels would not discharge garbage into the Caribbean Sea prior to entering the Gulf of Mexico. The MEPC approved the special area designation in 1991.

[13]The number and extent of designated special areas has grown, posing increased challenges for maritime operators. But special areas are unlikely to proliferate without restraint in the near term, because such designations may limit navigational freedoms significantly, and the 1982 Law of the Sea Convention recognized the importance of balancing protection of these freedoms (for the benefit of international commerce) with interests in protecting coastal ecosystems. Nevertheless, pressure to extend special area protections is likely to mount.

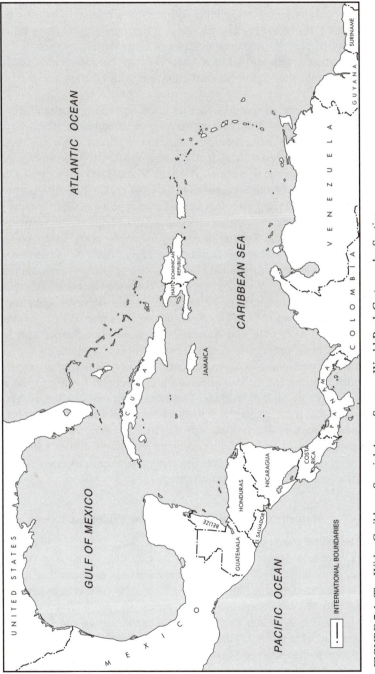

FIGURE 7-1 The Wider Caribbean Special Area. Source: World Bank Cartography Section.

egy provide for zero-discharge capability for vessels operating in special areas, and that the strategy include measures to assure adequate port reception facilities—not only in U.S. ports (a topic addressed in Chapter 5), but also bordering the Wider Caribbean special area. An international emphasis on building the capability of ports to implement Annex V is important, because the stringent rules associated with special area status will not be enforceable in the Wider Caribbean until the region has sufficient numbers of adequate port reception facilities.

Entry into force of special area rules in the Caribbean, as elsewhere, depends on a determination by IMO that sufficient port reception facilities are in place. However, IMO has not established definitive criteria as yet for determining whether this condition has been met. Clearly, MARPOL does not require that all adjacent states become signatories of Annex V, and IMO has indicated that there is no need for all nations adjacent to the special area to establish port reception facilities. But IMO recognizes that vessels cannot be expected to comply with the stringent special area restrictions unless there are adequate and relatively convenient opportunities for disposal of garbage in nearby ports. In the Wider Caribbean, adequate port reception facilities probably would be needed only in key littoral states—such as the United States, Cuba, and Mexico—in order for vessels to comply with Annex V without a great deal of inconvenience. Once this occurs, the special area designation may become enforceable. However, Cuba and Mexico are not now parties to MARPOL. Negotiations are under way to encourage their ratification of MARPOL and Annex V and to secure adequate port reception facilities in these nations.

It is no small undertaking to assure that all vessels that need zero-discharge capability achieve it, and that adequate port reception facilities exist near sea areas bordered by multiple nations. These are significant challenges that affect a number of maritime sectors and demand the involvement of multiple federal agencies. (Chapters 4 and 5 identified technical obstacles to achieving zero-discharge capability and providing adequate reception facilities.) Strong national leadership will be required to meet these challenges and develop and execute an effective Annex V implementation strategy.

Implementation of Annex V in the Wider Caribbean

The Wider Caribbean poses a greater challenge for U.S. implementation of Annex V than does any other region. The Coast Guard has reported a greater number of violations in the Gulf of Mexico and the Caribbean than in other U.S. waters, because ports there are "frequented more by vessels from nations that are not party to MARPOL and manned by crews who are unaware of the requirements" (Eastern Research Group, 1992). The severity of the Caribbean debris problem was what prompted the United States to petition IMO for a special area designation for the Gulf of Mexico. (To gain regional support for the initiative, the United States also launched diplomatic efforts in the United Nations Environment Program's Regional Seas Program for the Wider Caribbean Region.) After obtaining the designation, the U.S. government turned its attention to assuring that the prerequisites were met: Adjacent nations had to provide adequate reception facilities.

As a next step, the United States led the effort to gain World Bank support for an assessment of the need for waste reception facilities in the region's ports. Officials at the World Bank's Global Environmental Facility agreed to provide funding for such a study. The study determined that, in many of the island nations, the problem of handling vessel-generated garbage could not be separated from the larger issue of management of solid wastes produced on land; a comprehensive waste management strategy was needed for the islands and the region. The World Bank then developed a package of grants and loans to help address the waste management needs of the eastern Caribbean states.[14] In addition, recognizing that a region-wide program for financial and technical assistance is needed, the World Bank has initiated the Wider Caribbean Initiative on Ship-Generated Waste in Support of the MARPOL 73/78 Convention (World Bank/Global Environmental Facility, 1994).

The World Bank project for the eastern Caribbean states serves as an example of a project designed to produce comprehensive solutions through a waste management strategy that does not merely shift pollution from one place to another. The project also illustrates that finding solutions for other areas of the Caribbean and the globe will be neither inexpensive nor easy. Limited resources are available to aid nations lacking the domestic capacity to handle their own wastes as well as those generated by vessels entering their ports. The administrative and legal infrastructures needed to implement stringent environmental standards are beyond reach of many nations. Regional cooperation may be the only way to surmount these limitations. The United States could continue to exert leadership in promoting the development of cooperative and collaborative programs. One mechanism would be a regional memorandum of understanding that sets terms for the sharing of enforcement assets, training programs, and other resources.

Another way to promote international implementation of Annex V, and thereby assist in U.S. implementation efforts, would be to identify and overcome obstacles hindering participation in MARPOL by Caribbean nations. Not enough has been done to analyze how Caribbean states might carry out responsibilities for control of pollution from vessels cost effectively, either by ratifying MARPOL

---

[14]Members of the Organization of Eastern Caribbean States (OECS) are Antigua and Barbuda, Dominica, Grenada, St. Kitts and Nevis, St. Lucia, St. Vincent and the Grenadines, and Montserrat. Montserrat is not a member of the World Bank group. A workshop was held in 1993 at which OECS members were informed about the World Bank project.

or enacting domestic legislation. New approaches for assisting these states could be considered. For example, to encourage participation in MARPOL, provisions allowing qualifying states to defer certain obligations under annexes I and II may be useful. Another possibility is development of independent, regional agreements incorporating the obligations of Annex V without the burdens of annexes I and II.

SUMMARY

Implementation of Annex V will require attention to three overarching issues. One is the need for national leadership; many opportunities for improving Annex V implementation require the cooperation of multiple agencies and organizations and diverse maritime sectors. There are four ways to provide leadership: maintain the status quo, assign the task to one agency, establish an interagency task force, or create a national commission. The commission concept may offer the most potential benefits and entail the fewest drawbacks. A commission not only could coordinate the efforts of federal agencies, but also could serve as a high-level focal point for U.S. leadership in guiding the global maritime community toward increased standards of performance.

The second issue is enforcement of Annex V. Efforts are under way to improve prosecution of foreign violators. A number of other steps also might be taken to enhance the effectiveness of enforcement. Government authorities could seek to clarify the extent of and fully exercise port state control; issue tickets in civil cases involving Annex V violations, particularly in the fisheries and recreational boating sectors; require that ports provide receipts for garbage off-loaded into their reception facilities, and then compare the receipts to vessel logs; enlist the assistance of the NMFS, MMS, and state marine police in reporting Annex V violations; encourage ship operators to report inadequate garbage reception facilities at ports; and conduct public awareness campaigns urging citizens to report illegal garbage disposal.

The third issue is devising an Annex V implementation strategy that takes special areas into account. This issue has both domestic and international aspects. On the domestic side, vessels operating in special areas ultimately need to achieve zero-garbage-discharge capability, and port reception facilities bordering special areas need to be adequate. On the international side, the United States needs to find new and improved ways to assist with the development and improvement of vessel garbage control mechanisms in neighboring nations. One option would be to explore the formulation of memoranda of understanding for the sharing of information, enforcement assets, and other resources. There is a particular need for mechanisms to reduce the administrative burdens on developing countries. The United States also could seek means of increasing the numbers of adequate port reception facilities in special areas.

REFERENCES

Eastern Research Group (ERG), Inc. 1992. Report to Congress on Compliance with the Marine Plastic Pollution Research and Control Act of 1987. Report prepared for the U.S. Coast Guard by ERG, Arlington, Mass. (now Lexington, Mass.). June 24.

International Maritime Organization (IMO). 1991. Guidelines for the Designation of Special Areas and the Identification of Particularly Sensitive Sea Areas. IMO Resolution A.720(17). Adopted Nov. 6, 1991. Available from IMO, 4 Albert Embankment, London, SE1 7SR.

North, R.C. 1993. Prepared statement of Capt. Robert C. North, chief, Office of the Marine Safety, Security, and Environmental Protection, U.S. Coast Guard, before the Subcommittee on Superfund, Ocean, and Water Protection of the Committee on Environment and Public Works, U.S. Senate, 102nd Congress, Second Session, Washington, D.C., Sept. 17, 1992. Pp. 146-149 in Implementation of the Marine Plastic Pollution Research and Control Act. S. Hrg. 102–984. Washington, D.C.: U.S. Government Printing Office.

Podlich, M. 1993. Prepared Statement of Margaret Podlich, project director, Pollution Prevention Program, Center for Marine Conservation, for the Subcommittee on Superfund, Ocean, and Water Protection of the Committee on Environment and Public Works, U.S. Senate, 102nd Congress, Second Session, Washington, D.C., Sept. 17, 1992. Pp. 49-56 in Implementation of the Marine Plastic Pollution Research and Control Act. S. Hrg. 102–984. Washington, D.C.: U.S. Government Printing Office.

Schrinner, J.E. 1992. Pollution of Seas—Disposal of Waste. Paper presented to the Caribbean Shipping Association Semi-Annual General and Group Meetings, Grand Cayman, Cayman Islands, May 25–26, 1992.

United States. 1992. Enforcement of Pollution Conventions: MARPOL Annex V Violations. MEPC/INF.44. Document submitted to the 33rd session of the Marine Environment Protection Committee of the International Maritime Organization (IMO). Available from IMO, 4 Albert Embankment, London SE1 7SR. Sept. 9.

U.S. Department of Justice. 1993. News Release. United States Attorney, Southern District of Florida, Miami, Fla. April 15.

U.S. Environmental Protection Agency (EPA). 1988. Citizen Volunteers in Environmental Monitoring: Summary Proceedings of a National Workshop. EPA 503/9-89-001. EPA Office of Water, Washington, D.C. September.

Weikart, H. 1993. Presentation by Heather Weikart, National Marine Fisheries Service Observers Program, to the Committee on Shipborne Wastes of the National Research Council, Red Lion Inn, Seattle, Wash., July 15, 1993.

World Bank/Global Environmental Facility. 1994. Developing Countries of the Wider Caribbean Region: Wider Caribbean Initiative for Ship-Generated Waste. Report No. 12868LAC. Washington, D.C.: World Bank. June 30.

8

Measuring Progress in Implementation of Annex V

R egardless of what steps are taken to improve implementation of Annex V, it is important to be able to measure any resulting progress, not only to determine which interventions are effective but also to enhance scientific understanding of the oceans. A progress assessment would have two primary components. The first would involve record keeping to gauge Annex V compliance rates among vessels. The second component would involve environmental monitoring to determine whether the flux of vessel garbage to the marine environment is being reduced. To provide the data needed for these two types of evaluations, appropriate record-keeping practices and environmental monitoring techniques would need to be developed and employed.

This chapter examines possible approaches to record keeping and environmental monitoring from the perspective of measuring progress in implementation of Annex V. The first half of the chapter addresses the collection of records on compliance. There are numerous opportunities for routine collection of such information (as suggested at various points earlier in this report), but few data are compiled or analyzed systematically. The second half of the chapter outlines the options for environmental monitoring. It should be noted that this approach cannot be employed as the sole measure of progress in Annex V implementation, because vessel garbage is only one source of debris in the marine environment. Thus, although the committee concentrated on the problem of vessel garbage, this aspect of the analysis focuses on the broader problem of marine debris.

RECORD KEEPING AS A MEASURE OF IMPLEMENTATION

The committee's work has revealed that information is available that could be used to measure Annex V compliance but, for a variety of reasons, it is not yet put to that use. Progress in U.S. implementation of Annex V could be measured in a straightforward manner if comprehensive data were collected over time on numbers of vessels discharging garbage at ports, amounts of garbage discharged, numbers of complaints about garbage reception facilities, and numbers of repeat violations by vessels and ports. Such data would enable the federal government to conduct meaningful analyses of compliance that are not now possible. Such information also would support strategic planning and program evaluation showing, for example, the statistical relationship between educational programs and Annex V compliance, and between the status of port reception facilities and local levels of marine debris. Moreover, the agencies involved in Annex V implementation could identify weak spots where resources should be directed and gain access to useful data collected by other departments.

If a comprehensive record-keeping system were desired, then it would be necessary to develop a government-wide format for Annex V data, collect systematically various types of information from myriad sources, and then combine it all in an electronic database. The Coast Guard could input information from vessel garbage logs, Annex V enforcement reports, and the Certificate of Adequacy program. The Animal and Plant Health Inspection Service (APHIS) could input the data it collects on vessel compliance and amounts of garbage offloaded. Similarly, the Minerals Management Service, the National Marine Fisheries Service, the Department of State, and all other agencies involved with Annex V could collect and input their own data.

Such a task would be enormous. The work involved could not be justified for years—until enough data had been collected to enable meaningful analysis. In addition, there is the question of who would oversee such an interagency effort. If a national commission were established to oversee Annex V implementation (as suggested in Chapter 7), perhaps it could coordinate the development of a comprehensive database. In the meantime, a smaller-scale record-keeping regime might be feasible, particularly if it made use of records already available.

The most easily implemented and potentially most useful system might be a combined Coast Guard/APHIS record-keeping program on vessel garbage handling. APHIS retains but apparently makes little use of records of vessel boardings and garbage off-loading. One research team (Hollin and Liffman, 1993) had to collect *manually* the information recorded on more than 1,500 vessel boarding cards in order to identify an apparent trend in use of shipboard equipment to comply with Annex V. This type of information could be logged into a unified system. The Coast Guard and APHIS would have to agree to cooperate, establish a common reporting format, convert their data into electronic form, and input it into a database. Apart from providing benchmarks for measuring Annex V implementation, this strategy could be extremely useful in suggesting where the two agencies' monitoring and enforcement efforts should be directed.

In selecting the types of data to be recorded, it would be important to go beyond numbers of violations, prosecutions, or permits and attempt to document the *process* of building a permanent Annex V implementation regime—that is, to collect data reflecting *why* compliance problems arise. For instance, did a vessel fail to off-load garbage in a particular port because the reception facilities were full or not available? Did the crew discharge plastics overboard because they didn't know this practice was illegal? Routine collection of information about mariners' attitudes and behaviors would be useful in identifying where interventions were needed, and in satisfying the need (documented in Chapter 6) for evaluation of Annex V education programs.

The data bank could be enhanced further if cargo and cruise ships were required to off-load garbage at all U.S. port calls, and these discharges were recorded. (Surveys might be a more effective tool for small vessels.) At present, few ports are recording information on total weight of debris and usage of dumpsters. While neither Annex V nor the Marine Plastics Pollution Research and Control Act require vessels to off-load garbage, some other nations do mandate it upon both arrival and departure.

Another approach would be to model and then monitor vessel–port garbage transactions. The Environmental Protection Agency (EPA) maintains a computer model for solid waste management and might adapt it for ships or ports. The amounts of garbage off-loaded from ships in ports could be sampled or audited, and these data could be entered into the model to provide, over time, some indication as to whether the amounts were consistent based on days at sea, crew size, and vessel type. This approach would need to be applied to all sizes and types of ports, including small piers and marinas. In fact, it would be more important to conduct such studies in smaller ports, where there are no other methods for examining garbage disposal (such as routine Coast Guard and APHIS boardings and inspections).

Assessing Annex V Implementation Internationally

Both the International Maritime Organization and the U.S. government have mandated that potential polluters document their actions. Recently, some governments also have obliged waste management companies to "manifest" garbage shipments just as shippers keep records of cargos and shipping transactions. Such data could be useful in measuring Annex V implementation internationally. However, the history of international agreements shows that reporting—even when mandated—is generally poor, casting doubt on the effectiveness of such an approach.

In 1991, the U.S. General Accounting Office (GAO) was asked by the Congress to assess compliance with reporting requirements in a number of international agreements. The purpose was to determine whether international environmental agreements are effective, and whether nations are living up to their obligations. What the GAO discovered was reflected in the title selected for the report, *International Agreements Are Not Well Monitored* (U.S. General Accounting Office, 1992).

Reporting mandates often are placed in international treaties. The purpose of such mandates is to prompt compliance: Nations may risk international disapproval and retaliation if evidence reveals numerous violations and weak enforcement responses. Reporting requirements also give the international community a way to quantify over time any trends in compliance (i.e., the overall efficacy of multilateral arrangements). But, as the GAO report points out, compliance with reporting requirements is not a reliable indicator of compliance with international standards. Similarly, failure to report does not indicate the nation is violating the substantive obligations of the agreement. Record keeping to fulfill treaty reporting requirements may be beyond the administrative capacity of a government for a variety of reasons.

While the reports submitted to international secretariats are not a perfect measure of the efficacy of international agreements, they are the sole evidence of what is actually happening worldwide in fulfillment of treaty obligations. Equally importantly, these reports are indicative of the practical limits of government surveillance of, and control over, the behavior of seafarers while at sea.

The GAO studied the following eight agreements: the Montreal Protocol (which addresses ozone depletion), the Nitrogen Oxides Protocol (acid rain), the Basel Convention (transport of hazardous wastes), the London Dumping Convention, MARPOL (Annex I only), the Convention on International Trade in Endangered Species (CITES), the International Whaling Convention, and the International Tropical Timber Agreement (ITTA).

Seven of the eight currently require that members report annually on implementation (although the information requested usually is limited to numbers of permits issued, violations detected, or inspections conducted). The GAO found that reporting fell far short of what was mandated, and the few reports that were submitted often were incomplete and late. The GAO determined that, in most cases, the respective secretariat is ill equipped to press for better performance. Furthermore, because of their small size, lack of authority, and scant resources, secretariats are equally unable to assess implementation independently.

Most of the agreements examined provide measurable performance standards. CITES, for example, creates a permit system to ban trade in endangered species and control trade in threatened ones. The International Whaling Convention sets annual harvest quotas. MARPOL Annex I establishes specific limits on the amount of oil that can be discharged. Yet high rates of reporting on compliance were found for only three of the conventions studied: the Montreal Protocol, the Nitrogen Oxides Protocol, and the International Whaling Convention. Less than half the membership of CITES, the London Dumping Convention, MARPOL, and ITTA filed their required reports.

A mere 13 of the then-57 signatories to MARPOL had provided the secretariat with the required information on violations and penalties imposed. Only 59 percent of the parties had reported on the availability of (Annex I) oily waste reception facilities as mandated. With regard to CITES, only 25 of the 104 parties had delivered their annual reports containing information on trade in listed species. The GAO stated that "most parties either submit reports that are late, incomplete, or in the wrong format, or do not submit any report at all."

The GAO noted that secretariats are limited in authority and ability to assess compliance independently. CITES stands as the only agreement that specifically grants the secretariat the role of assessing compliance. Through a contract with a private organization, the Wildlife Trade Monitoring Unit, CITES data are analyzed and the violations summarized in a report. The secretariat then can recommend trade sanctions. Most secretariats are not positioned to verify information received from member governments; rather, they act as facilitators and information clearinghouses. The GAO study also concluded that secretariats are typically small with very limited funding and lacking in the resources to undertake more systematic monitoring.

No matter who does it, monitoring of international agreements is a major assignment. The Committee on Shipborne Wastes certainly has been challenged by the task of assembling the information needed to report on Annex V implementation across all fleets in the United States. To do so on an annual basis would require a level of organization and effort that does not now exist anywhere for collecting data on any international agreement.

ENVIRONMENTAL MONITORING

Environmental monitoring is an important aspect of environmental management. A monitoring system involves not only field assessments and data analysis but also integrated and coordinated activities with "the specified goal of producing predefined management information; it is the sensory component of environmental management" (National Research Council, 1990). Of great significance in the present context is the high cost of *not* monitoring; failure to monitor adequately poses a serious impediment to efforts to protect marine environmental quality (National Research Council, 1990).

Surveys of Beach Debris

Progress in implementation of Annex V could be measured most directly by changes in the flux of vessel garbage to beaches and the sea floor. Obtaining precise data is difficult. Two criteria govern the validity and utility of such measurement. First, the materials surveyed must be identifiable as vessel-generated garbage. Second, quality assurance and quality control practices are essential to assure scientifically valid results.

There are few, if any, surveillance programs designed to test the effectiveness of Annex V implementation. One that has some relevance is a monitoring program on a remote island in the South Atlantic, Inaccessible Island of the Tristan da Cuna group, where an exponential increase in the amount of beach litter was noted between 1984 and 1990 (Ryan and Moloney, 1993). Eighty percent of the debris was plastic, with most items having a source in South America, more than 3,000 kilometers away. The amount of debris originating from vessels was not ascertained. (Even when debris can be traced, it is difficult to use this information to determine whether a violation of Annex V occurred [Amos, 1993].)

While current surveillance programs are not oriented specifically to Annex V, extensive activities are devoted to studying the types, amounts, and sources of debris on coastal beaches and to heightening awareness of the marine debris problem. Much of the data has been gathered by the Center for Marine Conservation (CMC), which launched a beach cleanup campaign in 1986 in Texas. The effort has evolved into the annual International Coastal Cleanup Campaign, which relies on a network of state and country coordinators to organize thousands of citizens.[1]

The purpose of the event is not only to clean the beaches but also to collect data on the types and amounts of debris. The CMC produces an annual report, which provides data broken down at the national, state, and local levels. Reports of wildlife entangled or otherwise affected by debris are compiled. While identifying sources of debris is difficult even for trained experts, citizens have provided useful information, such as findings of debris traceable to cruise lines based on company names on product labels.

The use of volunteers to gather data is attractive from both an economic and a social perspective. However, whether volunteers can gather scientifically sound data is subject to debate. Amos (1993) noted a marked difference between beach surveys done by volunteers and those by scientists. In this single experiment, the volunteers appeared to under-count debris items by about 50 percent. A similar problem was reported at Padre Island (Miller, 1993). If volunteers are to be used to gather data for scientific purposes, then they need to be trained in data collection techniques. There also needs to be scientific oversight to assure adherence to research protocols (U.S. Environmental Protection Agency, 1988).

In recent years, there has been increasing recognition of the need to standardize monitoring methods. Without such standards, there is no baseline to which new data can be compared, and data cannot be shared among the various monitor-

---

[1]In 1992, this one-day event involved more than 160,000 volunteers in 33 countries (Hodge et al., 1993). Since then, the effort has expanded to include more than 222,000 volunteers in 40 countries.

ing groups. In 1989, the National Oceanic and Atmospheric Administration (NOAA) entered into an agreement with the National Park Service to conduct a five-year pilot study using a standard methodology for marine debris surveys[2], tested on beaches within nine national seashores. Recently, the EPA has been leading an effort to improve on the methodology. The EPA is working with NOAA, the National Park Service, the CMC, the Coast Guard, the Marine Mammal Commission, and selected scientists to establish a method for determining inputs of debris from specific ocean- and land-based sources and identifying trends. The EPA methodology has been tested at pilot sites in Maryland and New Jersey. A draft methodology has been developed and reviewed by all federal agencies that monitor marine debris, and final approval was expected by the end of 1994. A long-term marine debris sampling program, carried out by trained volunteers, is to be implemented at selected U.S. beaches in 1995.

Monitoring Trends in Biological Impacts

Another approach to measuring progress in Annex V implementation would be to monitor for trends in ecological effects, such as injury or mortality among species of wildlife. As discussed in Chapter 2, available information on the impacts of debris on marine organisms consists primarily of baseline studies. Trends might be determined if long-term studies were initiated focusing on groups and populations of marine species. However, despite widespread observations of marine debris, only a few animal populations are monitored so closely that the effects of such debris could be discerned among all the other influences on the population.

The potential for using this type of research to measure Annex V implementation is suggested by the ongoing northern fur seal studies, which provide a continuing census of a legally protected marine species. Through close and repeated observations, researchers are able to record information on the effects of fishing debris on seal colonies. A recent assessment notes a 50 percent decrease between 1981 and 1989 in the number of seals reported entangled in trawl webbing, possibly due to a reduction in the amount of net fragments discarded by fishing vessels (Fowler and Baba, 1991). That data set, initiated long before Annex V came into force, provides a record of the harm caused by uncontrolled vessel garbage. Continued collection of such data—particularly if the researchers were asked specifically to also record debris entanglements—might provide an

---

[2]The methodology was based on early drafts of a marine debris survey manual developed with the support of the Marine Entanglement Research Program. The manual (Ribic et al., 1992) was adopted for publication in 1993 by the Intergovernmental Oceanographic Commission's Working Committee on the Global Investigation of Pollution in the Marine Environment, which had launched the initial standardization effort in 1986 by agreeing to develop such a guide.

indication of whether the harm is abating. It is important to remember, however, that the fur seals are studied on land only, so the results may not reflect the total effects of marine debris, and that the case for population-level effects on these animals, while the strongest data available, is only circumstantial.

One group of researchers has recommended that all future studies of wildlife interactions with debris include statistically adequate sampling schemes designed to test hypotheses that the prevalence of debris is either increasing or decreasing in given areas or for specific taxa (Sileo, 1990).

Monitoring Plastics in the Marine Environment

As discussed in Chapter 2, plastics are the most abundant and most harmful type of marine debris. Their persistence in the marine environment is virtually infinite, according to some environmental scientists, and the solids can cause considerable harm in addition to aesthetic insults. Plastics can kill marine animals through ingestion or entanglement and inflict costly damage to vessel operations through fouling of propellers, water intake pipes, and fishing gear. Another threat may lie in the accumulation of plastics on the sea floor. Although plastics are buoyant when introduced to the marine environment, they quickly sink to the bottom, where they may inhibit gas exchange between the overlying waters and the pore waters of the sediments. Hypoxia or anoxia could result.

Therefore, for purely ecological reasons, it would be advisable to conduct long-term monitoring programs to measure amounts of plastics in the marine environment, both on beaches and on the coastal seafloor. The data also could provide a measure of progress in Annex V implementation, because plastic is the one material for which all overboard discharge is banned. The committee, drawing on the personal experience of several members and relying heavily on Ribic et al. (1992) and Amos (1993), devised a basic monitoring strategy that would be useful from both an Annex V and a scientific standpoint and therefore make the best possible use of resources. The strategy borrows from the basic methodology of the EPA's planned marine debris monitoring program but is different in three important respects: The committee's model focuses on plastics rather than all marine debris, attempts to isolate vessel garbage from land-source debris, and includes both beach and benthic surveys.

The goal would be to determine the fluxes of plastics through the marine environment as a function of time. The focus could be expanded to include other particularly harmful and problematic debris items, such as fishing gear. It might be appropriate to incorporate such an effort into NOAA's Status and Trends Program, which has been described as "the closest current approach to a standardized national assessment of marine pollution" (National Research Council, 1990). The NOAA program measures contaminants such as metals and chlorinated hydrocarbons at over 100 sites on an annual basis. Bottom-feeding fish, mussels, oysters, and sediments are collected. The goal is "to create, maintain, and assess a long-term record of contaminant concentrations and biological responses to contamination in the coastal and estuarine waters of the United States" (National Oceanic and Atmospheric Administration, 1988).

Beach Surveys of Plastic Debris

To date, most studies of plastics accumulating on beaches have two deficiencies with respect to pinpointing the flux of materials regulated by Annex V. First, these surveys may not be conducted often enough, in that the residence time of debris on beaches appears to be only a matter of months or, in some cases, days. If these estimates are accurate, then the results drawn from less-frequent sampling probably underestimate the true fluxes. Second, these studies record the incidence of *all* debris on a beach and may include plastic discards from non-ship sources such as storm drains, recreational activities, and sewers, thereby confusing the results.

To assure uniformity in data gathering, a dedicated collection team could be employed.[3] An alternative would be to train volunteers to identify debris items in a uniform manner, perhaps by using a manual such as the *Pocket Guide to Marine Debris* (Center for Marine Conservation, 1993). As data were collected, all debris would be removed from each sampling site. Materials to be counted would include all plastics and, in some areas, non-plastic debris such as waste from fishing activities.

For the program to be thorough, all U.S. coasts would have to be monitored. Monitoring sites might be designated on each coast and the Gulf of Alaska, where large amounts of debris from fishing activities accumulate. The collection team could survey each site on a regular basis. Because Annex V regulates only vessel garbage, monitoring sites could be sought that receive minimal discards from land sources. Perhaps uninhabited offshore islands would provide the most reasonable monitoring sites[4]; another possibility would be beaches closed to public use due to their association with active or abandoned naval target ranges.

Benthic Surveys of Plastic Debris

The objectives of benthic surveys would be to measure the amounts and

---

[3]Such a strategy was employed during the EPA-sponsored National Mussel Watch from 1976 to 1978, in which sentinel organisms were collected at over 100 stations on the East, West and Gulf coasts (Goldberg et al., 1983). Two scientists acted as a dedicated collection team. The program has been continued and expanded under NOAA's Status and Trends Program.

[4]Data collected on Sable Island provides ample evidence of the transport of human-generated garbage across vast expanses of water onto a sparsely inhabited, windswept island (Lucas, 1992). Copious amounts of debris from ships also have washed up on remote Hawaiian island beaches (Marine Mammal Commission, 1992).

types of plastic and other debris on the sea floor, the area covered by such materials, and any changes with time. Surveys can be conducted in a variety of ways, using trawls, submersibles, divers, side-scan sonar imaging, or underwater cameras. All these strategies are expensive and can cover but a small area of the ocean bottom. Trawl surveys appear to be the preferred as well as the least expensive strategy (Ribic et al., 1992), although the cost and efficiency of trawl and electronic surveys have not been assessed. Clearly, the trawl surveys would best be made in conjunction with the beach surveys at each site.

Ribic et al. (1992) identified the variables to be considered in trawl surveys. One variable is vessel capability to tow effectively. The mesh size of the net governs the sizes of particles captured. Fluctuations in survey depth provide a sense of whether the trawl is following the bottom.

Sampling Sites and Frequencies

Because all plastic material within a given stretch of beach is to be both counted and collected, a beach site must be both short enough that a survey can be executed and long enough to provide suitable statistics. Amos (1993) suggests a minimum length of one kilometer. Whatever the length, a site needs to encompass the total beach area so there are no difficulties with lateral transport of debris.

Sampling frequencies would be developed in line with quality control and quality assurance criteria. Quality control relates to the quality of the data itself, usually defined by statistical parameters, standard deviations, and precision. Quality assurance relates to the adequacy of the data to satisfy the goal of the project (i.e., whether the data reflect statistically valid changes with time in the flux of plastics from vessel discards to the coastal zone).

Sampling frequency would depend in part on how often the physical oceanographic properties of a site change. In this as well as other aspects of sampling design, a statistician is crucial, as emphasized by both Ribic et al. (1992) and Amos (1993). The former asserts that "a statistician should be consulted at the onset of survey planning and be involved through the completion of the study."

Data Collection and Management

The survey team would employ multiple data units, such as site-by-site volume, weight, and number of debris articles. In addition, other information would be collected with each site visit, including current patterns, weather, and some measure of vessels transiting nearby shipping lanes. At certain sites, measures of commercial and recreational fishing intensity also could be important.

Plastic containers often are imprinted with the year and even month of manufacture, country of origin, and manufacturer. Such information is extremely useful in associating the debris with a given source, such as a vessel as opposed to a land-based source. What is more, this information can establish recent use, suggesting the approximate time of discard (e.g., after ratification of Annex V).

Because the plastics survey project would be narrow in scope, the data gathered might be of only limited value. Still, this type of data, given appropriate quality control parameters, could be useful to national and international agencies implementing Annex V. Therefore, the data might be stored in a readily accessible computer for use in other marine debris research programs.

SUMMARY

Progress in implementation of Annex V could be measured through record keeping reflecting vessel compliance and, as a supplementary measure, environmental monitoring.

The most easily implemented record-keeping program might be a combined Coast Guard/APHIS system on vessel garbage handling, making use of existing APHIS records of vessel boardings and garbage off-loading, and information from Coast Guard enforcement reports and vessels' garbage logs. Apart from providing benchmarks for measuring Annex V implementation, the database could be used to determine where the two agencies' monitoring and enforcement resources should be directed. Both the data-gathering and enforcement efforts also could benefit if cargo and cruise ships were required to off-load garbage at all of their U.S. port calls.

An environmental monitoring program could be designed to determine the fluxes of plastics through the marine environment as a function of time. Such an effort might be incorporated into NOAA's Status and Trends Program. A collection team could collect plastic debris from selected beach sites on all U.S. coasts, in conjunction with trawl or electronic surveys of the coastal sea floor. The EPA could have some involvement, in order to capitalize on the experience and expertise gained in developing its beach monitoring program.

REFERENCES

Amos, A.F. 1993. Technical Assistance for the Development of Beach Debris Data Collection Methods. Final Report submitted to U.S. Environmental Protection Agency, Gulf of Mexico Program, New Orleans, La. TR/93-002. May 31.

Center for Marine Conservation (CMC). 1993. Pocket Guide to Marine Debris. Washington, D.C.: CMC.

Fowler, C.W. and N. Baba. 1991. Entanglement studies, St. Paul Island, 1990 Juvenile Male Northern Fur Seals. AFSC Processed Report 91-01. Available from the Marine Entanglement Research Program of the National Marine Fisheries Service (National Oceanic and Atmospheric Administration), Seattle, Wash.

Goldberg, E.D., M. Koyde, V. Hodge, A.R. Flegal, and J. Martin. 1983. U.S. Mussel Watch: 1977–1978 results on trace metals and radio nuclides. Estuarine, Coastal and Shelf Science (U.K.) 16:69-83.

Hodge, K., J. Glen, and D. Lewis. 1993. 1992 International Coastal Cleanup Results. Washington, D.C.: Center for Marine Conservation.

Hollin, D. and M. Liffman. 1993. Survey of Gulf of Mexico Marine Operations and Recreational Interests: Monitoring of MARPOL Annex V Compliance Trends. Report to the U.S. Environmental Protection Agency, Region 6, Gulf of Mexico Program by Dewayne Hollin, Texas A&M University Sea Grant College Program, and Michael Liffman, Louisiana State University Sea Grant College Program.

Lucas, Z. 1992. Monitoring persistent litter in the marine environment on Sable Island, Nova Scotia. Marine Pollution Bulletin 24(4):192-199). April.

Marine Mammal Commission (MMC). 1992. Annual Report of the Marine Mammal Commission, Calendar Year 1991, a report to Congress. Washington, D.C.:MMC. Jan. 31.

Miller, J. 1993. Marine Debris Investigation: Padre Island National Seashore, Texas. Corpus Christi, Tex.: National Park Service. December.

National Oceanic and Atmospheric Administration (NOAA). 1988. National Marine Pollution Program: Federal Plan for Ocean Pollution Research, Development, and Monitoring Fiscal Years 1988–1992. Rockville, Md.: NOAA. Cited in National Research Council (NRC). 1990. Managing Troubled Waters: The Role of Marine Environmental Monitoring. Marine Board, NRC. Washington, D.C.: National Academy Press.

National Research Council (NRC). 1990. Managing Troubled Waters: The Role of Marine Environmental Monitoring. Marine Board, NRC. Washington, D.C.: National Academy Press.

Ribic, C.A., T.R. Dixon, and I. Vining. 1992. Marine Debris Survey Manual. NOAA Technical Report NMFS 108. Available from the Marine Entanglement Research Program of the National Marine Fisheries Service (National Oceanic and Atmospheric Administration), Seattle, Wash.

Ryan, P.G. and C.L. Moloney. 1993. Marine litter keeps increasing. Nature 361:23. Jan. 7.

Sileo, L. (chair). 1990. Report of the working group on ingestion. Pp. 1226-1231 in Proc. of the Second International Conference on Marine Debris, 2–7 April, 1989, Honolulu, Hawaii (Vol. II), R.S. Shomura and M.L. Godfrey, eds. NOAA-TM-NMFS-SWFSC-154. Available from the Marine Entanglement Research Program of the National Marine Fisheries Service (National Oceanic and Atmospheric Administration), Seattle, Wash.

U.S. Environmental Protection Agency (EPA). 1988. Citizen Volunteers in Environmental Monitoring: Summary Proceedings of a Workshop. EPA 503/9-89-001. Washington, D.C.: EPA Office of Water. September.

U.S. General Accounting Office (GAO). 1992. International Environment: International Agreements are Not Well Monitored. GAO/RCED-92-43. Washington, D.C.: GAO Resources, Community, and Economic Development Division. Jan. 27.

9

National Strategy

Previous chapters have examined the problem of vessel garbage from a variety of perspectives, by addressing scientific understanding of marine debris, the legal requirements of MARPOL Annex V and the related U.S. law, and characteristics of the maritime sectors that must comply with these mandates. The report has identified a variety of barriers to compliance as well as potential solutions and factors complicating those solutions. The report also has explored strategic issues and limitations generated by considerations of importance to government agencies and the regulated communities.

The task now is to integrate all these elements into a coherent strategy that will enhance implementation of Annex V. In the committee's judgment, such a strategy needs to be tailored to practical realities, not only in terms of the needs and characteristics of each maritime sector but also in the context of the integrated solid waste management system (ISWMS) in place for land-generated waste. That is, the strategy should target problems and opportunities specific to each sector, and it should serve to integrate the handling of vessel garbage into the ISWMS, taking into account both the trends and the shortcomings of that system.

This approach suggests that progressive changes in the handling of land-generated garbage should be encouraged in the maritime world. Recycling, for instance, is now standard in many homes and offices. Residents in many parts of the country are accustomed to separating, cleaning, storing, and setting at curbside a variety of recyclable waste materials. It is therefore plausible that fisheries personnel could become accustomed to returning used nets and lines to port for recycling. While not specifically required by Annex V, recycling would foster compliance by reducing a source of marine debris. At the same time, it is important to recognize that the infrastructure for recycling plastic materials used by mariners is not well developed. Even where markets exist for recycled materials and products, there is seldom a convenient and cost-effective arrangement for converting the collected waste materials into products. But as this and other aspects of the ISWMS are improved, new opportunities will be created to improve management of vessel garbage.

This chapter takes such considerations into account in identifying, for each maritime sector, a set of strategic objectives that should serve as milestones in working toward the overall goal of Annex V implementation. In addition, specific actions are recommended or suggested that would foster attainment of these objectives. In combination, these sets of objectives and tactics constitute the foundation for a national Annex V implementation strategy. Federal actions needed to help execute this strategy across all fleets are described in Chapter 10.

The committee wishes to emphasize that an objective is something to be pursued, rather than an absolute requirement (as would be established by law), and that existing obstacles to Annex V compliance, however onerous, should not serve as justification for abandoning an objective.

The following introduction outlines the committee's approach to identifying priorities for each sector.

IDENTIFYING AND EVALUATING STRATEGIES AND TACTICS

The starting point for developing the sector-by-sector implementation strategy is the set of interventions identified in Chapter 4. The matrices in that chapter illustrate the options the committee considers worthy of serious consideration. While any of those interventions might yield some benefits, the committee believes certain objectives and actions to be compulsory if full implementation of Annex V is to be achieved. This chapter outlines these essential elements, which were identified based on the analysis presented in Chapters 4 through 8 and the collective judgment and expertise of the committee. The proposed interventions may be neither easy to execute nor rapidly achieved, but they are critical elements of a national Annex V implementation strategy.

As a guide in identifying the priorities, the committee established a set of criteria, which were employed to screen possible interventions. The committee relied on its collective judgment, rather than formal analysis, to determine whether an alternative met the criteria. (Formal analysis may be impossible, in any case, due to the paucity of data on marine debris and the difficulty of measuring debris levels.) *Authorities implementing Annex V should continue to employ these criteria consistently but informally, without elaborate analyses, in evaluating the effectiveness of any actions proposed here that are pursued.* The committee believes the implementation program would be strongest if these few criteria were applied informally to all activities, as opposed to a more complicated approach. The committee also believes the continuing evaluation process should retain the benefit of the direct observations and experiences of individuals engaged in implementing Annex V. In their daily work, the members of the various maritime sectors know far better than any outside observers what succeeds in their arena.

The following criteria were developed and used by the committee:

• **Effectiveness.** An intervention must be likely to reduce, or provide essential data for reducing the environmental hazard posed by vessel garbage, by either reducing the amount of material or improving handling of the material, in ways that undeniably can show trends in waste entry to the marine environment.

• **Cost Effectiveness.** An intervention must be effective enough to justify its cost. The committee did not examine costs of the various options in detail but believes the proposed actions would be effective enough to justify the expenses incurred. The most expensive proposals might have to be evaluated independently by those who would implement them. Other, less expensive proposals may be desirable in the short term.

• **Efficiency.** The interventions must interfere as little as possible with ongoing activities and must be affordable in terms of time and resources to the maritime sector(s) and government regulators involved.

• **Timeliness of Results.** The actions must allow for some reasonable level of preparation and control and yield improvement within an acceptable time frame.

• **Equity.** The interventions must provide remedies where most needed or in ways that distribute the implementation effort both within and among the maritime sectors.

• **Sustainability.** The actions must help build a permanent Annex V implementation regime and foster the mariner's capability to sustain compliance.

In using these criteria to identify priority objectives and tactics for each maritime sector, the committee did not attempt to rank the proposals. However, two biases emerged in the analysis that serve to emphasize certain types of proposals. *First, the committee placed priority on actions that are upstream (toward the left) in the hazard evolution model described in Chapters 3 and 4.* Logic dictates that these actions would tend to be the most beneficial environmentally (although not necessarily in terms of cost and social advantages) because they address the problem in its earliest stages. Waste reduction is an example of such an approach. *Second, the committee emphasized the need to achieve zero-discharge capability, where appropriate.* This is a legal mandate for vessels that operate in special areas (where only food waste may be discharged). It is also an appropriate objective for vessels dedicated to day trips, because zero discharge should be easy to achieve in this sector and the International Maritime Organization (IMO) guidelines for Annex V implementation recommend use of port reception facilities "whenever practicable." Furthermore, federal law supports the concept of zero discharge. (The Federal Water Pollution Control Act, Title I, Section 101 (1), states that "it is the national goal that the discharge of pollutants into the navigable waters be eliminated . . .")

Following are the strategic objectives and tactics identified for each maritime sector. The order of presentation reflects only the sequence in which sectors and topics were introduced in the preceding chapters.

STRATEGY FOR EACH MARITIME SECTOR

Recreational Boats and Their Marinas

Objective: Achieve zero-discharge capability

Because recreational boaters generally remain within 12 nautical miles of shore, they usually are prohibited from discharging any garbage overboard (unless the vessel is equipped with a comminuter). This situation, combined with the fact that most boaters take day trips, makes zero-discharge capability an objective for this sector. It should be fairly easy to store all garbage on board for disposal ashore. Even so, innovative measures may be needed to attain this objective, because boats tend to be small (with little storage space) and many boaters are unaccustomed to planning for proper garbage handling.

An obvious tactic for boaters would be to reduce use of disposable materials. In addition, convenient garbage storage bins should be incorporated into the design of new boats, and small commercial trash compactors should be installed on boats capable of extended voyages.

Objective: Assure adequacy of port reception facilities

Although reception facilities at marinas generally are not deficient, recreational boats may come ashore at a variety of simple docks and ramps. While small landing areas are not required by the Coast Guard to have reception facilities, it is important to assure that waste receptacles are available and easily accessible. "Clean marina" programs should be established by state licensing

---

Objectives for Recreational Boating Sector

- Achieve zero-discharge capability
- Assure adequacy of port reception facilities
- Assure that boaters are provided with appropriate Annex V information and education

agencies and trade or recreational associations to certify that landing areas meet established criteria for garbage reception facilities.

Objective: Assure that boaters are provided with appropriate Annex V information and education

Because implementation of Annex V depends heavily on responsible personal behavior, it is important that boaters receive the information needed to make the right decisions. Existing communication channels, including signs, the recreational media, and radio, should be employed for this purpose. Annex V information should be distributed at boat races, fishing derbies, and other activities, including contacts with the Sea Grant Marine Advisory Service. This information also should be included as part of state boater registration processes and Coast Guard inspections. In addition, boaters should be encouraged to participate in beach cleanups. There is a particular need for education concerning the problems caused by improper disposal of monofilament fishing line.

In addition, international channels should be created for distributing information about Annex V and compliance strategies. Effective strategies should be promoted and shared through racing associations and/or United Nations groups. International educational events should be sponsored for boaters. Boaters who undertake international voyages should be given Annex V information so they can inform foreign ports about their disposal needs.

To support all these efforts, Coast Guard, Customs, state marine police, and other officials who interact with boaters should be trained in how to persuade boaters to comply with Annex V.

Commercial Fisheries and Their Fleet Ports

Objective: Achieve zero-discharge capability for fishing vessels that operate as day boats

The vast majority of fishing vessels take day trips and should be able to refrain from discharging any garbage overboard. Although this objective is not reasonable for the minority of fishing vessels that take extended voyages, even they should be able to store most garbage on board for disposal in port.

Objective: Provide adequate port reception facilities

Port reception facilities in some remote areas are inadequate for receiving the garbage generated by fishing fleets. To encourage Annex V compliance by fisheries vessels, adequate garbage reception facilities should be provided at all fishing piers, not only for vessel-generated garbage and galley wastes but also for

---

Objectives for Commercial Fisheries

- Achieve zero-discharge capability for fishing vessels that operate as day boats
- Provide adequate port reception facilities
- Assure access to appropriate on-board garbage handling and treatment technologies
- Provide comprehensive vessel garbage management system
- Assure that seagoing and management personnel are provided with appropriate Annex V information, education, and training
- Improve Annex V enforcement
- Extend U.S. cooperation to encourage compliance by foreign-flag vessels

---

debris caught in fishing nets. State authorities who regulate state-numbered fishing vessels should be engaged in establishing reception facilities.

Objective: Assure access to appropriate on-board garbage handling and treatment technologies

Fishing vessels that undertake extended voyages may require installation of garbage handling and treatment technologies in order to achieve compliance with Annex V. Special efforts should be mounted to demonstrate and foster adoption of technologies appropriate to vessel size and operations, in both new and existing vessels. The National Marine Fisheries Service (NMFS) should offer grants to foster development and installation of integrated waste management systems for fishing vessels.

Objective: Provide comprehensive vessel garbage management system

Beyond providing reception facilities and on-board technologies, it is important to strengthen the overall vessel garbage management system. Fishing ports (especially those in remote areas) should be incorporated into the regional ISWMS. The NMFS should discourage abandonment of fishing gear, especially in heavily fished areas.

In addition, a national system for recycling fishing gear should be developed based on successful existing pilot programs, and the system should be integrated into the chemical industry (which produces the materials used in nets and lines). Because this is a unique waste stream that has not been recycled on a wide scale previously, it may be helpful to offer financial incentives to encourage fishermen to return their gear. For example, industry or the NMFS could require deposits on all monofilament lines and nets. Fishermen could collect this money when returning their old gear; unclaimed deposits could be used to help defray costs of establishing the recycling system.

Objective: Assure that seagoing and management personnel are provided with appropriate Annex V information, education, and training

Due to the lack of direct regulatory oversight of the fisheries sector, it is important to encourage voluntary compliance through education. Existing channels, including the Sea Grant Marine Advisory Service, can be used for this purpose. Annex V information should be included in processes for fishing license renewal and boat registration, and marine debris issues should be raised at regional fisheries forums. In addition, while most fishing vessels are uninspected, the Coast Guard's voluntary examination program should be exploited as an avenue for distributing Annex V information.

New approaches for distributing information also should be devised. Newsletters soliciting innovative educational and technological ideas should be developed and disseminated throughout the fisheries community, as is done in the agricultural population. In addition, because fishing is often a family business, families should be educated as a means of influencing their seagoing members.

Educational efforts should address, among other things, opportunities for recycling and uses for recycled plastics and other materials. Reports on gear lost in the oceans should be circulated to persuade fishermen of the potential reduction in fish stocks caused by ghost fishing.

Objective: Improve Annex V enforcement

The fisheries fleet is the one maritime sector where routine enforcement is needed and can be cost effective in assuring Annex V compliance. Where appropriate and feasible, fisheries observers should be enlisted to monitor garbage disposal practices. In addition, fishing nets could be labeled or imprinted with the name of the vessel using them, so vessel operators that lose or discard nets could be identified. Although it would be difficult to distinguish between illegal discards and accidental losses, the NMFS could keep track of the identifications on recovered nets and use the information to identify fisheries where special educational, monitoring, and possibly enforcement efforts are needed.

Fisheries councils also should require reporting of lost gear, both to collect information on this problem and to identify where additional measures to prevent such losses are needed. The IMO guidelines for Annex V implementation recommend that such records be kept and encourage development and deployment of such measures.

Objective: Extend U.S. cooperation to encourage compliance by foreign-flag vessels

Because garbage discharged outside U.S. waters can drift toward the coast, it is important to consider means of fostering Annex V implementation by foreign fishing fleets operating nearby. Such implementation should be a condition of any joint fishing ventures or possibly trade agreements with other nations. Other types of international agreements can serve as mechanisms for this purpose as well. For example, the NMFS scientific agreement with Mexico could encourage or require Annex V compliance by the Mexican shrimp industry, which is blamed in part for the debris in the Gulf of Mexico.

Cargo Ships and Their Itinerary Ports

Objective: Improve access to on-board garbage handling and treatment technologies

To reduce the amounts of garbage that must be discarded, vessel operators should install, maintain, and use on-board compactors, thermal processors, pulpers, and incinerators. These technologies should be retrofitted where feasible and appropriate and integrated into all new construction.

Objective: Provide comprehensive vessel garbage management system, including adequate port reception facilities

Numerous steps can and should be taken to improve the garbage management system for cargo ships. Improvements are needed in three general areas: monitoring of on-board garbage handling by both U.S.-flag and foreign-flag ships; port reception facilities; and handling of Animal and Plant Health Inspection Service (APHIS) waste.

All ocean-going, U.S.-flag ships of 12.2 meters (about 40 feet) or more in length are required to maintain a log documenting the volume, date, time, and location of each discharge of garbage. To provide for greater accountability, the Coast Guard should require all cargo ships (except those with comprehensive on-board waste management systems) to off-load Annex V garbage at every U.S. port call. (Such requirements are in place in the North Sea and other foreign waters; in these areas, record keeping is mandated by port states and applies to all

---

Objectives for Cargo Ships

- Improve access to on-board garbage handling and treatment technologies
- Provide comprehensive vessel garbage management system, including adequate port reception facilities
- Assure that seagoing and management personnel are provided with appropriate Annex V information, education, and training
- Fully exercise U.S. authority to improve compliance by foreign-flag vessels and by all vessels in foreign waters

vessels entering ports.) Vessel logs and on-board garbage handling and treatment technology should be examined during routine Coast Guard inspections.

To help improve the port side of the vessel garbage management system, state agencies should require adequate reception facilities as a condition of issuing permits to ports and should assure that garbage disposal is integrated with regional ISWMS. The Coast Guard should require a port to have the appropriate state permits as a condition of granting a Certificate of Adequacy (COA). Port and terminal operators also should assume expanded roles in overseeing the adequacy of reception facilities and assuring customer satisfaction with services. Cost issues need to be addressed in the permitting process. Ports should be able to recover disposal costs from users, but fees paid by ships should be in line with charges for disposal of land-based garbage. Alternatively, port tariffs or related user fees could be increased to cover garbage disposal.

In addition, the U.S. Department of Agriculture should work to integrate the APHIS program more fully with the Annex V regime to minimize compliance difficulties. Cargo ships should be required to off-load APHIS garbage at every U.S. port call (as is required of aircraft), and ship operators should be educated about the types of garbage subject to quarantine.

Objective: Assure that seagoing and management personnel are provided with appropriate Annex V information, education, and training

As with other fleets, it is important that merchant mariners be given sufficient information and training to enable compliance with Annex V. The need for such training extends throughout each company, from the chief executive officer, who controls the corporate culture, down to the employees who order supplies and personally handle the garbage.

Requirements for employee training in proper waste management should be enacted and enforced throughout this sector. In addition, employees responsible for vessel provisioning should receive training in how to reduce amounts of packaging taken on board and how to emphasize use of recyclable materials.

Objective: Fully exercise U.S. authority to improve compliance by foreign-flag vessels and by all vessels in foreign waters

Because most cargo vessels are foreign flag, it is imperative that special efforts be made to improve Annex V compliance by foreign-flag vessels transiting U.S. waters. The Coast Guard should continue to step up its enforcement activities targeting foreign vessels. The garbage log requirement should be extended to foreign-flag vessels, through either international agreement or unilateral U.S. action in accordance with its port state authorities, and violators should be punished.

Negotiations will be required in various international forums to improve garbage handling in foreign ports, not only because U.S.-flag vessels call at these ports but also because improper at-sea garbage disposal near the U.S. coastline can have adverse effects in the U.S. Exclusive Economic Zone and territorial waters. More specifically, steps must be taken to address the need for adequate port reception facilities in special areas. U.S. authorities should work with the International Maritime Organization (IMO) and other forums to develop clear international criteria and guidelines for port/vessel interfaces. To improve Annex V implementation in nations with scarce resources, the United States should explore the use of regional memoranda of understanding (MOUs) to enable the sharing of enforcement assets and other resources.

Passenger Day Boats, Ferries, and Their Terminals

Objective: Achieve zero-discharge capability (for plastics, glass, cans, and paper), integrating the handling of vessel garbage into local solid waste management systems

Due to the short duration of voyages by these vessels (some casino ships don't move at all) and the resulting ease of returning all garbage to shore, zero-discharge capability should be the objective in this sector. This may have been achieved already, but simple steps can be taken to assure success.

Vessel operators should strive to reduce use of packaging, particularly items that could be blown overboard by the wind. They also should cover Annex V in public announcements to passengers and provide numerous on-board MARPOL posters or placards and convenient trash cans. Ferry terminal operators should provide these informational services as well.

State governments should require ports serving day boats to have adequate waste receptacles as a condition of granting permits. Also, authorities should ensure that Annex V information is included in literature and guidelines directed at this sector (e.g., new IMO guidelines on roll-on/roll-off carriers).

Finally, ferries with international routes should be required to comply with Annex V as a condition of bilateral agreements signed by the nations involved.

Objective for Day Boat Sector

- Achieve zero-discharge capability, integrating the handling of vessel garbage into local solid waste management systems

---

Objectives for Small Public Vessels

- Improve on-board garbage handling and treatment technology
- Assure adequacy of port reception facilities
- Assure that seagoing and management personnel are provided with appropriate Annex V information, education, and training
- Develop model Annex V compliance program

Small Public Vessels and Their Home Ports

Objective: Improve on-board garbage handling and treatment technology

When on day trips, vessels in these fleets should be able to hold all garbage for proper disposal ashore. To reduce the amounts of garbage that must be stored in cramped quarters on longer voyages, advanced garbage handling and treatment technology should be incorporated into any new construction and, where feasible, retrofitted on older vessels. The Navy's technology development efforts should be expanded to include regional demonstration of a suite of on-board garbage treatment equipment for small vessels. The private sector might be encouraged to participate through cooperative and grant and contract programs.

Objective: Assure adequacy of port reception facilities

As a user of all types of ports and the enforcement agent for Annex V, the Coast Guard should redouble its efforts to monitor port reception facilities, through the COA program, informal contacts with port operators, and formal reporting of inadequate facilities. The Navy should report to the Coast Guard any inadequate reception facilities encountered at commercial ports.

Objective: Assure that seagoing and management personnel are provided with appropriate Annex V information, education, and training

To ensure that all personnel have sufficient information to comply with Annex V, all agencies that operate small public vessels should take advantage of their command management structures to implement and integrate appropriate management and education initiatives.

Objective: Develop model Annex V implementation program

All federal agencies that operate small public vessels should develop Annex

V compliance programs that can serve as models for the private sector. Each service should develop, in coordination with the other agencies, an internal strategy for compliance, and each service should articulate that strategy and end reliance on temporary coping mechanisms. The Navy should continue to develop a separate scheme for its auxiliary fleet. All strategies should emphasize source reduction and the provision of adequate garbage reception facilities at home ports. Zero-discharge capability should be achieved for vessels that take short trips or transit special areas.

Offshore Platforms, Rigs, Supply Vessels, and Their Shore Bases

Objective: Achieve zero discharge at sea

With storage space on offshore oil and gas platforms and a continuous stream of supply boats able to shuttle garbage to shore, this sector should be able to refrain from contributing to the marine debris problem once several key problems are addressed.

First, steps should be taken to minimize losses of supplies and waste materials that fall off platforms in harsh conditions and contribute to the marine debris problem. In addition, the Coast Guard should examine garbage logs during its occasional inspections of platforms. These records also could be examined by the Minerals Management Service (MMS) as part of its routine rig inspections. The Coast Guard could pursue an MOU with the MMS as a mechanism for enabling the latter to enforce Annex V on oil platforms.

Objective: Assure comprehensive garbage management system, including adequate port reception facilities

Although the MMS oversees offshore platforms, the other segments of the industry—supply boats and shore bases—are not regulated as tightly. These weak links in the garbage management system need to be strengthened.

Supply boats and shore bases should be monitored in some fashion to assure proper garbage handling. Boats could be boarded by the Coast Guard, and their

---

Objectives for Offshore Industry

- Achieve zero discharge at sea
- Assure comprehensive garbage management system, including adequate port reception facilities
- Assure that seagoing and management personnel are provided with appropriate Annex V information, education, and training

activities assessed, based on reports of marine debris in the area. Shore bases, which are required to have reception facilities but not necessarily COAs, should be required by the states that license them to provide adequate reception facilities. Terminals could be required to obtain COAs, even if the boats they serve are smaller than the minimum size qualifying as a port for the program.

Objective: Assure that seagoing and management personnel are provided with appropriate Annex V information, education, and training

Annex V educational efforts should target all segments of the offshore industry. Management personnel should be given information covering the full spectrum of requirements for the handling of solid waste. The industry's voluntary ban on use of foamed plastic should be held up as an example of how to minimize or eliminate garbage. Supply boat operators in particular need information about Annex V. In addition, MMS officials engaged in routine overflights of offshore operations could be informed about Annex V so they can report violations as well as concentrations of marine debris on the water or shorelines.

Planning for educational programs should recognize that the offshore industry hires a continuous flow of new workers unfamiliar with Annex V, and that the companies involved have fewer resources and narrower expertise than in the past. The MMS should focus its limited resources on encouraging marginal independent operators to comply with Annex V, the approach used to minimize oil spills.

Specific messages need to be emphasized. Like other seafarers, offshore operators should be encouraged to reduce the use of packaging. They also should be urged to transport operational wastes to shore in a timely fashion, to minimize losses at sea. Overall, new attitudes concerning environmental protection should be encouraged, so that industry personnel voluntarily refrain from tossing anything overboard.

Navy Surface Combatant Vessels and Their Home Ports

Objective: Develop plans for full Annex V compliance, including capability to achieve zero discharge in special areas, making the best use of existing technologies and strategies

While it must contend with special burdens in developing a plan for full Annex V compliance, the Navy also has unique opportunities due to the large sums of money that have been appropriated for research and development and its effective command and control organization that can implement successful strategies on a fleetwide basis. It is important to make the most of these assets.

To that end, the Navy should reconsider its decisions to abandon on-board garbage treatment technologies—specifically compactors and incinerators—employed successfully on large ships in other fleets (and, in fact, on some Navy

---

Objectives for Navy Surface Combatant Vessels

- Develop plans for full Annex V compliance, including capability to achieve zero discharge in special areas, making the best use of existing technologies and strategies
- Develop model Annex V implementation program

---

ships). The Navy already has devoted considerable time and resources to these technologies, and state-of-the-art units are available. Compactors are a basic element of compliance strategies in most other fleets, including Coast Guard ships that remain at sea for months at a time. Incinerators are standard on passenger cruise ships. Designed and used properly according to IMO guidelines, incinerators can eliminate garbage almost entirely—a significant benefit in that wastes need not be either stored or discharged overboard. The Navy should evaluate the possible use of incinerators that meet or exceed IMO guidelines and make a new decision based on rigorous scientific and engineering tests.

The Navy also should seek out and heed other lessons gained from experiences in other maritime sectors. For example, recycling programs—another standard practice on cruise ships—can help reduce waste streams. The Navy's shipboard recycling effort varies by operating unit. Even when on-board garbage treatment technology is installed, metal cans, glass, cardboard, and paper will continued to be discharged into the water as permitted by Annex V. The Navy should encourage its crews to reclaim and recycle ferrous and non-ferrous food and beverage containers for which a market and suitable on-board storage space exist. The Navy also should explore the feasibility of returning glass to shore for recycling or disposal. The Navy also should conduct a critical review of its food service system and provide leadership in source reduction and development of packaging systems that would reduce use of ferrous and glass containers.

While space shortages and fire hazard concerns preclude extended on-board storage of cardboard and paper wastes, the Navy has the option of using its pulpers or shredders to reduce the cellulosic material to particles less than 25 millimeters in size. The failure to obtain legislation allowing use of pulpers and shredders in special areas should not preclude the installation of this equipment. The Navy should consider installing pulpers and shredders for use where permitted, to eliminate discharge of floating debris.

To prepare for the entry into force of special areas such as the Mediterranean, where operations are extensive, the Navy must develop a capability to achieve zero discharge. Proposals are being solicited from industry for mature technologies suitable for shipboard use, and a separate National Research Council study is examining the Navy's compliance efforts. If no appropriate systems (including compactors and incinerators) can be developed and deployed, then the Navy should consider other alternatives, perhaps using the hazard evolution model (see Chapter 4) to identify "upstream" options.

Objective: Develop model Annex V implementation program

As the authority responsible for assuring U.S. compliance with Annex V, and as an international leader in IMO and in global environmental protection, the federal government should set an example through its own fleets for private and foreign vessels. It is especially important that the Navy not only satisfy the mandates of Annex V, but also, as the largest federal fleet, provide a model compliance program. As time passes, it is increasingly difficult to justify heavy fines against commercial ship operators for illegal garbage discharges, when similar actions carried out by the Navy are tolerated.

The Navy should make a top-level commitment to planning for and achieving full compliance. Priority should be placed on information exchange, both within the fleet and between the Navy and other maritime sectors. Successful technologies and strategies should be shared and deployed. To foster recycling and reduce volumes of garbage that must be discharged in port reception facilities, the Navy should establish comprehensive fleetwide recycling practices and explore marketing the metal and glass wastes it now collects and separates. State-of-the-art reception facilities should be provided in home ports, and commercial and foreign ports of call should be encouraged to provide such facilities as well. Foreign ports have economic motivations to comply in order to attract and retain naval business.

Passenger Cruise Ships and Their Itinerary Ports

Objective: Increase use of on-board garbage handling and treatment technologies

To reduce the amounts of garbage that must be stored on cruise ships for

---

Objectives for Cruise Ship Sector

- Increase use of on-board garbage handling and treatment technologies
- Assure comprehensive vessel garbage management system, including adequate port reception facilities
- Assure that seagoing and management personnel are provided with appropriate Annex V information, education, and training
- Exploit U.S. authority to improve compliance by foreign-flag vessels and by all vessels in foreign waters

disposal in port, modern garbage handling and treatment technologies should be integrated into new construction. The growing popularity of cruises that emphasize ecological knowledge and environmental pursuits offers an opportunity to test innovations in waste management aboard cruise ships with willing populations.

Objective: Assure comprehensive vessel garbage management system, including adequate port reception facilities

All measures proposed to improve the garbage management system for cargo vessels also apply to the cruise ship sector, because many of the same problems plague both fleets. Cruise ships should be required to off-load both Annex V and APHIS garbage at U.S. port calls. States and port operators should help ensure that reception facilities in U.S. ports are adequate to handle cruise ship garbage.

In addition, cruise ships should be required to provide Annex V educational programs (perhaps through videos, such as the safety presentations shown on airlines) for passengers and crews as a condition of access to U.S. ports, and violators should be punished.

Objective: Assure that seagoing and management personnel are provided with appropriate Annex V information, education, and training

Due to the large volumes of garbage generated on cruise ships, the rapid growth of the industry, and inability to monitor such large populations effectively, educational efforts targeting this sector, particularly crews and passengers, need to be expanded. Vessel operators should be encouraged to reduce amounts of packaging brought on board. Crews need to be trained in proper garbage handling practices. Passengers must be persuaded to respect the environment. Preservation of the ocean environment should be promoted as a basis for preserving cruise itineraries in unique and fragile locations (the standard should be the same regardless of the itinerary).

Objective: Exploit U.S. authority to improve compliance by foreign-flag vessels and by all vessels in foreign waters

All measures proposed to improve compliance by foreign-flag cargo ships also apply to the cruise ship sector. The Coast Guard should continue to step up its enforcement activities targeting foreign vessels. The garbage log requirement should be extended to foreign-flag vessels, and violators should be punished.

In addition, U.S. authorities should encourage islands on cruise ship itineraries to assist in implementation of Annex V by providing adequate garbage disposal services, because these islands derive economic benefits from the cruise trade. Particularly important in this respect is the World Bank's search for a regional mechanism that will improve waste management in the Caribbean; a solution will go a long way toward meeting the needs of cruise vessels operating in that region.

Research Vessels and Their Ports of Call

Objective: Provide model Annex V compliance program

Because research vessels visit pristine areas, are dedicated to the study and preservation of the marine environment, and often are supported by the federal government, this fleet should strive to provide a model Annex V compliance program. That means vessels operating in special areas should achieve zero-discharge capability. Vessel operators should consider all possible ways of reducing overboard discharges, including reducing the use of packaging. In addition, the Department of State should resolve, through IMO or other avenues, the procedural obstacles that block garbage off-loading at some foreign ports.

Objective: Improve on-board garbage handling and treatment technology

As they address other aspects of marine science, research vessel personnel should provide leadership in development and demonstration of garbage handling, treatment, and recycling technologies. Government agencies that sponsor marine research could draw the private sector into development of shipboard technology through cooperative and grant and contract programs. To ensure that operating funds are not depleted to cover the costs of garbage handling and treatment, funds should be earmarked for equipment to enable Annex V compliance.

Objective: Assure that seagoing and management personnel are provided with appropriate Annex V information, education, and training

Operators of research vessels have an obligation to educate not only their own crews and visitors but also, due to the nature of their work, the general public. Visiting scientists should be informed about Annex V, as they may be

---

Objectives for Research Vessel Sector

- Provide model Annex V compliance program
- Improve on-board garbage handling and treatment technology
- Assure that seagoing and management personnel are provided with appropriate Annex V information, education, and training

oblivious to shipboard rules and practices. In addition, vessel operators should hold open houses and laboratories to educate the public and other fleets about proper garbage handling and treatment methods. Researchers also should promote recognition of the marine debris problem at scientific research forums.

10

Federal Action to Improve Implementation of Annex V

A s the preceding chapters demonstrate, there are many opportunities for action to improve U.S. implementation of MARPOL Annex V. Although many specific actions need to be taken by mariners, ports, and private companies, there is also a critical need for sustained, directed, national leadership to establish nationwide information networks, standards, rules, and regulations. This chapter synthesizes the many components of the committee's analysis to draw overall conclusions and provide recommendations for federal action to improve implementation of Annex V across all fleets. Such action is needed because the U.S. government ratified Annex V without developing a detailed implementation plan.

The presentation is organized into six sections, based on themes drawn from Chapter 2, which identified scientific needs, and Chapter 9, which built on Chapters 3–8 to establish objectives and recommend specific tactics for each maritime sector. Chapter 2 demonstrated the need for improved scientific monitoring of the marine environment. Chapter 9 identified a number of topics requiring attention in many if not all maritime sectors: the vessel/shore interface; on-board technology; Annex V enforcement, education and training; and national leadership of Annex V implementation.

These six themes provide the framework for the committee's proposed Annex V implementation program. For each thematic area, the committee identified objectives (which are embedded in the conclusions) and the federal agencies that should lead the effort or provide support. The rationale for the selection of the designated agencies is provided. The committee also identified areas where the states, local governments, and private organizations should provide assistance.

SCIENTIFIC MONITORING

Environmental monitoring is a way of providing feedback for improving environmental management. A number of illuminating studies and surveys have been conducted on the fates and effects of marine debris, but there has not been any comprehensive, long-term research. Improved collection and analysis of data on marine debris not only would fill the numerous gaps in the existing scientific knowledge base but also would provide means for assessing Annex V and progress in its implementation. Reliable data would provide a rational basis for timely shifts in management programs to improve Annex V compliance. This type of monitoring is by nature long term and demands organizational commitment.

To expand understanding of the fates of marine debris, the *committee concludes that statistically valid long-term programs are needed to monitor the flux of plastics in the oceans and assess the rates of accumulation of debris in the benthos.* Research on the fate and transport of plastics in the global oceans would provide a basis for evaluating whether Annex V, as currently written and internationally implemented, is providing adequate protection. Plastic would be the logical target because it is the most prevalent and harmful type of debris and its overboard discharge is prohibited by Annex V. In addition, regular surveys to measure accumulation rates of plastic on beaches and the coastal sea floor would provide a measure of the current pollution problem and a benchmark for evaluating compliance with Annex V. It would be important to share the data with national and international agencies responsible for Annex V implementation. In addition, because it is difficult to obtain such data without a systematic, worldwide effort involving the cooperation of other maritime nations, it might be helpful to draw attention to the need for this type of monitoring through international forums, such as the International Maritime Organization (IMO) and the Intergovernmental Oceanographic Commission (IOC).

To expand understanding of the effects of marine debris, the *committee concludes that statistically valid long-term programs are needed to monitor interactions of marine species with debris in the oceans and the impact of debris on pristine areas.* Existing studies could be expanded and extended. New data on wildlife interactions (e.g., entanglements with and ingestion of debris) is needed to verify the ecological effects of debris that have been suggested by previous reports and surveys.

Standardized reporting forms, centralized data analysis, and information exchange are essential. It may be feasible to adapt existing research on non-Annex V topics, such as analyses of fish stomach contents, to also record the incidence of plastics and other debris. Another approach would be to conduct regular necropsies on dead stranded marine mammals and other animals. Research on the impact of debris in areas minimally affected by land-based sources would help assess progress in implementation of Annex V and the overall effectiveness of the mandate.

The committee further concludes that the National Oceanic and Atmospheric Administration (NOAA) is best equipped of all federal agencies to lead the monitoring effort, because its Marine Entanglement Research Program (MERP) has collected much of the existing knowledge on marine debris and its Status and Trends Program could be expanded readily to monitor plastic debris. NOAA could obtain assistance from the Environmental Protection Agency (EPA), which has considerable experience collecting data on land-based sources of debris and debris in urban waterfronts and has developed a beach monitoring program. NOAA also could obtain information from agencies such as the National Park Service, which routinely observes debris at national seashores, and coastal states that monitor beaches. The committee therefore recommends

NOAA, with the assistance of EPA, should establish statistically valid, long-term monitoring programs to gather data on the flux of marine debris, the physical transport and fate of marine debris, accumulation of plastic on beaches and in the benthos, wildlife interactions with debris, and the impact of debris on pristine areas. NOAA also should assure that the results of its monitoring programs are communicated to other agencies responsible for Annex V implementation and enforcement.

The U.S. government should draw attention to the need for an international data collection effort through IMO and the IOC.

VESSEL/SHORE INTERFACE

The most prevalent problem across the various maritime sectors is inadequate port reception facilities. This is a result of the lack of planning for Annex V implementation and is a major obstacle to full implementation; far-reaching changes and strong leadership and coordination will be required to overcome this problem.

As a first step toward improving the vessel/shore interface, the *committee concludes that vessel garbage management must be viewed as a system that includes port reception facilities, and this system needs to be combined with the integrated solid waste management system (ISWMS) for land-generated waste.* The ISWMS recognizes the diverse needs for waste treatment to accommodate the many materials generated in the nation's homes and industries, and vessel garbage can be integrated into that system. For a system to function efficiently, there must be a coherent overall management scheme and technical standards. Vessel operators can do much more to reduce shipborne waste and to return the residual to shore, but they need to be assured access to affordable reception facilities that meet their needs, and the garbage must be disposed of safely and efficiently. In addition, technical standards are needed to help operators of all types of ports, from large commercial ports to recreational marinas, satisfy the Annex V mandate for provision of "adequate" garbage reception facilities.

To encourage use of port reception facilities, the question of who should pay for garbage services, and how, needs to be addressed. Because port management is decentralized in the United States, the federal government may have to initiate discussions on this topic. As part of the process, port operators may need to cooperate in finding a rational basis for setting disposal fees, which now vary regionally. One option would be to require that fees paid by ships be comparable to local charges for disposal of land-generated garbage. Alternatively, port tariffs or related user fees could be increased to cover garbage disposal.

The committee also concludes that the handling of Animal and Plant Health Inspection Service (APHIS) waste needs to be integrated as fully as possible with the Annex V regime and the system for managing land-generated waste. The APHIS program, administered by the U.S. Department of Agriculture (USDA), historically has been separate from other waste management efforts, but the need for an efficient and effective overall system demands that the APHIS system be integrated into the ISWMS. The aim is to make compliance with both Annex V and APHIS regimes as easy as possible for vessel operators.

The committee further concludes that there is a need to assure accountability of both vessel operators and port operators. This need will be addressed in part by the Coast Guard requirement that operators of ocean-going, U.S.-flag commercial vessels over 12.2 meters (about 40 feet) in length maintain logs of garbage disposal practices. However, it would be difficult and time consuming to verify the accuracy of the logs in any way other than through spot checks. Accountability could be strengthened if ports issued receipts for garbage discharged into their reception facilities. (Knowing the size of the crew and the duration of the voyage since the last port call, the Coast Guard could estimate the amount of garbage that should be discharged at a specific port.) In addition, to assure that ports meet vessel needs for handling of garbage (including APHIS waste), vessel operators could be required to report any inadequate reception facilities using the IMO forms. Such reports would need to be followed up by the Coast Guard, to assure that the necessary improvements were made.

In keeping with trends in ISWMS, and based on the effectiveness of small-scale marina recycling projects, the *committee also concludes that recycling of vessel garbage needs to be promoted.* Materials that have been recycled include plastics, metal cans, and fishing nets. There are needs for infrastructure mechanisms for transporting the materials to processing centers, public awareness efforts to promote recycling, and widespread provision of port reception facilities for returned materials. There is a particular need to establish a recycling system for fishing nets, which are not now recycled but could be.

Finally, the *committee concludes that EPA is the logical agency to establish the overall framework for improving the vessel/shore facility interface, due to its expertise in and authority for national management of land-generated waste.* The EPA has the expertise to set minimum technical standards appropriate for reception facilities at each type of port. The EPA also has the authority to assure, through the states, that reception facilities meet the standards. The EPA can require that garbage from vessels docked at any port be included in the states' solid waste management plans, which are authorized by the Solid Waste Disposal Act, as amended by the Resource Conservation and Recovery Act. A congressional directive may be required, however, because the EPA has taken the position that this function is outside its purview.

Ultimately, this approach may obviate the need for the Certificate of Adequacy (COA) program run by the Coast Guard, which, realistically, has neither the expertise nor the resources to assess and monitor garbage reception facilities and, moreover, monitors only a limited number of ports. Unless and until the new system is in place, however, the COA program must continue to provide a check on the adequacy of port reception facilities.

The EPA can be assisted in improving the vessel/shore interface by the Coast Guard, which runs the COA program and enforces Annex V; the states, which develop solid waste management plans and issue permits to ports, docks, and piers; port and terminal operators, which could assume an expanded role in overseeing the adequacy of reception facilities and assuring customer satisfaction with services; the private sector (e.g., the Solid Waste Association of North America, the Center for Marine Conservation [CMC], professional societies, and industry trade associations), which can help promote recycling and Annex V compliance; and the various maritime sectors, which can communicate their needs and suggest solutions. The EPA also can make use of the forthcoming IMO manual on reception facilities. The committee therefore recommends

To improve management of vessel garbage and meet U.S. national and international commitments to implement Annex V, the Congress should direct EPA to use its current resources to establish an overall framework that (1) incorporates the vessel garbage management system into the ISWMS for land-generated waste, (2) requires states to include in their solid waste management plans the disposal of garbage from vessels docked at their ports, (3) establishes technical standards for reception facilities appropriate to each type of port, (4) provides for accountability by requiring commercial ports to issue receipts for garbage discharged at their facilities, and by assuring that states follow up reports of inadequate port reception facilities, and (5) promotes recycling of vessel garbage. The EPA should obtain assistance from the Coast Guard, the states, port and terminal operators, the private sector, and the maritime

sectors and should make use of the forthcoming IMO manual on reception facilities.

In developing their solid waste management plans, states should assure that vessel garbage disposal fees are set on some rational basis, and that a mechanism for collecting the fees is established. Port operators should consider cooperating in setting fees, which should be comparable to local fees for disposal of land-generated garbage.

The USDA should make any changes necessary to integrate the APHIS regime into the Annex V compliance program and the ISWMS as fully as possible.

The Coast Guard should require vessel operators to report inadequate reception facilities using the IMO forms and should follow up these reports. And, if ports are required to issue receipts for garbage discharged into their reception facilities, then the Coast Guard should examine these receipts when reviewing vessel garbage logs.

Unless and until the COA program is merged with the EPA program, the Coast Guard should incorporate into the program requirements that port reception facilities meet EPA technical standards and have any requisite state and EPA approvals.

ON-BOARD TECHNOLOGIES

At least some vessels in all fleets will require installation of appropriately sized and reliable compactors, pulpers, shredders, incinerators, or other technologies in order to minimize garbage for disposal in port. Although some equipment is available, it does not meet all the needs of all fleets, even the U.S. Navy, which has an extensive research and development (R&D) program dedicated to developing and demonstrating on-board garbage handling and treatment technologies. The cruise ship industry works with equipment vendors and engineers to meet individual needs, but the potential markets for many technologies, such as those needed for fisheries fleets, have not attracted commercial developers.

The committee concludes that new and improved on-board garbage handling and treatment technologies are needed, a problem that may be resolved in part by adapting commercial equipment used in homes, retail establishments, and industry. The difficulty of developing appropriate on-board equipment is illustrated by the experience of the Navy, which has been working on this problem since the early 1980s and does not expect to bring its surface fleets into compliance until the turn of the century. Other fleets do not have direct access to the Navy's expertise; in some cases, they can purchase commercial equipment off the shelf, but more often individual alterations or entirely new technology is needed. This is an opportunity for the federal government to work toward two of its goals: development of dual-use technology and protection of the environment. Development, testing, and evaluation are needed to make available a suite of appropriately sized and configured equipment for all maritime sectors.

To support and foster the wide use of new and improved on-board technologies, *the committee concludes that demonstration projects, research on operations and maintenance issues, and information exchange are needed.* Demonstration projects are important not only to gain experience with equipment but also to display it to the wider community and gain acceptance. The diverse equipment requirements of the various fleets could be met through small projects carried out through government grants or contracts with the private sector. New equipment could be demonstrated on various types of vessels in different fleets. Research also is needed to address operations and maintenance issues, such as human factors, safety, and reliability. Finally, exchange of technical information among the various maritime sectors is essential to maximize the return on R&D investments and avoid duplication of effort. Information about the Navy's equipment developments, for example, still needs to be shared with other government fleets and the private sector.

The committee also concludes that steps must be taken to resolve issues that may be impeding safe and efficient garbage storage and expanded use of compactors and incinerators. Guidelines on shipboard sanitation may need to be developed for fleets other than cruise ships and these fleets offered technical assistance to ensure that on-board storage procedures are safe and efficient. To foster expanded use of compactors, APHIS could develop standards based on compacted garbage. U.S. standards for on-board incinerators also are needed if use of this technology is to be expanded.

Finally, the committee concludes that economic issues—including the cost of technologies to vessel operators and the tradeoffs with garbage disposal fees— need to be addressed. Economic considerations will determine whether on-board garbage handling and treatment technologies actually are used. Vessel operators will weigh the costs of these technologies against port fees for disposal of waste "as is" and, perhaps, the possibility of being fined or losing business for violating Annex V. Therefore, technologies must be not only affordable but also cost-competitive with other garbage handling options. Operators of fisheries fleets may need a source of capital to enable the development, purchase, and installation of technology. One resource may be the National Marine Fisheries Service (NMFS) financial assistance programs for improvements in fisheries fleets. The Capital Construction Fund Program may be an appropriate source if the NMFS is willing to provide the funds for pollution-abatement equipment and waive the minimum cost requirements.

To accomplish all the activities necessary to develop and deploy on-board technologies to enable Annex V compliance, *the committee concludes that the Maritime Administration (MARAD) is the logical lead agency, due to its ongoing, broad-based marine technology assessment and development efforts.* MARAD could obtain technical assistance from the Navy and maintain contact with the various fleets through NOAA's Sea Grant Marine Advisory Service. To help execute narrow projects to meet the needs of small fleets, federal agencies could award grants and contracts to private companies. The R&D effort needs to be responsive to the needs of the Coast Guard, NOAA, and other government fleets, as well as the private sector. The committee therefore recommends

> **MARAD should develop and execute an on-board garbage treatment technology R&D program that addresses the needs for new equipment; alteration of commercial equipment; technology demonstration and information exchange; and operational, maintenance, and cost issues. MARAD should obtain technical support from the Navy and maintain contact with the various fleets through NOAA's Sea Grant Marine Advisory Service and the NMFS. The program should be responsive to the needs of the Coast Guard, NOAA, and other government fleets, as well as the private sector.**

> **The federal government should take steps to resolve issues that may be impeding safe garbage storage and expanded use of compactors and incinerators. To ensure that on-board storage procedures are safe and efficient, the government should examine the need for sanitation guidelines and related technical assistance for fleets other than cruise ships. APHIS should consider developing standards based on compacted garbage. The EPA should adopt IMO standards for shipboard incinerators.**

> **The NMFS should offer financial assistance to fisheries fleets investing in on-board garbage handling and treatment technology. The NMFS should waive policy conditions, such as minimum cost requirements, that limit access to these programs.**

ENFORCEMENT

This section addresses enforcement of Annex V standards at sea only (enforcement in ports is addressed in the previous section on the Vessel/Shore Interface).

Although voluntary compliance by seafarers is the linchpin of Annex V implementation, effective enforcement provides an extra impetus for compliance, an additional means of control over certain fleets, and some confidence that violators, once prosecuted, will not repeat their actions. At the same time, it is important to make enforcement as efficient as possible by targeting problem fleets, because limited resources and the vast expanse of the oceans combine to preclude comprehensive enforcement.

As a fundamental step toward strengthening Annex V enforcement among seafarers, *the committee concludes that enforcement action must be taken and followed up in every case where the United States can assert jurisdiction, even when the violator is a foreign-flag vessel.* The Coast Guard is making progress in this area by pursuing direct action against foreign-flag vessels that violate Annex V within the U.S. Exclusive Economic Zone. It will be important to work through IMO to establish clear procedures for exercising port state enforcement authorities. In addition, fines or penalties for violating Annex V need to be sufficiently high to serve as deterrents.

The committee also concludes that the Coast Guard needs to take additional steps to enhance enforcement where it is most needed. To provide additional means for enforcing Annex V among foreign-flag cargo and cruise ships particularly, the requirement for garbage logs could be extended to foreign-flag vessels. Recreational boaters, fishing fleets, and the offshore oil and gas industry also pose special challenges in implementation of Annex V. The Coast Guard could issue "tickets" in civil cases involving Annex V violations, particularly in the fisheries and recreational boating sectors, if the pilot projects using this type of streamlined approach to enforcing other laws are shown to be successful. The Coast Guard also could encourage violation reports by other federal officials engaged in surveillance of fisheries fleets and the offshore industry, as well as state marine police, who routinely come into contact with boaters. These agencies could provide additional eyes for enforcement at no extra cost. The Coast Guard also could pursue vigorously its planned public awareness campaign urging citizens to report illegal garbage disposal.

The committee concludes that, to make the best use of existing information and enforcement assets, systematic government record keeping and analysis is needed. While a comprehensive Annex V record-keeping system involving all relevant federal agencies is probably not feasible, the Coast Guard and APHIS could collaborate to develop and maintain a computerized database on vessel garbage handling. APHIS records of vessel boardings and garbage off-loading could be converted to electronic form and logged into the shared database. The Coast Guard could input information from vessel logs and enforcement reports. Data analyses could be used as a basis for determining where the two agencies' enforcement resources should be directed. The data bank would be most meaningful if cargo and cruise ships were required to off-load all garbage at every U.S. port call, and if ports issued receipts for all garbage discharged into their facilities.

The committee concludes that the Coast Guard, which already is legally responsible for Annex V enforcement, is the appropriate agency to lead the expanded enforcement effort. Support could be obtained from the NMFS, Minerals Management Service (MMS), and state marine police. The committee therefore recommends

> **The Coast Guard, together with the Department of State and Department of Justice, should continue to enforce Annex V aggressively against foreign-flag violators, consistent with the nation's international obligations, and should work through IMO to resolve ambiguities concerning the extent of port state authority in this regard. The requirement for garbage logs should be extended to foreign-flag vessels. The Coast Guard also should adopt a policy of issuing tickets in civil cases if pilot projects show this streamlined enforcement approach to be successful. In addition, the Coast Guard should request the assistance of the NMFS, MMS, and state marine police in reporting Annex V violations. Finally, the agency should pursue vigorously its campaign to encourage public reports of violations.**

> **The Coast Guard and APHIS should collaborate to develop, maintain, and use for enforcement purposes an Annex V record-keeping system incorporating information from vessel boardings, garbage logs, enforcement reports, and, if a receipt system is instituted, port receipts for offloaded garbage.**

> **The Coast Guard should issue a periodic report listing Annex V enforcement actions and the assistance provided by other federal agencies and marine police units in the states. Analyses of data from the Coast Guard/APHIS record-keeping system should be included. Such reports would allow the Congress to evaluate the adequacy of appropriations for Annex V implementation projects and enforcement.**

EDUCATION AND TRAINING

Education and training efforts targeting all levels of seafaring and management personnel as well as the general public are critical in establishing a sense of personal responsibility on the part of individuals and a high level of voluntary Annex V compliance.

Therefore, *the committee concludes that a sustained national program of Annex V education and training is needed that reaches all levels of all maritime sectors as well as non-traditional target groups, such as the packaging industry and government officials, and provides for information exchange, both domestically and internationally.* The program must include research, to develop a solid base of knowledge concerning how to package the message; execution, to carry the message to all levels of personnel and management in all sectors; and evaluation, to gather evidence to justify program expenditures. The program must make use of existing knowledge about effective teaching methods and build on successful past or ongoing educational efforts, notably those carried out by NOAA's MERP and Sea Grant programs and the CMC. Innovative strategies must be sought to reach and persuade mariners known to have poor records of compliance. Also essential is development of national and international channels, such as newsletters, for exchange of information across fleets about Annex V compliance strategies, including education and training programs and on-board garbage treatment equipment.

To assure leadership, stable funding, and innovation, *the committee concludes that a publicly chartered, independent foundation offers the most promise for coordinating and enhancing a successful education and training program over the long term.* There is considerable precedent for this approach to coordinating national programs. The National Boating Safety Advisory Council is an example. The Annex V foundation would award grants to private industry and associations, academic institutions, and public agencies to develop, test, and carry out education and training projects, with an emphasis on innovative concepts. The foundation also would develop information exchange strategies. Funding could be provided through modest congressional appropriations and industry support; oversight could be provided by a national commission (described and recommended in the following section on national leadership).

The committee therefore recommends

The Congress should charter and endow a foundation to coordinate a sustained, long-term, national program that would assure development and execution of Annex V education and training programs for all maritime sectors as well as non-traditional target groups and provide for domestic and international exchange of information on Annex V compliance strategies. The program should include research, execution, and evaluation components and should promote innovation. To develop and carry out projects, the foundation should award grants to private industry and associations, academic institutions, public agencies, and nonprofit organizations.

NATIONAL LEADERSHIP

Because many federal agencies are involved in implementing Annex V and the Marine Plastics Pollution Research and Control Act (MPPRCA), there is no clear leader or centralized coordination of all aspects of this complex effort. Yet the inherent scope and importance of this task demands leadership.

As a first step toward providing leadership, *the committee concludes that U.S. government and government-supported fleets, to set an example, need to work systematically to comply with Annex V, upgrade crew training and provisioning practices, and encourage transfer of successful experiences to other fleets.* Clearly, it would be difficult for the federal government to justify enforcing rules that its own fleets do not make every effort to observe. Zero discharge is required by law for vessels operating in special areas where the discharge rules are in force, and it is also an appropriate objective for vessels making day trips. The committee wishes to emphasize that an objective is something to strive for, rather than an absolute requirement as established by law. The committee recognizes that government fleets face serious and continuing difficulties in obtaining funds for Annex V implementation projects. The proposed Annex V foundation could be the mechanism for development of education and training materials and transfer of technologies and strategies among maritime sectors.

Furthermore, *the committee concludes that centralized oversight, direction, and coordination of Annex V implementation is needed.* Evidence of the need is documented throughout this report. In absence of such leadership, important data on debris accumulation and garbage disposal practices have not been gathered, the adequacy of port reception facilities has been given only cursory consideration, key educational and technology development projects have not been pursued, and information about successful programs and technologies have not been disseminated widely. Leadership is needed if comprehensive national implementation of Annex V is to be achieved.

The committee concludes that the United States needs to continue to take a leadership role in the international community with respect to Annex V implementation. Because U.S. implementation of Annex V is affected by the compliance levels of foreign-flag vessels, the United States needs to push for increased standards of performance worldwide. The nation could assist in the dissemination of Annex V information and technology to foreign maritime users through a variety of regional forums, including the United Nations Environment Programme's Regional Seas program, regional and bilateral fisheries agreements, the North Atlantic Treaty Organization, international oceanographic organizations, and tourism and yachting associations. The United States also needs to find ways to help assure the adequacy of port reception facilities in the Wider Caribbean special area, perhaps through the development of memoranda of understanding (MOUs) for the sharing of enforcement assets and other resources. The United States could assist in identifying and overcoming obstacles hindering Caribbean nations from adopting the provisions of MARPOL, either through ratification of the convention or national legislation.

To provide consistent, independent, expert oversight and coordination of Annex V and MPPRCA implementation, as well as international leadership, the committee concludes that a permanent national commission is needed. There is considerable precedent for the commission approach. The Congress has established a number of commissions to focus on specific, narrow issues and problems of major domestic and international concern. A commission would have greater flexibility than would federal agencies in working with the private sector to promote Annex V implementation and would be well positioned to promote U.S. leadership in the global maritime community. Furthermore, no single agency has all the requisite expertise and authority to fill a comprehensive leadership role.

To be effective, a national commission addressing Annex V implementation would require a clear legislative mandate establishing its overview authority and outlining its responsibilities, which could include (1) reviewing information on the sources, amounts, effects, and control of vessel garbage, (2) working with federal agencies to assure they carry out their roles and responsibilities and exchange relevant information, (3) making recommendations to agencies on actions or policies related to identification and control of sources of vessel garbage, (4) providing support for research, regulatory, and policy analyses, (5) providing the Congress with periodic reports on the state of the problem, progress in research and management measures, and factors limiting the success of implementation, (6) overseeing the Annex V educational foundation, and (7) overseeing international aspects of Annex V implementation. The legislation also would need to authorize funding sufficient for the commission to carry out its duties.

Finally, to carry out Annex V implementation efforts requiring the expertise and resources of multiple agencies, *the committee concludes that MOUs between relevant agencies need to be negotiated and observed.* These agreements would spell out specific roles and responsibilities and help assure that the work is accomplished.

The committee therefore recommends

The Congress should require that federal and federally supported fleets, to set an example, work systematically toward full Annex V compliance, upgrade crew training and provisioning practices, and encourage transfer of successful experiences to commercial fleets.

The Congress should establish a permanent national commission with a clear legislative mandate establishing its authority to oversee the national Annex V and MPPRCA implementation effort. The panel should be modeled on other national commissions, such as the Marine Mammal Commission, established to address major issues of concern. The legislation should outline the commission's responsibilities and authorize funding sufficient for execution of its duties.

The commission should (1) review information on the sources, amounts, effects, and control of vessel garbage, (2) work with federal agencies to assure they carry out their roles and responsibilities and share relevant information, (3) assure that MOUs for Annex V implementation are negotiated and observed, (4) make recommendations to federal agencies

on actions or policies related to identification and control of sources of vessel garbage, (5) provide support for research, regulatory, and policy analyses, (6) provide the Congress with periodic reports on the state of the problem, progress in research and management measures, and factors limiting the effectiveness of implementation, (7) oversee the Annex V educational foundation, and (8) oversee international aspects of Annex V implementation.

In closing, the committee observes that many of its conclusions and recommendations may be applicable to the problem of marine debris in general as well as the more specific problem of vessel garbage, and that the Annex V educational foundation and national commission may be useful mechanisms for implementing all components of MARPOL. The broad utility of the committee's recommendations may provide additional justification for implementing them.

APPENDIXES

A
Committee on Shipborne Wastes
Biographical Information

William R. Murden, Jr. (NAE), Chairman, is a principal of Murden Marine, Inc. a consulting engineering firm he established. He is nationally and internationally recognized as an authority on marine port issues. Mr. Murden built his technical career within the U.S. Army Corps of Engineers, eventually becoming chief of the Dredging Division of the Office of Chief of Engineers. In that position, he was responsible for managing all aspects of the $400 million U.S. dredging program and for the design and construction of the dredges, derrick boats, towboats, and other small craft in the Corps' floating plant. He has written numerous technical papers on dredging technology and marine engineering. He is a former member of the Marine Board (1988 to 1991). Mr. Murden attended the Citadel but interrupted his studies to serve as a command pilot during World War II. He later earned a B.S. degree in Mechanical Engineering from Elizabethtown College and an M.B.A. from Heed University in Florida.

Anthony Frank Amos is a research associate at the University of Texas Marine Science Institute in Port Aransas. In that position, he has gained considerable experience with oceanographic expeditions in remote locations, including the Antarctic. Mr. Amos is known widely for having introduced scientific rigor to the study of beach litter along the Gulf of Mexico coastline. Long before marine debris was a popular concern and the focus of regulations, he conducted surveys of litter on Texas beaches, developing a methodology to quantify and categorize the phenomenon and note its harmful effects. His work has provided the most complete long-term data and scientific observations available on marine debris and has formed the basis for identification of pollutant sources and remedies. Mr.

Amos also has become a prominent advocate for change in laws and attitudes to eliminate marine debris at its sources. His awards include the Texas Marine Educator's Association Award in 1989. Mr. Amos is the author of more than 50 scientific documents and writes a weekly "Island Observer" column for his local newspaper. A British citizen, he is a permanent resident of the United States. He was educated at the Glyn School in Surrey, U.K.

Anne D. Aylward is a member of the Marine Board. She served as executive director of the National Commission on Intermodal Transportation and was formerly the maritime director of the Massachusetts Port Authority, where she was responsible for the development, marketing, and operation of the Port of Boston. She has served as chairman of the North Atlantic Port Conference, vice chairman of the Boston Harbor Association, a member of the Board of Governors for the Boston Shipping Association, and past chairman of the Board and U.S. Delegation for the American Association of Port Authorities. She is a member of the Executive Committee of the Marine Board and a member of the Women's Transportation Seminar, Boston Chapter. Ms. Aylward received her A.B. degree from Radcliffe College and her M.A. in City Planning from Massachusetts Institute of Technology.

James Ellis is vice president of the Boat Owners Association of the United States (BOAT/U.S.), an association with more than 400,000 dues-paying members. He is the executive director of the BOAT/U.S. Foundation for Boating Safety and in 1989-1990 served as president of the National Safe Boating Council. In the latter position, he directed the council's activities, including the National Safe Boating Week Campaign (an outreach program that delivers safety information to more than 20,000 boating clubs). Mr. Ellis is an accomplished sailor who has directed an offshore sailing school for 2,000 students and has raced nationally and internationally for most of his adult life. He owns four recreational vessels. He received a 1991 Rolex Navigators Award and is a national honorary member of the U.S. Power Squadron.

Edward D. Goldberg (NAS) is an eminent professor of chemistry at the Scripps Institute of Oceanography in La Jolla, California. He has written widely on subjects central to the understanding of the well-being of the oceans; his scholarly publications have addressed marine pollution, the composition of sea water, sediments and marine organisms, and environmental management. He directed the 1975 NAS study *Assessing Potential Ocean Pollutants*, which prepared a widely cited estimate of garbage pollution in the ocean. His oceanographic work has been recognized through numerous awards and fellowships, including a Guggenheim Fellowship in Berne, Switzerland and a NATO Fellowship in Brussels, Belgium. Dr. Goldberg earned his Ph.D. at the University of Chicago.

William G. Gordon is a fisheries expert, recently retired from the New Jersey Marine Sciences Consortium, where he served for four years as vice president for programs and Sea Grant director. From 1981 to 1986 he headed the National Marine Fisheries Service, where he was responsible for managing national fisheries programs and coordinating these activities with other federal agencies and foreign governments. In that role, Mr. Gordon was recognized for his effectiveness in representing the interests of fisheries and fishermen while negotiating numerous international fisheries agreements. Prior to serving as director he held numerous positions in which he directed efforts to strengthen research capabilities, develop new fisheries, encourage international programs, and manage fisheries, including recreational fisheries. In 1989, Mr. Gordon served as chairman of the technology working group at the International Marine Debris Symposium and presented a report on technical trials of thermal reprocessing of fishing net materials. He served as vice chairman of the Marine Board's 1990-1991 study on fishing vessel safety. Mr. Gordon earned an M.S. in Fisheries at the University of Michigan, where he also pursued post-graduate studies.

Michael P. Huerta (resigned) is the executive director of the Port of San Francisco, which encompasses diverse facilities ranging from heavy industrial cargo operations to recreational waterfronts, including Fisherman's Wharf. His professional accomplishments emphasize economic development, trade expansion, and development of organizational capabilities to create the infrastructure needed to support economic development. Mr. Huerta previously worked as the commissioner of the City of New York Department of Ports, International Trade and Commerce, where he was responsible for administering 578 miles of waterfront operations, including construction. In addition, he worked through the Agency for International Development to encourage employment and investment in the eastern Caribbean nation of St. Christopher (St. Kitts) and Nevis. Mr. Huerta earned his M.P.A. in International Relations and Policy Analysis from the Woodrow Wilson School, Princeton University.

Shirley Laska is vice chancellor for research and professor of sociology at the University of New Orleans. She is the founder and former director of the Environmental Social Science Research Institute. Her research focuses on how communities are affected by both natural disasters and human interventions in the environment. She recently has studied the impacts of offshore oil and gas extraction on coastal communities, management of coastal wetlands, and environmental attitudes of coastal users, including attitudes toward marine debris and beach litter. She is the author of 27 publications, including recent works on environmental controversies surrounding the use of solid waste incinerators and the risk communication content of print and broadcast reports of a natural hazard. Dr. Laska earned a Ph.D. in Sociology from Tulane University.

Stephen A. Nielsen is vice president, Marine Operations, for Princess Cruises. He has extensive experience managing the spectrum of cruise ship operations: itinerary planning; logistics and shore tours; passenger services; security; and diverse tourism, hospitality, and protocol arrangements. He has served as a consultant to a number of ports during the remodeling or construction of cruise ship terminals and located and planned the development of Princess Cruises' two private islands in the Caribbean. He was a founder and remains a senior officer of L.A. Cruiseship Terminals, Inc., a consortium of seven cruise companies formed to work with the Port of Los Angeles in the design, construction, and operation of the port's World Cruise Center. In addition to his personal expertise, Mr. Nielsen is able to call upon the extensive marine experience and resources of Princess Cruises' parent company, the Peninsular & Oriental Steam Navigation Co., including its U.K.-based technical consultancy, which has conducted shipboard garbage studies and designed new equipment for the fleet's use. Mr. Nielsen is a former member of the International Committee of Passenger Lines' subcommittee on the U.S. Public Health Service Vessel Sanitation Inspection Program and a current member of the Florida Caribbean Cruise Association and the Northwest Cruise Ship Association.

Kathryn J. O'Hara is director of the Pollution Prevention Program at the Center for Marine Conservation (CMC), which is recognized as the lead membership-based environmental organization in the drive to reduce marine debris and the environmental harm which it causes. Her work focuses on education and government, industry, and citizen cooperation. Educational materials developed by CMC aim to increase public awareness and inform seafaring communities and related industries of means to reduce sources of marine debris. Ms. O'Hara directs the center's International Coastal Cleanup Program, an annual event that has grown over 7 years to include 220,000 volunteers in 35 states and 40 foreign countries. In 1988, CMC initiated the Marine Debris Database Program, using volunteers to collect data on beach litter. Ms. O'Hara devised standardized forms for data gathering suitable for use by volunteers, thereby improved the utility of the data to both researchers and regulators. In her focused attention to reducing marine debris, she has demonstrated an ability to interact with a wide range of industry, government, and grassroots groups and has become a key source for information about marine debris and Annex V implementation activities in diverse local settings. Ms. O'Hara earned her B.S. degree in Zoology from Duke University and her M.S. degree in Marine Biology at the College of Charleston, South Carolina.

Joseph D. Porricelli (deceased) was a co-founder and managing principal of ECO, Inc., where he worked on projects relating to liquified natural gas transportation, deep-water ports, Very Large Crude Carrier operations, mobile offshore drilling units, and port operations. Several projects involved the adaptation of waste handling technologies to marine systems. He received a B.S. degree from the U.S. Coast Guard Academy and an M.S.E. degree in Naval Architecture and Marine Engineering from the University of Michigan. Mr. Porricelli was a life member of the Society of Naval Architects and Marine Engineers, a member of numerous professional organizations, and participated in many international marine technical and safety forums. He was a former member of the Marine Board.

Richard J. Satava is a senior superintendent for Sea-Land Service's Ship Management group and is responsible for the day-to-day operations of vessels in Sea-Land's Pacific Northwest fleet. In this position, he has been responsible for implementing Annex V on the company's vessels and providing shore-based support for those efforts. He is a master mariner with 15 years in the maritime industry and has had a broad range of experience on vessels of all types, including chemical and oil tankers, freighters, container ships, and bulk carriers. Mr. Satava has been a member of several industry associations, including the American Institute of Merchant Shipping, Pacific Merchant Shipping Association, and the Puget Sound Steamship Operators Association. He also was a partner in an operating shellfish farm, for which he developed the pre-market shellfish purification and packaging standards and procedures. He is a graduate of the U.S. Merchant Marine Academy and holds a current Master's license.

N.C. Vasuki is the general manager and chief executive officer of the Delaware Solid Waste Authority and the immediate past international president of the Solid Waste Association of North America (SWANA). His professional abilities are put to use at many levels of government, from the local jurisdictions of Delaware to the international plane and the transboundary domains of the SWANA membership. He has earned a reputation for effective implementation of solid waste handling strategies and is a technical leader in the government response to U.S. solid waste disposal problems. Earlier in his career, Mr. Vasuki was responsible for administering Delaware's environmental protection programs. He has served as president of the Chesapeake Water Pollution Control Association, a member of the steering committee for the Governor's Environmental Legacy Program, and a member of the Governor's Committee on Oil Transportation. He is a diplomate of the American Academy of Environmental Engineers and the author of more than 30 technical publications and one reference book. Mr. Vasuki earned a B.S. degree in Civil Engineering at the National Institute of Engineering in India and an M.S. in Civil Engineering from the University of Delaware.

Miranda S. Wecker serves as counsel to the Center for International Environmental Law-U.S., a public interest law organization advocating the development and use of international law to protect the global environment. She also directs a consulting company that provides advice on environmental law and policy. From 1985 to 1991, Ms. Wecker served as associate director and director of policy studies for the Council on Ocean Law (COL), an organization founded to promote U.S. adherence to the Third United Nations Convention on the Law of the Sea. She regularly served on the U.S. delegation to meetings of the United Nations Environment Program for the Wider Caribbean Region and edited a monthly newsletter on ocean law developments. Ms. Wecker earned her J.D. and an L.L.M. degree in Marine Affairs and Law from the University of Washington in Seattle.

B

Annex V of MARPOL 73/78
Regulations for the Prevention of Pollution
by Garbage from Ships

with attachments:
Guidelines for the Implementation of Annex V of MARPOL 73/78
Standard Specification for Shipboard Incinerators

Regulation 1
Definitions

For the purposes of this Annex:

(1) *Garbage* means all kinds of victual, domestic and operational waste excluding fresh fish and parts thereof, generated during the normal operation of the ship and liable to be disposed of continuously or periodically except those substances which are defined or listed in other Annexes to the present Convention.

(2) *Nearest land.* The term "from the nearest land" means from the baseline from which the territorial sea of the territory in question is established in accordance with international law except that, for the purposes of the present Convention, "from the nearest land" off the north-eastern coast of Australia shall mean from a line drawn from a point on the coast of Australia in

latitude 11°00′ S, longitude 142°08′ E
to a point in latitude 10°35′ S, longitude 141°55′ E,
thence to a point latitude 10°00′ S, longitude 142°00′ E,
thence to a point latitude 9°10′ S, longitude 143°52′ E,
thence to a point latitude 9°00′ S, longitude 144°30′ E,
thence to a point latitude 13°00′ S, longitude 144°00′ E,
thence to a point latitude 15°00′ S, longitude 146°00′ E,
thence to a point latitude 18°00′ S, longitude 147°00′ E,
thence to a point latitude 21°00′ S, longitude 153°00′ E,
thence to a point on the coast of Australia in
latitude 24°42′ S, longitude 153°15′ E.

(3) *Special area* means a sea area where for recognized technical reasons in relation to its oceanographical and ecological condition and to the particular character of its traffic the adoption of special mandatory methods for the prevention of sea pollution by garbage is required. Special areas shall include those listed in regulation 5 of this Annex.

Regulation 2
Application

The provisions of this Annex shall apply to all ships.

Regulation 3
Disposal of garbage outside special areas

(1)   Subject to the provisions of regulations 4, 5 and 6 of this Annex:

   (a)   the disposal into the sea of all plastics, including but not limited to synthetic ropes, synthetic fishing nets and plastic garbage bags, is prohibited;

   (b)   the disposal into the sea of the following garbage shall be made as far as practicable from the nearest land but in any case is prohibited if the distance from the nearest land is less than:

   (i)   25 nautical miles for dunnage, lining and packing materials which will float;

   (ii)   12 nautical miles for food wastes and all other garbage including paper products, rags, glass, metal, bottles, crockery and similar refuse;

   (c)   disposal into the sea of garbage specified in subparagraph (b)(ii) of this regulation may be permitted when it has passed through a comminuter or grinder and made as far as practicable from the nearest land but in any case is prohibited if the distance from the nearest land is less than 3 nautical miles. Such comminuted or ground garbage shall be capable of passing through a screen with openings no greater than 25 millimetres.

(2)   When the garbage is mixed with other discharges having different disposal or discharge requirements the more stringent requirements shall apply.

Regulation 4
Special requirements for disposal of garbage

(1)   Subject to the provisions of paragraph (2) of this regulation, the disposal of any materials regulated by this Annex is prohibited from fixed or floating platforms engaged in the exploration, exploitation and associated offshore processing of sea-bed mineral resources, and from all other ships when alongside or within 500 metres of such platforms.

(2)   The disposal into the sea of food wastes may be permitted when they have been passed through a comminuter or grinder from such fixed or floating platforms located more than 12 nautical miles from land and all other

ships when alongside or within 500 metres of such platforms. Such comminuted or ground food wastes shall be capable of passing through a screen with openings no greater than 25 millimetres.

Regulation 5
Disposal of garbage within special areas

(1) For the purposes of this Annex the special areas are the Mediterranean Sea area, the Baltic Sea area, the Black Sea area, the Red Sea area, the "Gulfs area", the North Sea area, the Antarctic area and the Wider Caribbean Region, including the Gulf of Mexico and the Caribbean Sea, which are defined as follows:

(a) The *Mediterranean Sea area* means the Mediterranean Sea proper including the gulfs and seas therein with the boundary between the Mediterranean and the Black Sea constituted by the 41° N parallel and bounded to the west by the Straits of Gibraltar at the meridian 5°36′ W.

(b) The *Baltic Sea area* means the Baltic Sea proper with the Gulf of Bothnia and the Gulf of Finland and the entrance to the Baltic Sea bounded by the parallel of the Skaw in the Skagerrak at 57°44.8′ N.

(c) The *Black Sea area* means the Black Sea proper with the boundary between the Mediterranean and the Black Sea constituted by the parallel 41° N.

(d) The *Red Sea area* means the Red Sea proper including the Gulfs of Suez and Aqaba bounded at the south by the rhumb line between Ras si Ane (12°8.5′ N, 43°19.6′ E) and Husn Murad (12°40.4′ N, 43°30.2′ E).

(e) The *Gulfs area* means the sea area located north-west of the rhumb line between Ras al Hadd (22°30′ N, 59°48′ E) and Ras al Fasteh (25°04′ N, 61°25′ E).

(f) The *North Sea area** means the North Sea proper including seas therein with the boundary between:

  (i) the North Sea southwards of latitude 62° N and eastwards of longitude 4° W;

  (ii) the Skagerrak, the southern limit of which is determined east of the Skaw by latitude 57°44.8′ N; and

  (iii) the English Channel and its approaches eastwards of longitude 5° W and northwards of latitude 48°30′ N.

---

* Regulation 5(1)(f) was adopted by the MEPC at its twenty-eighth session and entered into force on 18 April 1991.

(g)   The Antarctic area* means the sea area south of latitude 60° S.

(h)   The Wider Caribbean Region**, as defined in article 2, paragraph
      1 of the Convention for the Protection and Development of the
      Marine Environment of the Wider Caribbean Region (Cartagena de
      Indias, 1983), means the Gulf of Mexico and Caribbean Sea proper
      including the bays and seas therein and that portion of the Atlantic
      Ocean within the boundary constituted by the 30° N parallel from
      Florida eastward to 77°30′ W meridian, thence a rhumb line to the
      intersection of 20° N parallel and 59° W meridian, thence a rhumb
      line to the intersection of 7°20′ N parallel and 50° W meridian, thence
      a rhumb line drawn south-westerly to the eastern boundary of French
      Guiana.

(2)   Subject to the provisions of regulation 6 of this Annex:

      (a)   disposal into the sea of the following is prohibited:

            (i)   all plastics, including but not limited to synthetic ropes,
                   synthetic fishing nets and plastic garbage bags; and

            (ii)   all other garbage, including paper products, rags, glass, metal,
                   bottles, crockery, dunnage, lining and packing materials;

      (b)   except as provided in subparagraph (c) of this paragraph,*** disposal
            into the sea of food wastes shall be made as far as practicable from
            land, but in any case not less than 12 nautical miles from the nearest
            land;

      (c)   disposal into the Wider Caribbean Region of food wastes which have
            been passed through a comminuter or grinder shall be made as far
            as practicable from land, but in any case not subject to regulation 4
            not less than 3 nautical miles from the nearest land. Such
            comminuted or ground food wastes shall be capable of passing
            through a screen with openings no greater than 25 millimetres.***

(3)   When the garbage is mixed with other discharges having different disposal
      or discharge requirements the more stringent requirements shall apply.

(4)   Reception facilities within special areas:

      (a)   The Government of each Party to the Convention, the coastline of
            which borders a special area, undertakes to ensure that as soon as
            possible in all ports within a special area adequate reception facilities
            are provided in accordance with regulation 7 of this Annex, taking
            into account the special needs of ships operating in these areas.

---

* Regulation 5(1)(g) was adopted by the MEPC at its thirtieth session and is expected to enter into
force on 17 March 1992.
** Regulation 5(1)(h) was adopted by the MEPC at its thirty-first session and is expected to enter
into force on 4 April 1993.
*** These amendments were adopted by the MEPC at its thirty-first session and are expected to
enter into force on 4 April 1993.

(b) The Government of each Party concerned shall notify the Organization of the measures taken pursuant to subparagraph (a) of this regulation. Upon receipt of sufficient notifications the Organization shall establish a date from which the requirements of this regulation in respect of the area in question shall take effect. The Organization shall notify all Parties of the date so established no less than twelve months in advance of that date.

(c) After the date so established, ships calling also at ports in these special areas where such facilities are not yet available, shall fully comply with the requirements of this regulation.

(5)* Notwithstanding paragraph 4 of this regulation, the following rules apply to the Antarctic area:

(a) The Government of each Party to the Convention at whose ports ships depart *en route* to or arrive from the Antarctic area undertakes to ensure that as soon as practicable adequate facilities are provided for the reception of all garbage from all ships, without causing undue delay, and according to the needs of the ships using them.

(b) The Government of each Party to the Convention shall ensure that all ships entitled to fly its flag, before entering the Antarctic area, have sufficient capacity on board for the retention of all garbage while operating in the area and have concluded arrangements to discharge such garbage at a reception facility after leaving the area.

Regulation 6
Exceptions

Regulations 3, 4 and 5 of this Annex shall not apply to:

(a) the disposal of garbage from a ship necessary for the purpose of securing the safety of a ship and those on board or saving life at sea; or

(b) the escape of garbage resulting from damage to a ship or its equipment provided all reasonable precautions have been taken before and after the occurrence of the damage, for the purpose of preventing or minimizing the escape; or

(c) the accidental loss of synthetic fishing nets, provided that all reasonable precautions have been taken to prevent such loss.

---

* This amendment was adopted by the MEPC at its thirtieth session and is expected to enter into force on 17 March 1992.

Regulation 7
Reception facilities

(1) The Government of each Party to the Convention undertakes to ensure the provision of facilities at ports and terminals for the reception of garbage, without causing undue delay to ships, and according to the needs of the ships using them.

(2) The Government of each Party shall notify the Organization for transmission to the Parties concerned of all cases where the facilities provided under this regulation are alleged to be inadequate.

Guidelines for the implementation of Annex V of MARPOL 73/78

Preface

The main objectives of these guidelines are to (1) assist governments in developing and enacting domestic laws which give force to and implement Annex V, (2) assist vessel operators in complying with the requirements set forth in Annex V and domestic laws and, (3) assist port and terminal operators in assessing the need for, and providing, adequate reception facilities for garbage generated on different types of ships. Part IV (Garbage) of the Organization's *Guidelines on the Provision of Adequate Reception Facilities in Ports*, June 1978, has been modified and incorporated in this publication to consolidate all Annex V related guidelines. In the interest of uniformity, governments are requested to refer to these guidelines when preparing appropriate national regulations.

1   Introduction and definitions

1.1   These guidelines have been developed taking into account the regulations embodied in Annex V, the articles and resolutions of the International Convention for the Prevention of Pollution from Ships, 1973, as modified by the Protocol of 1978 relating thereto (MARPOL 73/78) (hereinafter referred to as the "Convention"). Their purpose is to provide guidance to countries which have ratified Annex V and are in the process of implementing the Annex. The guidelines are divided into seven categories that provide a general framework upon which governments will be able to formulate programmes for education and training of seafarers and others to comply with the regulations; methods of reducing shipboard generation of garbage; shipboard garbage handling and storage procedures; shipboard equipment for processing garbage; estimation of the amounts of ship-generated garbage delivered to port; and actions to ensure compliance with the regulations.

1.2   Recognizing that Annex V regulations promote waste management systems for ships, and that ships vary tremendously in size, mission, complement and capability, these guidelines include a range of waste management options that may be combined in many ways to facilitate compliance with Annex V. Further, recognizing that waste management technology for ships is in an early stage of development, it is recommended that governments and the Organization continue to gather information and review these guidelines periodically.

1.3   Although Annex V permits the discharge of a range of garbage into the sea, it is recommended that whenever practicable ships use, as a primary means, port reception facilities.

1.4   Governments should stimulate the provision and use of port reception facilities for garbage from ships, as outlined in section 7.2 of these guidelines.

1.5   The Convention provides definitions for terms used throughout these guidelines which establish the scope of Annex V requirements. These definitions are incorporated in section 1 of these guidelines and in regulation 1 of Annex V. Definitions taken directly from the Convention are listed in section 1.6, and are followed by other definitions which are useful.

1.6   *Definitions from the Convention*

1.6.1   *Regulations* means the regulations contained in the annexes to the Convention.

1.6.2   *Harmful substance* means any substance which, if introduced into the sea, is liable to create hazards to human health, harm living resources and marine life, damage amenities or interfere with other legitimate uses of the sea, and includes any substance subject to control by the Convention.

1.6.3   *Discharge*, in relation to harmful substances or effluents containing such substances, means any release, howsoever caused, from a ship and includes any escape, disposal, spilling, leaking, pumping, emitting or emptying.

1.6.3.1   *Discharge* does not include:

   (i)   dumping, within the meaning of the Convention on the Prevention of Marine Pollution by Dumping of Wastes and Other Matter, done at London on 13 November 1972; or

   (ii)   release of harmful substances directly arising from the exploration, exploitation and associated offshore processing of sea-bed mineral resources; or

   (iii) release of harmful substances for purposes of legitimate scientific research into pollution abatement or control.

1.6.4   *Ship* means a vessel of any type whatsoever operating in the marine environment and includes hydrofoil boats, air-cushion vehicles, submersibles, floating craft and fixed or floating platforms.

1.6.5   *Incident* means an event involving the actual or probable discharge into the sea of a harmful substance, or effluents containing such a substance.

1.6.6   *Organization* means the International Maritime Organization.

1.7   *Other definitions*

1.7.1   *Wastes* means useless, unneeded or superfluous matter which is to be discarded.

1.7.2 *Food wastes* are any spoiled or unspoiled victual substances, such as fruits, vegetables, dairy products, poultry, meat products, food scraps, food particles, and all other materials contaminated by such wastes, generated aboard ship, principally in the galley and dining areas.

1.7.3 *Plastic* means a solid material which contains as an essential ingredient one or more synthetic organic high polymers and which is formed (shaped) during either manufacture of the polymer or the fabrication into a finished product by heat and/or pressure. Plastics have material properties ranging from hard and brittle to soft and elastic. Plastics are used for a variety of marine purposes including, but not limited to, packaging (vapour-proof barriers, bottles, containers, liners), ship construction (fibreglass and laminated structures, siding, piping, insulation, flooring, carpets, fabrics, paints and finishes, adhesives, electrical and electronic components), disposable eating utensils and cups, bags, sheeting, floats, fishing nets, strapping bands, rope and line.

1.7.4 *Domestic waste* means all types of food wastes and wastes generated in the living spaces on board the ship.

1.7.5 *Cargo-associated waste* means all materials which have become wastes as a result of use on board a ship for cargo stowage and handling. Cargo-associated waste includes but is not limited to dunnage, shoring, pallets, lining and packing materials, plywood, paper, cardboard, wire, and steel strapping.

1.7.6 *Maintenance waste* means materials collected by the engine department and the deck department while maintaining and operating the vessel, such as soot, machinery deposits, scraped paint, deck sweeping, wiping wastes, and rags, etc.

1.7.7 *Operational wastes* means all cargo-associated waste and maintenance waste, and cargo residues defined as garbage in 1.7.10.

1.7.8 *Dishwater* is the residue from the manual or automatic washing of dishes and cooking utensils which have been pre-cleaned to the extent that any food particles adhering to them would not normally interfere with the operation of automatic dishwashers. *Greywater* is drainage from dishwater, shower, laundry, bath and washbasin drains and does not include drainage from toilets, urinals, hospitals, and animal spaces, as defined in regulation 1(3) of Annex IV, as well as drainage from cargo spaces.

1.7.9 *Oily rags* are rags which have been saturated with oil as controlled in Annex I to the Convention. *Contaminated rags* are rags which have been saturated with a substance defined as a harmful substance in the other annexes to the Convention.

1.7.10 *Cargo residues* for the purposes of these guidelines are defined as the remnants of any cargo material on board that cannot be placed in proper cargo holds (loading excess and spillage) or which remain in cargo holds and elsewhere after unloading procedures are completed (unloading residual and spillage). However, cargo residues are expected to be in small quantities.

1.7.11 *Fishing gear* is defined as any physical device or part thereof or combination of items that may be placed on or in the water with the intended purpose of capturing, or controlling for subsequent capture, living marine or freshwater organisms.

1.7.12 *Seafarers* for the purposes of these guidelines means anyone who goes to sea in a ship for any purpose including, but not limited to transport of goods and services, exploration, exploitation and associated offshore processing of sea-bed mineral resources, fishing and recreation.

1.8   *Application*

1.8.1   Dishwater and greywater are not included as garbage in the context of Annex V.

1.8.2   *Ash* and *clinkers* from shipboard incinerators and coal-burning boilers are operational wastes in the meaning of Annex V, regulation 1(1) and therefore are included in the term *all other garbage* in the meaning of Annex V, regulations 3(1)(b)(ii) and 5(2)(a)(ii).

1.8.3   Cargo residues are to be treated as garbage under Annex V except when those residues are substances defined or listed under the other annexes to the Convention.

1.8.4   Cargo residues of all other substances are not explicitly excluded from disposal as garbage under the overall definition of garbage in Annex V. However, certain of these substances may pose harm to the marine environment and may not be suitable for disposal at reception facilities equipped to handle general garbage because of their possible safety hazards. The disposal of such cargo residues should be based on the physical, chemical and biological properties of the substance and may require special handling not normally provided by garbage reception facilities.

1.8.5   The release of small quantities of food wastes for the specific purpose of fish feeding in connection with fishing or tourist operations is not included as garbage in the context of Annex V.

2   Training, education and information

2.1   The definition of ships used in the Convention requires these guidelines to address not only the professional and commercial maritime community but also the non-commercial seafaring population as sources of pollution of the sea by garbage. The Committee recognized that uniform programmes in the field of training and education would make a valuable contribution to raising the level of the seafarers' compliance with Annex V, thereby ensuring compliance with the Convention. Accordingly, governments should develop and undertake training, education and public information programmes suited for all seafaring communities under their jurisdictions.

2.2 Governments may exchange and maintain information relevant to compliance with Annex V regulations through the Organization. Accordingly, governments are encouraged to provide the Organization with the following:

2.2.1 Technical information on shipboard waste management methods such as recycling, incineration, compaction, sorting and sanitation systems, packaging and provisioning methods;

2.2.2 Copies of current domestic laws and regulations relating to the prevention of pollution of the sea by garbage;

2.2.3 Educational materials developed to raise the level of compliance with Annex V. Contributions of this type might include printed materials, posters, brochures, photographs, audio and video tapes, and films as well as synopses of training programmes, seminars and formal curricula;

2.2.4 Information and reports on the nature and extent of marine debris found along beaches and in coastal waters under their respective jurisdictions. In order to assess the effectiveness of Annex V, these studies should provide details on amounts, distribution, sources and impacts of marine debris.

2.3 Governments are encouraged to amend their maritime certification examinations and requirements, as appropriate, to include a knowledge of duties imposed by national and international law regarding the control of pollution of the sea by garbage.

2.4 Governments are recommended to require all ships of their registry to permanently post a summary declaration stating the prohibition and restrictions for discharging garbage from ships under Annex V and the penalties for failure to comply. It is suggested this declaration be placed on a placard at least 12.5 cm by 20 cm, made of durable material and fixed in a conspicuous place in galley spaces, the mess deck, wardroom, bridge, main deck and other areas of the ship, as appropriate. The placard should be printed in the language or languages understood by the crew and passengers.

2.5 Governments are encouraged to have maritime colleges and technical institutes under their jurisdiction develop or augment curricula to include both the legal duties as well as the technical options available to professional seafarers for handling ship-generated garbage. These curricula should also include information on environmental impacts of garbage. A list of suggested topics to be included in the curriculum are listed below:

2.5.1 Garbage in the marine environment, sources, types and impacts;

2.5.2 National and international laws relating to, or impinging upon shipboard waste management;

2.5.3 Health and sanitation considerations related to the storage, handling and transfer of ship-generated garbage;

2.5.4 Current technology for on-board and shoreside processing of ship-generated garbage;

2.5.5 Provisioning options, materials and procedures to minimize the generation of garbage aboard ship.

2.6 Professional associations and societies of ship officers, engineers, naval architects, shipowners and managers, and seamen are encouraged to ensure their members' competency regarding the handling of ship-generated garbage.

2.6.1 Vessel and reception facility operators should establish training programmes for personnel operating and maintaining garbage reception or processing equipment. It is suggested that the programme include instruction on what constitutes garbage and the applicable regulations for handling and disposing of it. Such training should be reviewed annually.

2.7 Generalized public information programmes are needed to provide information to non-professional seafarers, and others concerned with the health and stability of the marine environment, regarding the impacts of garbage at sea. Governments and involved commercial organizations are encouraged to utilize the Organization's library and to exchange resources and materials, as appropriate, to initiate internal and external public awareness programmes.

2.7.1 Methods for delivering this information include radio and television, articles in periodicals and trade journals, voluntary public projects such as beach clean-up days and adopt-a-beach programmes, public statements by high government officials, posters, brochures, conferences and symposia, co-operative research and development, voluntary product labelling and teaching materials for public schools.

2.7.2 Audiences include recreational boaters and fishermen, port and terminal operators, coastal communities, ship supply industries, shipbuilders, waste management industries, plastic manufacturers and fabricators, trade associations, educators and governments.

2.7.3 The subjects addressed in these programmes are recommended to include the responsibilities of citizens under national and international law; options for handling garbage at sea and upon return to shore; known sources and types of garbage; impacts of plastic debris on sea-birds, fish, marine mammals, sea turtles and ship operations; impacts on coastal tourist trade; current actions by governments and private organizations, and sources of further information.

3   Minimizing the amount of potential garbage

3.1   All ship operators should minimize the taking aboard of potential garbage and on-board generation of garbage.

3.2 Domestic wastes may be minimized through proper provisioning practices. Ship operators and governments should encourage ships' suppliers and provisioners to consider their products in terms of the garbage they generate. Options available to decrease the amount of domestic waste generated aboard ship include the following:

3.2.1 Bulk packaging of consumable items may result in less waste being created. However, factors such as inadequate shelf-life once a container is open must be considered to avoid increasing wastes.

3.2.2 Reusable packaging and containers can decrease the amount of garbage being generated. Use of disposable cups, utensils, dishes, towels and rags and other convenience items should be limited and replaced by washable items when possible.

3.2.3 Where practical options exist, provisions packaged in or made of materials other than disposable plastic should be selected to replenish ship supplies unless a reusable plastic alternative is available.

3.3 Operational waste generation is specific to individual ship activities and cargoes. It is recommended that manufacturers, shippers, ship operators and governments consider the garbage associated with various categories of cargoes and take action as needed to minimize their generation. Suggested actions are listed below:

3.3.1 Consider replacing disposable plastic sheeting used for cargo protection with permanent, reusable covering material;

3.3.2 Consider stowage systems and methods that reuse coverings, dunnage, shoring, lining and packing materials;

3.3.3 Dunnage, lining and packaging materials generated in port during cargo discharge should preferably be disposed of to the port reception facilities and not retained on board for discharge at sea.

3.4 Cargo residues are created through inefficiencies in loading, unloading and on-board handling.

3.4.1 As cargo residues fall under the scope of these guidelines, it may, in certain cases, be difficult for port reception facilities to handle such residues. It is therefore recommended that cargo be unloaded as efficiently as possible in order to avoid or minimize cargo residues.

3.4.2 Spillage of the cargo during transfer operations should be carefully controlled, both on board and from dockside. Since this spillage typically occurs in port, it should be completely cleaned up prior to sailing and either delivered into the intended cargo space or into the port reception facility. Shipboard areas where spillage is most common should be protected such that the residues are easily recovered.

3.5   Fishing gear, once discharged, becomes a harmful substance. Fishing vessel operators, their organizations and their respective governments are encouraged to undertake such research, technology development and regulations as may be necessary to minimize the probability of loss, and maximize the probability of recovery of fishing gear from the ocean. It is recommended that fishing vessel operators record and report the loss and recovery of fishing gear. Techniques both to minimize the amount of fishing gear lost in the ocean and to maximize recovery of same are listed below.

3.5.1   Operators and associations of fishing vessels using untended, fixed or drifting gear are encouraged to develop information exchanges with such other ship traffic as may be necessary to minimize accidental encounters between ships and gear. Governments are encouraged to assist in the development of information systems where necessary.

3.5.2   Fishery managers are encouraged to consider the probability of encounters between ship traffic and fishing gear when establishing seasons, areas and gear-type regulations.

3.5.3   Fishery managers, fishing vessel operators and associations are encouraged to utilize gear identification systems which provide information such as vessel name, registration number and nationality, etc. Such systems may be useful to promote reporting, recovery and return of lost gear.

3.5.4   Fishing vessel operators are encouraged to document positions and reasons for loss of their gear. To reduce the potential of entanglement and "ghost fishing" (capture of marine life by discharged fishing gear), benthic traps, trawl and gill-nets could be designed to have degradable panels or sections made of natural fibre twine, wood or wire.

3.5.5   Governments are encouraged to consider the development of technology for more effective fishing gear identification systems.

3.6   Governments are encouraged to undertake research and technology development to minimize potential garbage and its impacts on the marine environment. Suggested areas for such study are listed below:

3.6.1   Development of recycling technology and systems for synthetic materials returned to shore as garbage;

3.6.2   Development of technology for degradable synthetic materials to replace current plastic products as appropriate. In this connection, governments should also study the impacts on the environment of the products of degradation of such new materials.

4   Shipboard garbage handling and storage procedures

4.1   Limitations on the discharge of garbage from ships as specified in Annex V are summarized in table 1. Although discharge at sea, except in special areas, of a wide range of ship-generated garbage is permitted outside specified distances from the nearest land, preference should be given to disposal at shore reception facilities.

Table 1 — Summary of at sea garbage disposal regulations

Garbage type	***All ships except platforms		***Offshore platforms
	Outside special areas	**In special areas	
Plastics – includes synthetic ropes and fishing nets and plastic garbage bags	Disposal prohibited	Disposal prohibited	Disposal prohibited
Floating dunnage, lining and packing materials	>25 miles offshore	Disposal prohibited	Disposal prohibited
Paper, rags, glass, metal, bottles, crockery and similar refuse	>12 miles	Disposal prohibited	Disposal prohibited
All other garbage including paper, rags, glass, etc. comminuted or ground	>3 miles	Disposal prohibited	Disposal prohibited
Food waste not comminuted or ground	>12 miles	>12 miles	Disposal prohibited
*Food waste comminuted or ground	>3 miles	>12 miles	>12 miles
Mixed refuse types	****	****	****

* Comminuted or ground garbage must be able to pass through a screen with mesh size no larger than 25 mm.
** Garbage disposal regulations for special areas shall take effect in accordance with regulation 5(4)(b) of Annex V.
*** Offshore platforms and associated ships include all fixed or floating platforms engaged in exploration or exploitation of sea-bed mineral resources, and all ships alongside or within 500 m of such platforms.
**** When garbage is mixed with other harmful substances having different disposal or discharge requirements, the more stringent disposal requirements shall apply.
Note: The Baltic Sea Special Area Disposal Regulations took effect on 1 October 1989.

4.1.1   Compliance with these limitations requires personnel, equipment and procedures for collecting, sorting, processing, storing and disposing of garbage. Economic and procedural considerations associated with these activities include storage space requirements, sanitation, equipment and personnel costs and in-port garbage service charges.

4.1.2   Compliance with the provisions of Annex V will require careful planning by the ship operator and proper execution by crew members as well as other seafarers. The most appropriate procedures for handling and storing garbage on ship will vary depending on factors such as the type and size of the ship, the area of operation (e.g. distance from nearest land), shipboard garbage processing equipment and storage space, crew size, duration of voyage, and regulations and reception facilities at ports of call. However, in view of the cost involved with the different ultimate disposal techniques, it may also be economically advantageous to keep garbage requiring special handling separate from other garbage. Proper handling and storage will minimize shipboard storage space requirements and enable efficient transfer of retained garbage to port reception facilities.

4.2   To ensure that the most effective and efficient handling and storage procedures are followed, it is recommended that vessel operators develop waste management plans that can be incorporated into crew and vessel operating manuals. Such manuals should identify crew responsibilities (including an environmental control officer) and procedures for all aspects of handling and storing garbage aboard the ship. Procedures for handling ship-generated garbage can be divided into four phases: collection, processing, storage, and disposal. A generalized waste management plan for handling and storing ship-generated garbage is presented in table 2. Specific procedures for each phase are discussed below.

4.3   *Collection*

Procedures for collecting garbage generated aboard ship should be based on consideration of what can and cannot be discarded overboard while *en route*. To reduce or avoid the need for sorting after collection, it is recommended that three categories of distinctively marked garbage receptacles be provided to receive garbage as it is generated. These separate receptacles (e.g. cans, bags, or bins) would receive (1) plastics and plastics mixed with non-plastic garbage; (2) food wastes (which includes materials contaminated by such wastes); and (3) other garbage which can be disposed of at sea. Receptacles for each of the three categories of garbage should be clearly marked and distinguishable by colour, graphics, shape, size, or location. These receptacles should be provided in appropriate spaces throughout the ship (e.g. the engine-room, mess deck, wardroom, galley, and other living or working spaces) and all crew members and passengers should be advised of what garbage should and should not be discarded in them. Crew responsibilities should be assigned for collecting or emptying these receptacles and taking the garbage to the appropriate processing or storage location. Use of such a system will facilitate subsequent shipboard processing and minimize the amount of garbage which must be stored aboard ship for return to port.

Table 2 — Options for shipboard handling and disposal of garbage

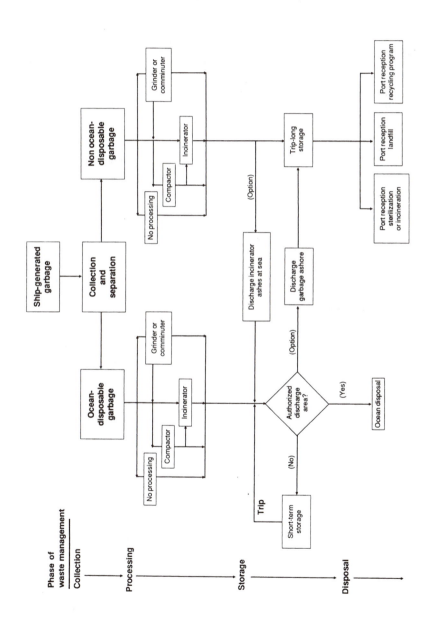

4.3.1   Plastics and plastics mixed with non-plastic garbage

Plastic garbage must be retained aboard ship for discharge at port reception facilities unless reduced to ash by incineration. When plastic garbage is not separated from other garbage, the mixture must be treated as if it were all plastic.

4.3.2   Food wastes

Some governments have regulations for controlling human, plant, and animal diseases that may be carried by foreign food wastes and materials that have been associated with them (e.g. food packaging and disposable eating utensils). These regulations may require incinerating, sterilizing, or other special treatment of garbage to destroy possible pest and disease organisms. Such garbage should be kept separate from other garbage and preferably retained for disposal in port in accordance with the laws of the receiving country. With regard to such garbage, governments are reminded of their obligation to assure the provision of adequate reception facilities. Precautions must be taken to ensure that plastics contaminated by food wastes (e.g. plastic food wrappers) are not discharged at sea with other food wastes.

4.3.3   Other garbage

Garbage in this category includes, but is not limited to, paper products, rags, glass, metal, bottles, crockery, dunnage, lining and packing materials. Vessels may find it desirable to separate dunnage, lining and packing material which will float since this material is subject to a different discharge limit than other garbage in this category (see table 1). Such garbage should be kept separate from other garbage and preferably retained for disposal in port.

4.3.4   Additional receptacles which might be useful

4.3.4.1   Separate cans or bags could be provided for receiving and storing glass, metal, plastics, paper or other items which can be recycled. To encourage crew members to deposit such items in receptacles provided, proceeds generated from their return might be added to a ship's recreational fund.

4.3.4.2   Synthetic fishing net and line scraps generated by the repair or operation of fishing gear may not be discarded at sea and should be collected in a manner that avoids its loss overboard. Such material may be incinerated, compacted, or stored along with other plastic waste or it may be preferable to keep it separate from other types of garbage if it has strong odour or great volume.

4.3.5   Recovery of garbage at sea

4.3.5.1   Fishermen and other seafarers who recover derelict fishing gear and other persistent garbage during routine operations are encouraged to retain this material for disposal on shore. If lost pots or traps are recovered and space is not available for storage, fishermen and other seafarers are encouraged to remove and transport any line and webbing to port for disposal and return the bare frames to the water, or minimally, to cut open the traps to keep them from continuing to trap marine life.

4.3.5.2 Seafarers are further encouraged to recover other persistent garbage from the sea as opportunities arise and prudent practice permits.

4.3.6 Oily rags and contaminated rags must be kept on board and discharged to a port reception facility or incinerated.

4.4 *Processing*

Depending on factors such as the type of ship, area of operation, size of crew, etc., ships may be equipped with incinerators, compactors, comminuters, or other devices for shipboard garbage processing (see section 5). Appropriate members of the crew should be assigned responsibility for operating this equipment on a schedule commensurate with ship needs. In selecting appropriate processing procedures, the following should be considered.

4.4.1 Use of compactors, incinerators, comminuters, and other such devices has a number of advantages, such as making it possible to discharge certain garbage at sea which otherwise might not be permitted, reducing shipboard space requirements for storing garbage, making it easier to off-load garbage in port, and enhancing assimilation of garbage discharged into the marine environment.

4.4.2 It should be noted that special rules on incineration may be established by authorities in some ports and may exist in some special areas. Incineration of the following items requires special precaution due to the potential environmental and health effects from combustion of by-products: hazardous materials (e.g. scraped paint, impregnated wood) and certain types of plastics (e.g. PVC-based plastics). The problems of combustion of by-products are discussed in 5.4.6.

4.4.3 Ships operating primarily in special areas or within 3 nautical miles from the nearest land should choose between storage of either compacted or uncompacted material for off-loading at port reception facilities or incineration with retention of ash and clinkers. This is the most restrictive situation in that no discharge is permitted. The type of ship and the expected volume and type of garbage generated will determine the suitability of compaction, incineration, or storage options.

4.4.4 Compactors make garbage easier to store, to transfer to port reception facilities, and to dispose of at sea when discharge limitations permit. In the latter case, compacted garbage may also aid in sinking, which would reduce aesthetic impacts in coastal waters and along beaches, and perhaps reduce the likelihood of marine life ingesting or otherwise interacting with discharged materials.

4.4.5 Ships operating primarily beyond 3 nautical miles from the nearest land are encouraged to install and use comminuters to grind food wastes to a particle size capable of passing through a screen with openings no larger than 25 mm. Although larger food scraps may be discharged beyond 12 nautical miles, it is recommended that comminuters be used even outside this limit because they hasten assimilation into the marine environment. Because food wastes comminuted with plastics cannot be discharged at sea, all plastic materials must be removed before food wastes are ground up.

4.5 Storage

Garbage collected from living and working areas throughout the ship should be delivered to designated processing or storage locations. Garbage that must be returned to port for disposal may require long-term storage depending on the length of the voyage or arrangements for off-loading (e.g. transferring garbage to an offshore vessel for incineration or subsequent transfer ashore). Garbage which may be discarded overboard may require short-term or no storage. In all cases, garbage should be stored in a manner which avoids health and safety hazards. The following points should be considered when selecting procedures for storing garbage:

4.5.1 Ships should use separate cans, drums, boxes, bags or other containers for short-term (disposable garbage) and trip-long (non-disposable garbage) storage. Short-term storage would be appropriate for holding otherwise disposable garbage while a ship is passing through a restricted discharge area.

4.5.2 Sufficient storage space and equipment (e.g. cans, drums, bags or other containers) should be provided. Where space is limited, vessel operators are encouraged to install compactors or incinerators. To the extent possible, all processed and unprocessed garbage which must be stored for any length of time should be in tight, securely covered containers.

4.5.3 Food wastes and associated garbage which are returned to port and which may carry diseases or pests should be stored in tightly covered containers and be kept separate from garbage which does not contain such food wastes. Both types of garbage should be stored in separate clearly marked containers to avoid incorrect disposal and treatment on land.

4.5.4 Storage of waste fishing gear on deck may be appropriate if materials have strong odours or if their size is too great to permit storage elsewhere on the ship. In cases where gear is fouled with marine growth or dead organisms, it may be reasonable to tow gear behind the vessel for a time to wash it out before storing. If it cannot be recovered by the vessel, the appropriate coastal State should be notified of its location.

4.5.5 Disinfection and both preventative and remedial pest control methods should be applied regularly in garbage storage areas.

4.6 Disposal

Although disposal is possible under Annex V, discharge of garbage to port reception facilities should be given first priority. Disposal of ship-generated garbage must be done in a manner consistent with the regulations summarized in table 1. When disposing of garbage, the following points should be considered:

4.6.1 Garbage which may be disposed of at sea can simply be discharged overboard. Disposal of uncompacted garbage is convenient, but results in a maximum number of floating objects which may reach shore even when discharged beyond 25 nautical miles from the nearest land. Compacted garbage is more likely to sink and thus less likely to pose aesthetic problems. If necessary and possible, weights should be added to promote sinking. Compacted bales of garbage should be discharged over deep water (50 m or more) to prevent rapid loss of their structural integrity due to wave action and currents.

4.6.2 Floating cargo-associated waste that is not plastic or otherwise regulated under other MARPOL annexes may be discharged beyond 25 nautical miles from the nearest land. Cargo-associated waste that will sink and is not plastic or otherwise regulated may be discharged beyond 12 nautical miles from the nearest land. Most cargo-associated waste may be generated during the loading and unloading process, usually at dock side. It is recommended that every effort be made to deliver these wastes to the nearest port reception facility system prior to the ship's departure.

4.6.3 Maintenance wastes are generated more or less steadily during the course of routine ship operations. In some cases, maintenance wastes may be contaminated with substances, such as oil or toxic chemicals, controlled under other annexes or other pollution control laws. In such cases, the more stringent disposal requirements take precedence.

4.6.4 To ensure timely transfer of large quantities of ship-generated garbage to port reception facilities, it is essential for ships or their agents to make arrangements well in advance for garbage reception. At the same time, disposal needs should be identified in order to make arrangements for garbage requiring special handling or other necessary arrangements. Special disposal needs might include off-loading food wastes and associated garbage which may carry certain disease or pest organisms, or unusually large, heavy, or odorous derelict fishing gear.

5 Shipboard equipment for processing garbage

5.1 The range of options for garbage handling aboard ships depends largely upon costs, personnel limitations, generation rate, capacity, vessel configuration and traffic patterns. The types of equipment available to address various facets of shipboard garbage handling include incinerators, compactors, comminuters and their associated hardware.

5.2 Grinding or comminution

When not in a special area, the discharge of comminuted food wastes and all other comminuted garbage (except plastics and floatable dunnage, lining and packing materials) may be permitted under regulation 3(1)(c) of Annex V beyond 3 nautical miles from the nearest land. Such comminuted or ground garbage must be capable of passing through a screen with openings no greater than 25 mm unless such comminuters or grinders comply with international or governmentally accepted standards which effectively accomplish this. It is recommended that garbage not be discharged into a ship's sewage treatment system unless it is approved for treating such garbage. Furthermore, garbage should not be stored in bottoms or tanks containing oily wastes. Such actions can result in faulty operation of sewage treatment or oily-water separator equipment and can cause sanitary problems for crew members and passengers. Options for grinding or comminution include the following:

5.2.1 A wide variety of food waste grinders are available in the market and are commonly fitted in most modern ships' galleys. These food waste grinders produce a slurry of food particles and water that washes easily through the required 25 mm screen. Output ranges from 10 to 250 litres per minute. It is recommended that the discharge from shipboard comminuters be directed into a holding tank when the vessel is operating within an area where discharge is prohibited.

5.2.2 Size reduction of certain other garbage items can be achieved by shredding or crushing and machines for carrying out this process are available for use on board ships.

5.2.3 Information on the development and use of comminuters for garbage aboard ships should be forwarded to the Organization.

5.3   Table 3 shows compaction options for various types of garbage.

5.3.1 Most garbage can be compacted; the exceptions include unground plastics, fibre and paper board, bulky cargo containers and thick metal items. Pressurized containers should not be compacted since they present an explosion hazard.

5.3.2 Compaction can reduce the volume of garbage into bags, boxes, or briquettes. When these compacted slugs are equally formed and structurally strong, they can be piled up in building block form; this permits the most efficient use of space in the storage compartments. The compaction ratio for normal mixed shipboard garbage may range as high as 12:1.

5.3.3 Some of the available compactors have options such as sanitizing, deodorizing, adjustable compaction ratios, bagging in plastic or paper, boxing in cardboard (with or without plastic or wax paper lining), baling, etc. Paper or cardboard tends to become soaked and weakened by moisture in the garbage during long periods of on-board storage. There have also been problems due to the generation of gas and pressure which can explode tight plastic bags.

5.3.4 If grinding machines are used prior to compaction, the compaction ratio can be increased and the storage space decreased.

5.3.5 A compactor should be installed in a compartment with adequate room for operating and maintaining the unit and storing trash to be processed. The compartment should be located adjacent to the areas of food processing and commissary · store-rooms. If not already required by regulations it is recommended that the space have freshwater washdown service, coamings, deck drains, adequate ventilation and hand or automatic fixed fire-fighting equipment.

5.3.6 Information on the development and use of shipboard compactors should be forwarded to the Organization.

Table 3 — Compaction options for shipboard-generated garbage

Typical examples	Special handling by vessel personnel before compaction	Compaction characteristics			On-board storage space
		Rate of alteration	Retainment of compacted form	Density of compacted form	
Metal, food and beverage containers, glass, small wood pieces	None	Very rapid	Almost 100%	High	Minimum
Comminuted plastics, fibre and paper board	Minor – reduce material to size for feed, minimal manual labour	Rapid	Approximately 80%	Medium	Minimum
Small metal drums, un-comminuted cargo packing, large pieces of wood	Moderate – longer manual labour time required to size material for feed	Slow	Approximately 50%	Relatively low	Moderate
Uncomminuted plastics	Major – very long manual labour time to size material for feed; usually impractical	Very slow	Less than 10%	Very low	Maximum
Bulky metal cargo containers, thick metal items	Impractical for shipboard compaction; not feasible	Not applicable	Not applicable	Not applicable	Maximum

5.4   In comparison with the technology of land-based incineration, the state of the art in marine incinerators is not highly advanced, primarily because the technology has not yet been subject to constraints on air emissions nor to the types of materials that could be incinerated. Marine incinerators in current use are predominantly designed for intermittent operation and hand stoking and typically do not include any provisions for air pollution control. Control of air pollution is normally required in many ports in the world. Prior to using an incinerator while in port, permission may be required from the port authority concerned. In general, the use of shipboard garbage incinerators in ports in or near urban areas should be discouraged as their use will add to possible air pollution in these areas. Special considerations for incinerators are listed below:

5.4.1   Table 4 presents options for incineration of garbage, including considerations for special handling by vessel personnel, combustibility, reduction of volume, residual materials, exhaust, and on-board storage space. Most garbage is amenable to incineration with the exception of metal and glass.

5.4.2   In contrast to land-based incinerators, shipboard incinerators must be as compact as practicable, and with operating personnel at a premium, automatic operation is desirable. Most shipboard incinerators are designed for intermittent operation: the waste is charged to the incinerator, firing is started, and combustion typically lasts for three to six hours.

5.4.3   Commercial marine incinerators currently available vary greatly in size, have natural or induced draught, and are hand fired. It should be noted that incinerator ratings are usually quoted on the basis of heat input rate rather than on a weight charged basis because of the variability of the heat content in the wastes. Some modern incinerators are designed for continuous firing, and can handle simultaneous disposal of nearly all shipboard waste.

5.4.4   Some of the advantages of the most advanced incinerators may include that they operate under negative pressure, they are highly reliable since they have few moving parts, they require minimal operator skill, they are low in weight, and they have low exhaust and external skin temperatures.

5.4.5   Some of the disadvantages of incinerators may include the possible hazardous nature of the ash or vapour, dirty operation, excessive labour required for charging, stoking and ash removal, and they may not meet air pollution regulations imposed in certain harbours. Some of these disadvantages can be remedied by automatic equipment for charging, stoking and ash discharge into the sea outside areas where such discharge is prohibited. The additional equipment to perform these automatic functions requires more installation space.

5.4.6   The incineration of predominantly plastic wastes, as might be considered under some circumstances in complying with Annex V, requires more air and much higher temperatures for complete destruction. If plastics are to be burnt in a safe manner, the incinerator should be suitable for the purpose, otherwise the following problems can result:

Table 4 — Incineration* options for shipboard-generated garbage

Typical examples	Special handling by vessel personnel before incineration	Incineration characteristics				On-board storage space
		Combust-ibility	Reduction of volume	Residual	Exhaust	
Paper packaging, food and beverage containers	Minor – easy to feed into hopper	High	Over 95%	Powder ash	Possibly smoky and not hazardous	Minimum
Fibre and paper board	Minor – reduce material to size for feed; minimum manual labour	High	Over 95%	Powder ash	Possibly smoky and not hazardous	Minimum
Plastic packaging, food and beverage containers, etc.	Minor – easy to feed into hopper	High	Over 95%	Powder ash	Possibly smoky and hazardous based on incineration design	Minimum
Plastic sheeting, netting, rope and bulk material	Moderate manual labour time for size reduction	High	Over 95%	Powder ash	Possibly smoky and hazardous based on incinerator design	Minimum
Rubber hoses and bulk pieces	Major manual labour time for size reduction	High	Over 95%	Powder ash	Possibly smoky and hazardous based on incinerator design	Minimum
Metal food and beverage containers, etc.	Minor – easy to feed into hopper	Low	Less 10%	Slag	Possibly smoky and not hazardous	Moderate
Metal cargo, bulky containers, thick metal items	Major manual labour time for size reduction (not easily incinerated)	Very low	Less 5%	Large metal fragments and slag	Possibly smoky and not hazardous	Maximum
Glass food and beverage containers, etc.	Minor – easy to feed into hopper	Low	Less 10%	Slag	Possibly smoky and not hazardous	Moderate
Wood, cargo containers and large wood scraps	Moderate manual labour time for size reduction	High	Over 95%	Powder ash	Possibly smoky and not hazardous	Minimum

* Check local rules for possible reductions.

5.4.6.1   Depending on the type of plastic and conditions of combustion, some toxic gases can be generated in the exhaust stream, including vaporized hydrochloric (HCl) and hydrocyanic (HCN) acids. These and other intermediary products of plastic combustion can be extremely dangerous.

5.4.6.2   The ash from the combustion of some plastic products may contain heavy metal or other residues which can be toxic and should therefore not be discharged into the sea. Such ashes should be retained on board, where possible, and discharged at port reception facilities.

5.4.6.3   The temperatures generated during incineration of primarily plastic wastes are high enough to possibly damage some garbage incinerators.

5.4.6.4   Plastic incineration requires three to ten times more combustion air than average municipal refuse. If the proper level of oxygen is not supplied, high levels of soot will be formed in the exhaust stream.

5.4.7   Certain ship classification societies have established requirements for the operation or construction of incinerators. The International Association of Classification Societies can provide information as to such requirements.

5.4.8   Information on the development and utilization of marine garbage incinerator systems for shipboard use should be forwarded to the Organization.

6   Port reception facilities for garbage

6.1   The methodology for determining the adequacy of a reception facility should be based on the needs of each type of ship, as well as the number and types of ships using the port. The size and location of a port should be considered in determining adequacy. Emphasis should also be made on calculating the quantities of garbage from ships which are not discharged to the sea in accordance with the provisions of regulations 3, 4 and 5 of Annex V.

6.2   It should be noted that, due to possibly existing different procedures for reception, port reception may require separation on board of:

6.2.1   food wastes (e.g. raw meat because of risk of animal diseases);

6.2.2   cargo-associated waste; and

6.2.3   domestic waste and maintenance waste.

6.3   *Estimates of quantities of garbage to be received*

6.3.1   Vessel, port and terminal operators should consider the following when determining quantities of garbage on a per ship basis:

6.3.1.1   type of garbage;

6.3.1.2   ship type and design;

6.3.1.3   ship operating route;

6.3.1.4   number of persons on board;

6.3.1.5   duration of voyage;

6.3.1.6   time spent in areas where discharge into the sea is prohibited or restricted; and

6.3.1.7   time spent in port.

6.3.2   Governments, in assessing the adequacy of reception facilities, should also consider the technological problems associated with the treatment and disposal of garbage received from ships. Although the establishment of waste management standards is not within the scope of the Convention, governments should take responsible actions within their national programmes to consider such standards.

6.3.2.1   The equipment for treatment and disposal of garbage is a significant factor in determining the adequacy of a reception facility. It not only provides a measure of the time required to complete the process, but it also is the primary means for ensuring that ultimate disposal of the garbage is environmentally safe.

6.3.2.2   Governments are urged to initiate, at the earliest opportunity, studies into the provision of reception facilities at ports in their respective countries. Governments should carry out the studies in close co-operation with port authorities and other local authorities responsible for garbage handling. Such studies should include information such as a port-by-port listing of available garbage reception facilities, the types of garbage they are equipped to handle (e.g. food wastes contaminated with foreign disease or pest organisms, large pieces of derelict fishing gear, or refuse and operational wastes only), their capacities and any special procedures required to use them. Governments should transmit the results of their studies to the Organization for inclusion in the Annex V library (see section 2.2).

6.3.2.3   While selecting the most appropriate type of reception facility for a particular port, consideration should be given to several alternative methods available. In this regard, floating plants for collection of garbage, such as barges or self-propelled ships, might be considered more effective in a particular location than land-based facilities.

6.3.3   The purpose of these guidelines will be attained if they can provide the necessary stimulus to governments to initiate, and continue studies of, reception facilities as well as treatment and disposal technology. Information on developments in this respect should be forwarded to the Organization.

7   Ensuring compliance with Annex V

Recognizing that direct enforcement of Annex V regulations, particularly at sea, is difficult to accomplish, governments are encouraged to consider not only restrictive and punitive measures but also the removal of any disincentives, creation of positive incentives, and the development of voluntary measures within the regulated community when developing programmes and domestic legislation to ensure compliance with Annex V.

7.1 *Enforcement*

7.1.1   Governments should encourage their flag vessels to advise them of ports in foreign countries Party to Annex V which do not have port reception facilities for garbage. This will provide a basis for advising responsible governments of possible problems and calling the Organization's attention to possible infractions. An acceptable reporting format is reproduced in the attached appendix.

7.1.2   Governments should establish a documentation system (e.g. letters or certificates) for ports and terminals under its jurisdiction, stating that adequate facilities are available for receiving ship-generated garbage. Periodic inspection of the reception facilities is recommended.

7.1.3   Governments should identify appropriate enforcement agencies, providing legal authority, adequate training, funding and equipment to incorporate the enforcement of Annex V regulations into their responsibilities. In those cases where customs or agricultural officials are responsible for receiving and inspecting garbage, governments should ensure that the necessary inspections are facilitated as much as possible.

7.1.4   Governments should consider, where applicable, the use of garbage discharge reporting systems (e.g. existing ship's deck log-book or record book) for ships. Such logs, at a minimum, should document the date, time, location by latitude and longitude, or name of port, type of garbage (e.g. food, refuse, cargo-associated waste or maintenance waste) and estimated amount of garbage discharged. Particular attention should be given to the reporting of:

7.1.4.1   the loss of fishing gear;

7.1.4.2   the discharge of cargo residues;

7.1.4.3   any discharge in special areas;

7.1.4.4   discharge at port reception facilities; and

7.1.4.5   discharge of garbage at sea.

7.1.5   The issue of documents or receipts by port reception facilities might also assist the reporting system.

7.2 *Compliance incentive systems*

7.2.1   The augmentation of port reception facilities to serve ship traffic without undue delay or inconvenience may require capital investment from port and terminal operators as well as the waste management companies serving those ports. Governments are encouraged to evaluate means within their authority to lessen this impact, thereby helping to ensure that garbage delivered to port is actually received and disposed of properly at reasonable cost or without charging special fees to individual ships. Such means include, but are not limited to:

7.2.1.1   tax incentives;

7.2.1.2   loan guarantees;

7.2.1.3   public vessel business preference;

7.2.1.4   special funds to assist in problem situations such as remote ports with no land-based waste management system in which to deliver ships' garbage;

7.2.1.5   government subsidies; and

7.2.1.6   special funds to help defray the cost of a bounty programme for lost, abandoned or discarded fishing gear or other persistent garbage. The programme would make appropriate payments to persons who retrieve such fishing gear, or other persistent garbage other than their own, from marine waters under the jurisdiction of government.

7.2.2   The installation of shipboard garbage processing equipment would facilitate compliance with Annex V and lessen the burden on port reception facilities to process garbage for disposal. Therefore, governments should consider actions to encourage certain types of garbage processing equipment to be installed on ships operating under its flag. For example, programmes to lessen costs to shipowners for purchasing and installing such equipment, or requirements for installing compactors, incinerators and comminuters during construction of new ships would be very helpful.

7.2.3   Governments are encouraged to consider the economic impacts of domestic regulations intended to force compliance with Annex V. Unrealistic regulations may lead to higher levels of non-compliance than an education programme without specific regulatory requirements beyond Annex V itself. Due to the highly variable nature of ship operations and configurations, it seems appropriate to maintain the highest possible level of flexibility in domestic regulations to permit ships the greatest range of options for complying with Annex V.

7.2.4   Governments are encouraged to support research and development of technology that will simplify compliance with Annex V regulations for ships and ports. This research should concentrate on:

7.2.4.1   shipboard waste handling systems;

7.2.4.2   ship provision innovations to minimize garbage generation;

7.2.4.3   loading and unloading technology to minimize dunnage, spillage and cargo residues; and

7.2.4.4   new ship construction design to facilitate garbage management and transfer.

7.2.5   Governments are encouraged to work within the Organization to develop port reception systems that simplify the transfer of garbage for international vessels.

7.3   *Voluntary measures*

7.3.1   Governments are encouraged to assist ship operators and seafarers' organizations in developing resolutions, by-laws and other internal mechanisms that will encourage compliance with Annex V regulations. Some of these groups include:

7.3.1.1   seamen's and officers' unions;

7.3.1.2   associations of shipowners and insurers, and classification societies; and

7.3.1.3   pilot associations, fishermen's organizations.

7.3.2   Governments are encouraged to assist and support, where possible, the development of internal systems to promote compliance with Annex V in port authorities and associations, terminal operators' organizations, stevedores' and longshoremen's unions and land-based waste management authorities.

Appendix

Form for reporting alleged inadequacy of port reception facilities for garbage

1. Country...
    Name of port or area..
    Location in the port (e.g. berth/terminal/jetty) ....
    Date of incident...

2. Type and amount of garbage for discharge to facility:

    a. Total amount:
        food waste ....m^3
        cargo-associated waste ....m^3
        maintenance waste ....m^3
        other ....m^3

    b. Amount not accepted by the facility:
        food waste ....m^3
        cargo-associated waste ....m^3
        maintenance waste ....m^3
        other ....m^3

3. Special problems encountered:

    Undue delay
    Inconvenient location of facilities
    Unreasonable charges for use of facilities
    Use of facility not technically possible
    Special national regulations
    Other

4. Remarks: (e.g. information received from port authorities or operators
             of reception facilities: reasons given concerning 2 above)

5. Ships's particulars ....
    Name of ship ....
    Owner or operator ....
    Distinctive number or letters.................................
    Port of registry ....
    Number of persons on board ....

....
Date of completion of form   *Signature of master*

Resolution MEPC.59(33)
(adopted on 30 October 1992)

Revised Guidelines for the Implementation of Annex V of MARPOL 73/78

THE MARINE ENVIRONMENT PROTECTION COMMITTEE,

RECALLING Article 38(c) of the Convention on the International Maritime Organization concerning the function of the Marine Environment Protection Committee,

RECOGNIZING that Annex V of the International Convention for the Prevention from Ships, 1973, as modified by the Protocol of 1978 relating thereto (MARPOL 73/78), provides regulations for the prevention of pollution by garbage from ships,

RECOGNIZING ALSO the necessity of providing guidelines to assist Governments in developing and enacting domestic laws and regulations which give effect to and implement Annex V of MARPOL 73/78,

NOTING that part IV (Garbage) of the Guidelines on the Provision of Adequate Reception Facilities developed by the Organization in 1978,

BEING AWARE that the Committee at its twenty-sixth session modified the above-mentioned guidelines for garbage and developed guidelines for the implementation of Annex V of MARPOL 73/78 which were incorporated in the publication MARPOL 73/78, Consolidated Edition, 1991,

BEING ALSO AWARE that the Assembly at its seventeenth session adopted resolution A.719(17) on prevention of air pollution from ships, and requested the Committee and the Maritime Safety Committee to develop environmentally based standards for incineration of garbage and other ship-generated waste,

HAVING CONSIDERED the recommendations of the Sub-Committee on Ship Design and Equipment at its thirty-fifth session and the Sub-Committee on Bulk Chemicals at its twenty-second session regarding the standard specification for shipboard incinerators,

1.   ADOPTS the Revised Guidelines for the Implementation of Annex V of MARPOL 73/78; and

2.   RECOMMENDS Governments to implement the provisions of Annex V of MARPOL 73/78 in accordance with the revised guidelines.

Annex

Revised Guidelines for the Implementation of Annex V of MARPOL 73/78

The guidelines contained in *MARPOL 73/78, Consolidated Edition, 1991,* are amended as set out hereunder.

Replace paragraph 5.4.7 with the following:

"5.4.7   Shipboard incinerators should be designed, constructed, operated and maintained in accordance with the Standard Specification for Shipboard Incinerators set out in appendix 2."

2   Re-number the present appendix as appendix 1.

3   Add the following as appendix 2:

Appendix 2
Standard specification for shipboard incinerators

1   *Scope*

1.1   This specification covers the design, manufacture, performance, operation, functioning, and testing of incinerators intended to incinerate garbage and other shipboard wastes generated during the ship's normal service (i.e. maintenance, operational, domestic and cargo-associated wastes, excluding cargo-associated wastes contaminated with Annex II and III substances as defined in the International Convention for the Prevention of Pollution from Ships, 1973, as modified by the Protocol of 1978 (MARPOL 73/78)).

1.2   This specification applies to those incinerator plants with capacities up to 1,160 kW per unit.

1.3   This specification does not apply to systems on special incinerator ships, e.g. for burning industrial wastes such as chemicals, manufacturing residues, etc.

1.4   This specification does not address the electrical supply to the unit, nor the foundation connections and stack connections.

1.5   This specification provides emission requirements in annex A1 and fire protection requirements in annex A2. Provisions for incinerators integrated with heat recovery units and provisions for flue gas temperature are given in annex A3 and annex A4, respectively.

1.6   This standard may involve hazardous materials, operations and equipment. This standard does not purport to address all of the safety problems associated with its use. It is the responsibility of the user of this standard to establish appropriate safety and health practices and determine the applicability of regulatory limitations prior to use, including possible port State limitations.

2   *Definitions*

2.1   *Ship* means a vessel of any type whatsoever operating in the marine environment and includes hydrofoil boats, air-cushioned vehicles, submersibles, floating craft and fixed or floating platforms.

2.2   *Incinerator* means shipboard facilities for incinerating solid wastes approximating in composition to household waste and liquid wastes arising from the operation of the ship, e.g. domestic waste, cargo-associated waste, maintenance waste, operational waste, cargo residues, and fishing gear, etc. These facilities may be designed to use or not to use the heat energy produced.

2.3   *Garbage* means all kinds of victual, domestic and operational waste excluding fresh fish and parts thereof, generated during normal operation of the ship as defined in Annex V to MARPOL 73/78.

2.4   *Waste* means useless, unneeded or superfluous matter which is to be discarded.

2.5   *Food wastes* are any spoiled or unspoiled victual substances, such as fruits, vegetables, dairy products, poultry, meat products, food scraps, food particles, and all other materials contaminated by such wastes, generated aboard ship, principally in the galley and dining areas.

2.6   *Plastic* means a solid material which contains as an essential ingredient one or more synthetic organic high polymers and which is formed (shaped) during either manufacture of the polymer or the fabrication into a finished product by heat and/or pressure. Plastics havematerial properties ranging from hard and brittle to soft and elastic. Plastics are used for a variety of marine purposes including, but not limited to, packaging (vapour-proof barriers, bottles, containers, liners), ship construction (fibreglass and laminated structures, siding, piping, insulation, flooring, carpets, fabrics, paints and finishes, adhesives, electrical and electronic components), disposable eating utensils and cups, bags, sheeting, floats, fishing nets, strapping bands, rope and line.

2.7   *Domestic waste* means all types of food wastes, sewage and wastes generated in the living spaces on board the ship.

2.8   *Cargo-associated* waste means all materials which have become wastes as a result of use on board a ship for cargo stowage and handling. Cargo-associated waste includes but is not limited to dunnage, shoring pallets, lining and packing materials, plywood, paper, cardboard, wire and steel strapping.

2.9   *Maintenance waste* means materials collected by the engine department and the deck department while maintaining and operating the vessel, such as soot, machinery deposits, scraped paint, deck sweeping, wiping wastes, oily rags, etc.

2.10   *Operational wastes* means all cargo-associated wastes and maintenance waste (including ash and clinkers), and cargo residues defined as garbage in 2.13.

2.11   *Sludge oil* means sludge from fuel and lubricating oil separators, waste lubricating oil from main and auxiliary machinery, waste oil from bilge water separators, drip trays, etc.

2.12   *Oily rags* are rags which have been saturated with oil as controlled in Annex I to the Convention. Contaminated rags are rags which have been saturated with a substance defined as a harmful substance in the other Annexes to MARPOL 73/78.

2.13   *Cargo residues* for the purposes of this standard are defined as the remnants of any cargo material on board that cannot be placed in proper cargo holds (loading excess and spillage) or which remains in cargo holds and elsewhere after unloading procedures are completed (unloading residual and spillage). However, cargo residues are expected to be in small quantities.

2.14   *Fishing gear* is defined as any physical device or part thereof or combination of items that may be placed on or in the water with the intended purpose of capturing, or controlling for subsequent capture, living marine or freshwater organisms.

3   *Materials and manufacture*

3.1   The materials used in the individual parts of the incinerator are to be suitable for the intended application with respect to heat resistance, mechanical properties, oxidation, corrosion, etc., as in other auxiliary marine equipment.

3.2   Piping for fuel and sludge oil should be seamless steel of adequate strength and to the satisfaction of the Administration. Short lengths of steel, or annealed copper nickel, nickel copper, or copper pipe and tubing may be used at the burners. The use of nonmetallic materials for fuel lines is prohibited. Valves and fittings may be threaded in sizes up to and including 60 mm OD (outside diameter), but threaded unions are not to be used on pressure lines in sizes 33 mm OD and over.

3.3   All rotating or moving mechanical and exposed electrical parts should be protected against accidental contact.

3.4   Incinerator walls are to be protected with insulated fire bricks/ refractory and a cooling system. The outside surface temperature of the incinerator casing being touched during normal operations should not exceed 20°C above ambient temperature.

3.5   Refractory should be resistant to thermal shocks and resistant to normal ship's vibration. The refractory design temperature should be equal to the combustion chamber design temperature plus 20%. (See 4.1).

3.6   Incinerating systems should be designed such that corrosion will be minimized on the inside of the systems.

3.7   In systems equipped for incinerating liquid wastes, safe ignition and maintenance of combustion must be ensured, e.g. by a supplementary burner.

3.8   The combustion chamber(s) should be designed for easy maintenance of all internal parts including the refractory and insulation.

3.9   The combustion process should take place under negative pressure, which means that the pressure in the furnace under all circumstances should be lower than the ambient pressure in the room where the incinerator is installed. A flue gas fan may be fitted to secure negative pressure.

3.10   The incinerating furnace may be charged with solid waste either by hand or automatically. In every case, fire dangers should be avoided and charging should be possible without danger to the operating personnel.

For instance, where charging is carried out by hand, a charging lock may be provided which ensures that the charging space is isolated from the fire box as long as the filling hatch is open.

Where charging is not effected through a charging lock, an interlock should be installed to prevent the charging door from opening while the incinerator is in operation with burning of garbage in progress or while the furnace temperature is above 220°C.

3.11   Incinerators equipped with a feeding sluice or system should ensure that the material charged will move to the combustion chamber. Such system should be designed such that both operator and environment are protected from hazardous exposure.

3.12   Interlocks should be installed to prevent ash removal doors from opening while burning is in progress or while the furnace temperature is above 220°C.

3.13   The incinerator should be provided with a safe observation port of the combustion chamber in order to provide visual control of the burning process and waste accumulation in the combustion chamber. Neither heat, flame nor particles should be able to pass through the observation port. An example of a safe observation port is high-temperature glass with a metal closure.

3.14   *Electrical requirements*

3.14.1 International Electrotechnical Commission (IEC) standards, particularly IEC Publication 92, *Electrical Installations in Ships and Mobile and Fixed Offshore Units*, are applicable for this equipment.

3.14.2 Electrical installation requirements should apply to all electrical equipment, including controls, safety devices, cables, and burners and incinerators.

3.14.2.1 A disconnecting means capable of being locked in the open position should be installed at an accessible location at the incinerator so that the incinerator can be disconnected from all sources of potential energy. This disconnecting means should be an integral part of the incinerator or adjacent to it. (See 5.1).

3.14.2.2 All uninsulated live metal parts should be guarded to avoid accidental contact.

3.14.2.3 The electrical equipment should be so arranged that failure of this equipment will cause the fuel supply to be shut off.

3.14.2.4 All electrical contacts of every safety device installed in the control circuit should be electrically connected in series. However, special consideration should be given to arrangements when certain devices are wired in parallel.

3.14.2.5 All electrical components and devices should have a voltage rating commensurate with the supply voltage of the control system.

3.14.2.6 All electrical devices and electric equipment exposed to the weather should be according to IEC Publication 92-201, table V.

3.14.2.7 All electrical and mechanical control devices should be of a type tested and accepted by a nationally recognized testing agency, according to international standards.

3.14.2.8 The design of the control circuits should be such that limit and primary safety controls should directly open a circuit that functions to interrupt the supply of fuel to combustion units.

3.14.3 Overcurrent protection

3.14.3.1 Conductors for interconnecting wiring that is smaller than the supply conductors should be provided with overcurrent protection based on the size of the smallest interconnecting conductors external to any control box, according to IEC rules.

3.14.3.2 Overcurrent protection for interconnecting wiring should be located at the point where the smaller conductors connect to the larger conductors. However, overall overcurrent protection is acceptable if it is sized on the basis of the smallest conductors of the interconnecting wiring, or according to IEC requirements.

3.14.3.3   Overcurrent protection devices should be accessible and their function should be identified.

3.14.4   Motors

3.14.4.1   All electric motors should have enclosures corresponding to the environment where they are located, at least IP 44 according to IEC Publication 529.

3.14.4.2   Motors should be provided with a corrosion-resistant nameplate specifying information in accordance with IEC Publication 92-301.

3.14.4.3   Motors should be provided with running protection by means of integral thermal protection, by overcurrent devices, or a combination of both, in accordance with manufacturer's instructions, which should be in accordance with IEC Publication 92-202.

3.14.4.4   Motors should be rated for continuous duty and should be designed for an ambient temperature of 45°C or higher.

3.14.4.5   All motors should be provided with terminal leads or terminal screws in terminal boxes integral with, or secured to, the motor frames.

3.14.5   Ignition system

3.14.5.1   When automatic electric ignition is provided, it should be accomplished by means of either a high-voltage electric spark, a high-energy electric spark or a glow coil.

3.14.5.2   Ignition transformers should have an enclosure corresponding to the environment where they are located, at least IP 44 according to IEC Publication 529.

3.14.5.3   Ignition cable should conform to the requirements of IEC Publication 92-503.

3.14.6   Wiring

3.14.6.1   All wiring for incinerators should be rated and selected in accordance with IEC Publication 92-352.

3.14.7   Bonding and grounding

3.14.7.1   Means should be provided for grounding the major metallic frame or assembly of the incinerators.

3.14.7.2   Non-current-carrying enclosures, frames and similar parts of all electrical components and devices should be bonded to the main frame or assembly of the incinerator. Electrical components that are bonded by their installation do not require a separate bonding conductor.

3.14.7.3 When an insulated conductor is used to bond electrical components and devices, it should show a continuous green colour, with or without a yellow stripe.

4   *Operating requirements*

4.1   The incinerator system should be designed and constructed for operation with the following conditions:

Maximum combustion chamber flue gas outlet temperature	1,200°C
Minimum combustion chamber flue gas outlet temperature	850°C
Pre-heat temperature of combustion chamber	650°C

For batch-loaded incinerators, there are no pre-heating requirements. However, the incinerator should be so designed that the temperature in the actual combustion space reaches 600°C within 5 min after start.

Pre-purge, before ignition:	at least four air changes in the chamber(s) and stack, but not less than 15 s
Time between restarts:	at least four air changes in the chamber(s) and stack, but not less than 15 s
Post-purge, after shutoff fuel oil:	not less than 15 s after the closing of the fuel oil valve
Incinerator discharge gases:	Minimum 6% O_2.

4.2   Outside surfaces of combustion chamber(s) should be shielded from contact such that people in normal work situations are not exposed to extreme heat (20°C above ambient temperature) or direct contact with surface temperatures exceeding 60°C. Examples of alternatives to accomplish this are a double jacket with an air flow in between or an expanded metal jacket.

4.3   Incinerating systems are to be operated with underpressure (negative pressure) in the combustion chamber such that no gases or smoke can leak out to the surrounding areas.

4.4   The incinerator should have warning plates attached in a prominent location on the unit, warning against unauthorized opening of doors to combustion chamber(s) during operation and against overloading the incinerator with garbage.

4.5   The incinerator should have instruction plate(s) attached in a prominent location on the unit that clearly addresses the following:

4.5.1   Cleaning ashes and slag from the combustion chamber(s) and cleaning of combustion air openings before starting the incinerator (where applicable).

4.5.2   Operating procedures and instructions. These should include proper start-up procedures, normal shutdown procedures, emergency shutdown procedures and procedures for loading garbage (where applicable).

4.6   To avoid the building up of dioxins, the flue gas should be shock-cooled to a maximum 350°C right after the incinerator.

5   Operating controls

5.1   The entire unit should be capable of being disconnected from all sources of electricity by means of one disconnect switch located near the incinerator. (See 3.14.2.1)

5.2   There should be an emergency stop switch located outside the compartment, which stops all power to the equipment. The emergency stop switch should also be able to stop all power to the fuel pumps. If the incinerator is equipped with a flue gas fan, the fan should be capable of being restarted independently of the other equipment on the incinerator.

5.3   The control equipment should be so designed that any failure of the following equipment will prevent continued operations and cause the fuel supply to be cut off.

5.3.1   Safety thermostat/draught failure

5.3.1.1 A flue gas temperature controller, with a sensor placed in the flue gas duct, should be provided that will shut down the burner if the flue gas temperature exceeds the temperature set by the manufacturer for the specific design.

5.3.1.2 A combustion temperature controller, with a sensor placed in the combustion chamber, should be provided that will shut down the burner if the combustion chamber temperature exceeds the maximum temperature.

5.3.1.3 A negative pressure switch should be provided to monitor the draught and the negative pressure in the combustion chamber. The purpose of this negative pressure switch is to ensure that there is sufficient draught/negative pressure in the incinerator during operations. The circuit to the program relay for the burner will be opened and an alarm activated before the negative pressure rises to atmospheric pressure.

5.3.2   Flame failure/fuel oil pressure

5.3.2.1 The incinerator should have a flame safeguard control consisting of a flame sensing element and associated equipment for shutdown of the unit in the event of ignition failure and flame failure during the firing cycle. The flame safeguard control should be so designed that the failure of any component will cause a safety shutdown.

5.3.2.2 The flame safeguard control should be capable of closing the fuel valves in not more than 4 s after a flame failure.

5.3.2.3 The flame safeguard control should provide a trial-for-ignition period of not more that 10 s during which fuel may be supplied to establish flame. If flame is not established within 10 s, the fuel supply to the burners should be immediately shut off automatically.

5.3.2.4 Whenever the flame safeguard control has operated because of failure of ignition, flame failure or failure of any component, only one automatic restart may be provided. If this is not successful then manual reset of the flame safeguard control should be required for restart.

5.3.2.5 Flame safeguard controls of the thermostatic type, such as stack switches and pyrostats operated by means of an open bimetallic helix, are prohibited.

5.3.2.6 If fuel oil pressure drops below that set by the manufacturer, a failure and lockout of the program relay should result. This also applies to sludge oil used as a fuel. (Applies where pressure is important for the combustion process or a pump is not an integral part of the burner.)

5.3.3   Loss of power

If there is a loss of power to the incinerator control/alarm panel (not remote alarm panel), the system should shut down.

5.4   Fuel supply

Two fuel control solenoid valves should be provided in series in the fuel supply line to each burner. On multiple burner units, a valve on the main fuel supply line and a valve at each burner will satisfy this requirement. The valves should be connected electrically in parallel so that both operate simultaneously.

5.5   Alarms

5.5.1   An outlet for an audible alarm should be provided for connection to a local alarm system or a central alarm system. When a failure occurs, a visible indicator should show what caused the failure. (The indicator may cover more than one fault condition.)

5.5.2   The visible indicators should be designed so that, where failure is a safety-related shutdown, manual reset is required.

5.6   After shutdown of the oil burner, provision should be made for the fire box to cool sufficiently. (As an example of how this may be accomplished, the exhaust fan or ejector could be designed to continue to operate. This would not apply in the case of an emergency manual trip.)

6   Other requirements

6.1   Documentation

A complete instruction and maintenance manual with drawings, electric diagrams, spare parts list, etc., should be furnished with each incinerator.

6.2   Installation

All devices and components should, as fitted in the ship, be designed to operate when the ship is upright and when inclined at any angle of list up to and including 15° either way under static conditions and 22.5° either way under dynamic conditions (rolling) and simultaneously inclined dynamically (pitching) 7.5° by bow or stern.

6.3   Incinerator

6.3.1   Incinerators are to be fitted with an energy source with sufficient energy to ensure a safe ignition and complete combustion. The combustion is to take place at sufficient negative pressure in the combustion chamber(s) to ensure no gases or smoke leak out to the surrounding areas. (See 5.3.1.3)

6.3.2   A drip tray is to be fitted under each burner and under any pumps, strainers, etc., that require occasional examination.

7   Tests

7.1   Prototype tests

An operating test for the prototype of each design should be conducted, with a test report completed indicating results of all tests. The tests should be conducted to ensure that all of the control components have been properly installed and that all parts of the incinerator, including controls and safety devices, are in satisfactory operating condition. Tests should include those described in section 7.3 below.

7.2   Factory tests

For each unit, if preassembled, an operating test should be conducted to ensure that all of the control components have been properly installed and that all parts of the incinerator, including controls and safety devices, are in satisfactory operating condition. Tests should include those described in 7.3 below.

7.3   Installation tests

An operating test after installation should be conducted to ensure that all of the control components have been properly installed and that all parts of the incinerator, including controls and safety devices, are in satisfactory operating condition.

7.3.1 *Flame safeguard.* The operation of the flame safeguard system should be verified by causing flame and ignition failures. Operation of the audible alarm (where applicable) and visible indicator should be verified. The shutdown times should be verified.

7.3.2 *Limit controls.* Shutdown due to the operation of the limit controls should be verified.

7.3.2.1 *Oil pressure limit control.* The lowering of the fuel oil pressure below the value required for safe combustion should initiate a safety shutdown.

7.3.2.2 *Other interlocks.* Other interlocks provided should be tested for proper operation as specified by the unit manufacturer.

7.3.3 *Combustion controls.* The combustion controls should be stable and operate smoothly.

7.3.4 *Programming controls.* Programming controls should be verified as controlling and cycling the unit in the intended manner. Proper pre-àpurge, ignition, post-purge and modulation should be verified. A stopwatch should be used for verifying intervals of time.

7.3.5 *Fuel supply controls.* The satisfactory operation of the two fuel control solenoid valves for all conditions of operation and shutdown should be verified.

7.3.6 *Low voltage test.* A low voltage test should be conducted to satisfactorily demonstrate that the fuel supply to the burners will be automatically shut off before an incinerator malfunction results from the reduced voltage.

7.3.7 *Switches.* All switches should be tested to verify proper operation.

8 *Certification*

8.1 Manufacturer's certification that an incinerator has been constructed in accordance with this standard should be provided (by letter or certificate or in the instruction manual).

9 *Marking*

9.1 Each incinerator should be permanently marked indicating:

9.1.1 Manufacturer's name or trademark.

9.1.2 Style, type, model or other manufacturer's designation for the incinerator.

9.1.3 Capacity – to be indicated by net designed heat release of the incinerator in heat units per timed period; for example, British Thermal Units per hour, megajoules per hour, kilocalories per hour.

10   *Quality assurance*

10.1 Incinerators should be designed, manufactured and tested in a manner that ensures they meet the requirements of this standard.

10.2 The incinerator manufacturer should have a quality system that meets ISO 9001, "Quality Systems – Model for Quality Assurance in Design/Development, Production, Installation and Servicing". The quality system should consist of elements necessary to ensure that the incinerators are designed, tested and marked in accordance with this standard. At no time should an incinerator be sold with this standard designation that does not meet the requirements herein (see "Certification").

Annex

A1 – *Emission standard for shipboard incinerators with capacities of up to 1,160 kW*

Minimum information to be provided

A1.1 An IMO Type Approval Certificate should be required for each shipboard incinerator. In order to obtain such certificate, the incinerator should be designed and built to an IMO approved standard. Each model should go through a specified type approval test operation at the factory or an approved test facility, and under the responsibility of the Administration.

A1.2 Type approval test should include measuring of the following parameters:

Max. capacity	kW or kcal/h kg/h of specified waste kg/h per burner
Pilot fuel consumption	kg/h per burner
O_2 average in combustion chamber/zone	%
CO average in flue gas	mg/MJ
Soot number average	Bacharach or Ringelman scale
Combustion chamber flue gas outlet temperature average	°C
Amount of unburned components in ashes	% by weight

A1.3   *Duration of test operation*

For sludge oil burning	6–8 hours
For solid waste burning	6–8 hours

A1.4   *Fuel/Waste specification for type approval test (% by weight)*

Sludge oil consisting of	75% sludge oil from heavy fuel oil 5% waste lubricating oil 20% emulsified water
Solid waste (class 2) consisting of	50% food waste 50% rubbish containing approx. 30% paper, approx. 40% cardboard, approx. 10% rags, approx. 20% plastic The mixture will have up to 50% moisture and 7% incombustible solids

*Classes of waste**

Class 0   Trash, a mixture of highly combustible waste such as paper, cardboard, wood boxes, and combustible floor sweepings, with up to 10% by weight of plastic bags, coated paper, laminated paper, treated corrugated cardboard, oily rags and plastic or rubber scraps. This type of waste contains up to 10% moisture, 5% incombustible solids and has a heating value of about 19,700 kJ/kg as fired.

Class 1   Rubbish, a mixture of combustible waste such as paper, cardboard cartons, wood scrap, foliage and combustible floor sweepings. The mixture contains up to 20% by weight of galley or cafeteria waste, but contains little or no treated papers, plastic or rubber wastes. This type of waste contains 25% moisture, 10% incombustible solids and has a heating value of about 15,100 kJ/kg as fired.

Class 2   Refuse, consisting of an approximately even mixture of rubbish and garbage by weight. This type of waste, common to passenger ship occupancy, consists of up to 50% moisture, 7% incombustible solids and has a heating value of about 10,000 kJ/kg as fired.

Class 3   Garbage, consisting of animal and vegetable wastes from restaurants, cafeterias, galleys, sick bays and like installations. This type of waste contains up to 70% moisture, up to 5% incombustible solids and has a heating value range of about 2,300 kJ/kg as fired.

---

* Reference: *Waste Classification*, Incinerator Institute of America.

Class 4　　　Aquatic life forms and animal remains, consisting of carcasses, organs and solid organic wastes from vessels carrying animal-type cargoes, consisting of up to 85% moisture, 5% incombustible solids and having a heating value range of about 2,300 kJ/kg as fired.

Class 5　　　By-product waste, liquid or semi-liquid, such as tar, paints, solvents, sludge, oil, waste oil, etc., from shipboard operations. BTU values must be determined by the individual materials to be destroyed.

Class 6　　　Solid by-product waste, such as rubber, plastics, wood waste, etc., from industrial operations. BTU values must be determined by the individual materials to be destroyed.

Calorific values	kcal/kg	kJ/kg
Vegetable and putrescibles	1,360	5,700
Paper	3,415	14,300
Rag	3,700	15,500
Plastics	8,600	36,000
Oil sludge	8,600	36,000
Sewage sludge	716	3,000

Densities	kg/m^3
Paper (loose)	50
Refuse (75% wet)	720
Dry rubbish	110
Scrap wood	190
Wood sawdust	220

Density of loose general waste generated on board ship will be about 130 kg/m^3.

A1.5　*Required emission standards to be verified by type approval test*

O_2 in combustion chamber　　　6 –12%

CO in flue gas maximum average　　200 mg/MJ

Soot number maximum average　　Bacharach 3 or Ringelman 1 (a higher soot number is acceptable only during very short periods such as starting up)

Unburned components in ash residues　　max. 10% by weight

Combustion chamber flue gas outlet temperature range　　900 –1,200°C

A high temperature in the actual combustion chamber/zone is an absolute requirement in order to obtain a *complete* and smoke-free incineration, including that of plastic and other synthetic materials while minimizing dioxin and VOC (volatile organic compounds) emissions.

A1.6 *Fuel-related emission*

A1.6.1 Even with good incineration technology the emission from an incinerator will depend on the type of material being incinerated. If for instance a vessel has bunkered a fuel with high sulphur content, then sludge oil from separators which is burned in the incinerator will lead to emission of SO_x. But again, the SO_x emission from the incinerator would only amount to less than one per cent of the SO_x discharged with the exhaust from main and auxiliary engines.

A1.6.2 Principal organic constituents (POC) cannot be measured on a continuous basis. Specifically, there are no instruments with provision for continuous time telemetry that measures POC, hydrogen chloride (HCl) or waste destruction efficiency to date. These measurements can only be made using grab sample approaches, where the sample is returned to a laboratory for analysis. In the case of organic constituents (undestroyed wastes), the laboratory work requires considerable time to complete. Thus, continuous emission control can only be assured by secondary measurements.

A1.6.3 *On-board operation/emission control*

For a shipboard incinerator with IMO type approval, emission control/ monitoring should be limited to the following:

.1 control/monitor O_2 content in combustion chamber (spot checks only);

.2 control/monitor temperature in combustion chamber flue gas outlet.

By continuous (auto) control of the incineration process, ensure that the above-mentioned two parameters are kept within the prescribed limits. This mode of operation will ensure that particulates and ash residue contain only traces of organic constituents.

A1.7 *Passenger/Cruise ships with incinerator installations having a total capacity of more than 1,160 kW*

A1.7.1 On board this type of vessel, the following conditions will probably exist:

.1 generation of huge amounts of burnable waste with a high content of plastic and synthetic materials;

.2 incinerating plant with a high capacity operating continuously over long periods;

.3 this type of vessel will often be operating in very sensitive coastal areas.

A1.7.2 In view of the fuel-related emission from a plant with such a high capacity, installation of a flue gas seawater scrubber should be considered. This installation can perform an efficient after-cleaning of the flue gases, thus minimizing the content of HCl, SO_x, particulate matter.

A1.7.3 Any restriction in nitrogen oxide (NO_x) should only be considered in connection with possible future regulations on pollution from the vessel's total pollution, i.e. main and auxiliary machinery, boilers, etc.

A2 – Fire protection requirements for incinerators and waste stowage spaces

For the purpose of construction, arrangement and insulation, incinerator spaces and waste stowage spaces should be treated as category A machinery spaces (SOLAS II-2/3.19) and service spaces (SOLAS II-2/3.12), respectively. To minimize the fire hazards these spaces represent, the following SOLAS requirements in chapter II-2 should be applied:

A2.1 For passenger vessels carrying more than 36 passengers:

   .1   regulation 26.2.2(12) should apply to incinerator and combined incinerator/waste storage spaces, and the flue uptakes from such spaces; and

   .2   regulation 26.2.2(13) should apply to waste storage spaces and garbage chutes connected thereto.

A2.2 For all other vessels, including passenger vessels carrying not more than 36 passengers:

   .1   regulation 44.2.2(6) should apply to incinerator and combined incinerator/waste spaces, and the flue uptakes from such spaces; and

   .2   regulation 44.2.2(9) should apply to waste storage spaces and garbage chutes connected thereto.

A2.3 Incinerators and waste stowage spaces located on weather decks (regulation II-2/3.17) need not meet the above requirements but should be located:

   .1   as far aft on the vessel as possible;

   .2   not less than 3 m from entrances, air inlets and openings to accommodations, service spaces and control stations;

   .3   not less than 5 m measured horizontally from the nearest hazardous area, or vent outlet from a hazardous area; and

   .4   not less than 2 m should separate the incinerator and the waste material storage area, unless physically separated by a structural fire barrier.

A2.4 A fixed fire detection and fire-extinguishing system should be installed in enclosed spaces containing incinerators, in combined incinerator/waste storage spaces and in any waste storage space in accordance with the following table:

	Automatic sprinkler system	Fixed fire-extinguishing system	Fixed fire detection system
Combined incinerator and waste storage space	X		
Incinerator space		X	X
Waste storage space	X		

A2.5 Where an incinerator or waste storage space is located on weather decks it must be accessible with two means of fire extinguishment: either fire hoses, semi-portable fire extinguishers, fire monitors or a combination of any two of these extinguishing devices. A fixed fire-extinguishing system is acceptable as one means of extinguishment.

A2.6 Flue uptake piping/ducting should be led independently to an appropriate terminus via a continuous funnel or trunk.

A3 – *Incinerators integrated with heat recovery units*

A3.1 The flue gas system, for incinerators where the flue gas is led through a heat recovery device, should be designed so that the incinerator can continue operation with the economizer coils dry. This may be accomplished with bypass dampers if needed.

A3.2 The incinerator unit should be equipped with a visual and an audible alarm in case of loss of feed-water.

A3.3 The gas side of the heat recovery device should have equipment for proper cleaning. Sufficient access should be provided for adequate inspection of external heating surfaces.

A4 – *Flue gas temperature*

A4.1 When deciding upon the type of incinerator, consideration should be given as to what the flue gas temperature will be. The flue gas temperature can be a determining factor in the selection of materials for fabricating the stack. Special high-temperature material may be required for use in fabricating the stack when the flue gas temperatures exceed 430°C.

Annex

**Form of IMO Type Approval Certificate
for shipboard incinerators with capacities
of up to 1,160 kW**

Certificate of Shipboard Incinerator

Name of Administration

Badge
or
Cypher

This is to certify that the shipboard incinerator listed has been examined
and tested in accordance with the requirement of the standard for
shipboard incinerators for disposing of ship-generated waste appended to
the Guidelines for the implementation of Annex V of MARPOL 73/78.

Incinerator manufactured by . . . .

Style, type or model of the incinerator* . . . .

Max. capacity . . . . kW or kcal/h
. . . . kg/h of specified waste
. . . . kg/h per burner

O_2 average in combustion chamber/zone . . . .%

CO average in flue gas . . . .
mg/MJ

Soot number average . . . . Bacharach or
Ringelman scale

Combustion chamber flue gas
outlet temperature average . . . . °C

Amount of unburned components
in ashes . . . .% by weight

A copy of this certificate should be carried on board a vessel fitted with
this equipment at all times.

Official stamp   Signed . . . .

Administration of
. . . .
Dated this . . . . . day of . . .   .

---

* Delete as appropriate.

APPENDIX

C

The International Law of the Sea: Implications for Annex V Implementation

By Miranda Wecker
Center for International Environmental Law
South Bend, Washington

International law governing the uses of the oceans provides the foundation for important U.S. authorities. International law underlies the right of U.S. commercial and military ships and aircraft to freely use the seas for navigation and overflight, supports the rights of scientific research vessels to conduct critical studies of oceanic and climatic processes, provides the basis for controlling pollution that would damage U.S. coastal environments, and underwrites dispute settlement procedures that resolve conflicts peacefully. But international legal principles also restrain the United States from unilaterally applying to foreign vessels standards as stringent as may be desired.

It is within this context of international law that the United States must fashion a strategy for protecting its waters and shorelines from marine debris and for promoting progress at the regional and global levels. To elucidate options available to the United States in advancing towards cleaner seas, it is important to understand both the opportunities offered by international law and the constraints on unilateral action that it imposes.

The earliest principles of international law embraced the notion of unimpeded and unrestricted use of the oceans. The freedom to transport goods was instrumental in the dramatic rise of international commerce and fundamental to stable relations among nations. Traditional high seas freedoms included the rights of navigation—both civil and military—and the right to freely take the sea's living and mineral resources. Throughout history, nations also have demanded mutual recognition of sovereign rights in waters adjacent to their soil (i.e., in territorial seas), in order to provide for national defense. Traditional claims to some more limited authorities also have been asserted in a broader band of sea, referred to as the contiguous zone.

The past several decades, however, have yielded an unprecedented level of conflict and competition among nations as the exploitation of ocean resources accelerated. Nations competed long distance for capture of fish and for access to minerals. They traded diplomatic protests over differences in pollution control standards and conservation of shared resources. These conflicts led to a call for a more detailed and definitive articulation of international principles of procedure, jurisdiction, and substantive obligations.

In the late 1960s, national leaders began the most complex and comprehensive treaty negotiation in world history, ultimately producing a universally agreed-upon set of rules governing uses of the oceans. After a full decade of conferences involving nearly 150 nations, the Third United Nations Convention on the Law of the Sea (UNCLOS III) was adopted in 1982. In addition to a multitude of different rules, UNCLOS III provides the ground rules for each nation's approach to controlling shipborne wastes, and, in particular, the extent to which another nation's right to establish its own approach must be respected.

Guyana's deposit of its ratification, on November 16, 1993, triggered the start of a one-year period following which UNCLOS III entered into force. Although most industrialized nations have not yet become parties to the treaty (pending imminent modification of the mining regime it establishes), there has long been near-universal acceptance of and support for its provisions relating to protection of the marine environment and the rights of navigation.

Regardless of whether the United States becomes a party to UNCLOS III, the convention, once it has entered into force, is expected to bolster significantly international efforts to protect the marine environment. This paper examines the implications of UNCLOS III for the control of shipborne wastes.

Entry into force will strengthen the legal stature of UNCLOS III's fundamental system of authorities and responsibilities for both coastal nations and nations with ships flying their flags. This system of clearly defined rights and responsibilities removes most ambiguities regarding the rights of port and coastal states to demand responsible conduct by foreign vessels and the right to take certain enforcement activities when a violation of rules occurs. The system also explicitly mandates environmentally responsible behavior by flag states with regard to regulation of their vessels.

The following provisions of UNCLOS III provide relevant authorities and obligations for the implementation of MARPOL Annex V. They serve to provide the United States with justification to control the discharge of marine debris and to demand that other nations act in accordance with their duties.

FLAG STATE RIGHTS AND RESPONSIBILITIES

A flag state has sovereign jurisdiction over vessels flying its flag[1]. Thus, under international law, the United States may require its flag vessels to comply with Annex V at all times, no matter where the vessels sail. In addition, pursuant to UNCLOS III, all flag states have a number of affirmative duties: to assure the compliance of their vessels with international standards (Article 217 (1)); to ensure that their vessels are seaworthy (Article 217 (2)); to periodically inspect and provide the requisite certificates to their ships (Article 217 (3)); to investigate all written complaints against their vessels and promptly institute proceedings where warranted (Article 217 (6)); to inform the relevant states and international organizations of enforcement proceedings (Article 217 (7)); and to provide sufficient penalties to discourage further violations (Article 217 (8)).

These general obligations reinforce the specific duties authorized by particular treaties such as MARPOL. A flag state also has the right to require that any legal actions against its vessels by other states be suspended and all records turned over, so that it can carry out the necessary legal remedies for pollution violations.[2]

COASTAL STATE AUTHORITIES

The rights and duties of coastal states varies in the different maritime zones recognized in UNCLOS III. In internal waters, the coastal state is recognized to be sovereign: It can place any condition on access to its ports, except in case of extreme emergency. Thus, the United States has the right to enforce Annex V with respect to any vessel that voluntarily enters its internal waters. Within its territorial sea, the coastal state has near-sovereign authority but lacks the right to hamper "innocent passage."[3] The United States, therefore, may take enforcement measures within its territorial waters, so long as they do not hamper innocent passage. Physical inspections within the territorial sea are authorized, but in order to avoid undue delay, most inspections take place in ports. In general, UNCLOS III affirms the rights of coastal states to adopt laws and regulations to

---

[1]Article 228 (3) affirms that in regard to vessels flying its flag, a State is not limited by UNCLOS III provisions governing the rights of coastal and port States. That is, the flag State can take any measures, including actions to impose penalties, irrespective of the prior proceedings of other States.

[2]Article 218 (4) addresses the suspension of actions brought by port States to allow flag State enforcement proceedings. Article 228 pertains to the suspension of proceedings by a coastal State for violations of international standards committed in the territorial sea.

[3]UNCLOS III defines certain acts as prejudicial to the peace, good order, or security of the coastal state and therefore not in keeping with the concept of "innocent passage." Among the actions considered prejudicial, listed in Article 19, are "(g) the loading or unloading of any commodity, currency, or person contrary to the customs, fiscal, immigration or sanitary laws and regulations of the coastal State" and "(h) any act of willful and serious pollution contrary to this Convention."

protect their shores, and to implement their customs, fiscal, immigration, or sanitary laws.

UNCLOS III provisions pertaining to the contiguous zone reaffirm the traditional principle that the coastal state has important interests but fewer authorities: Its jurisdiction is limited to the enforcement of fiscal, immigration, customs, and sanitary laws. However, under Part XII addressing obligations to protect the marine environment, UNCLOS III establishes the new jurisdictional principle that coastal states have a recognized interest in controlling pollution within 200 nautical miles of their shores. Under the concept of a 200-mile-wide Exclusive Economic Zone (EEZ), coastal states are allowed to claim control over the exploitation of living and non-living resources, as well as to exercise the following specific pollution-control rights and responsibilities.

If there are clear grounds to believe a violation took place in its territorial sea or EEZ, the coastal state has the right to seek information from the suspect vessel while it is in the EEZ. If the alleged violation resulted in a substantial discharge causing or threatening significant pollution of the marine environment, and the vessel refused to provide information, the coastal state has the right to physically inspect the suspect vessel while in the EEZ. If the evidence is clear and objective and the alleged violation resulted in an actual or threatened discharge causing major damage to the coastline or related interests, or to resources of the territorial sea or EEZ, then the coastal state may institute proceedings, including detention of the vessel.

In relation to U.S. responsibilities for implementation of MARPOL Annex V, these UNCLOS III provisions offer firm jurisdictional grounds for direct actions against violations occurring within 200 nautical miles of U.S. shores. A new U.S. policy based on this authority was initiated recently to allow the Justice Department to take direct action against a foreign vessel when there is evidence that an Annex V violation took place within the EEZ.

Although coastal states' rights clearly were expanded under the framework of UNCLOS III, there also appeared more definite articulations of the limits of prerogatives. In light of the importance of navigational freedoms and the sensitivity of nations regarding interference with their vessels, the right of the coastal state to unilaterally adopt laws affecting foreign ships was constrained explicitly in accordance with its diminishing authorities over more distant ocean space. Within the EEZ, coastal states may adopt only vessel pollution control laws that conform to "generally accepted international rules and standards established through the competent international organization." However, where there are "special circumstances" in "a particular, clearly defined area" requiring special mandatory measures due to "oceanographic and ecological condition," the coastal state may petition the competent international organization for special area designation. The coastal state then may propose the adoption of additional rules for approval by the international organization.

The significance of these constraints should not be underestimated. As an advocate of multilateral solutions and a key world leader on environmental issues, the United States often has been faced with accepting international standards and obligations that are less stringent than would be preferred. U.S. foreign policy is driven in part by its undeniable long-term interest in cooperating to raise international standards, and in part by its need to defend the environment. In relation to implementation of Annex V, the United States should pursue a cooperative multilateral approach that adheres strictly to the limits of U.S. authorities and emphasizes the identification of difficulties facing other countries and the provision of technical assistance to overcome obstacles. The United States also should aggressively use the recognized multilateral opportunities for accepting greater than minimal obligations, such as through special area designation.

PORT STATE CONTROL

An important advance in international law was made during the UNCLOS III negotiation with the detailed explication of port state authorities. If a vessel voluntarily enters a port (and thus either internal waters or an offshore terminal), then it is subject to the jurisdiction of the port state. Enforcement actions can be taken with respect to violations of applicable international standards. If, however, the violations were committed in the maritime zones of another state, the port state must receive requests for such legal proceedings from (1) the flag state, (2) the state in whose waters the violation was committed, or (3) a state damaged or threatened by the discharge violation. If the violation has caused or is likely to cause pollution in the zones of the port state, it also may proceed against a vessel in its port.

A port state has an affirmative duty to comply with the requests of nations that reasonably suspect a violation took place in any of their zones. Likewise, the port state has a duty to cooperate with flag states wishing to investigate the conduct of their vessels, irrespective of where alleged violations took place. After the port state has instituted proceedings, the coastal state in whose waters the violation took place may request all the relevant records and also may demand the suspension of proceedings undertaken by the port state. Legal actions taken by flag states take precedence over actions by either port or coastal states, unless the flag state has "repeatedly disregarded its obligation to enforce effectively the applicable international rules." Cases involving major damage to a coastal state also are an exception to the rule of flag state preeminence.

The rules regarding port state authorities set forth in the UNCLOS III treaty were seen widely as revolutionary: Once in force and operational, they set the stage for a new era in the enforcement of international law. The treaty's articulation of the active role that may be played by port state officials not only removed any jurisdictional questions, but also, and more importantly, bolstered the idea that port officials have an affirmative duty to inspect, investigate alleged violations, and institute proceedings. The very substantial opportunities for more effective implementation of Annex V through the creation and enhancement of collaborative port state control mechanisms are discussed in greater detail in the report of the full Committee on Shipborne Wastes. These opportunities are particularly worthy of exploration in relation to regional arrangements among developing countries that individually lack the infrastructures and assets needed for enforcement activities.

LIMITS ON NATIONAL AUTHORITIES: SAFEGUARDS FOR INTERNATIONAL SHIPPING

UNCLOS III established a number of important constraints on states with regard to their enforcement of international standards. These constraints serve to protect the legitimate interests of international shipping as well as the world community's common interest in the free flow of commerce. According to the treaty, nations may not discriminate "in form or in fact" against vessels of any other State; they may not cause undue delay; they may only apply monetary penalties for violations within the territorial sea unless the case involved "a wilful and serious act of pollution"; and they must notify the flag state and other affected states of any enforcement actions. With regard to implementation of Annex V, adherence to these constraints may be particularly important in light of the difficulties associated with monitoring compliance and the corollary need for voluntary commitment to compliance. Such cooperation is far more likely if the legitimate interests of the shipping community are respected.

The burdens of membership in international agreements controlling pollution inadvertently create disincentives to participation. To circumvent the tendency for ships of non-members to be held to lesser standards, many port states have adopted a policy of requiring that all ships entering their ports comply with international standards. In this way, no preference or economic advantage is given to ships of non-members. The UNCLOS III treaty recognizes and affirms that port states are empowered to require such compliance as a condition of entry. However, coastal states cannot place such conditions on access to the EEZ or territorial sea.

MORE GENERAL OBLIGATIONS

UNCLOS III contains a number of more general obligations that are relevant to Annex V implementation. These duties are not as clear and focused as those outlined previously, but they can be used to bolster the more specific obligations of other treaties and to motivate nations to move toward higher environmental standards of conduct.

In general, parties agree to protect the marine environment through regulating polluting activities under their jurisdiction and preventing trans-boundary damage to the environment of other states. Rare and fragile ecosystems as well as the habitat of endangered species must be given special protection. States are obligated to cooperate in the establishment of international environmental rules, standards, and recommended practices. States also commit to notify other nations of actual or imminent environmental threats, as well as to join in formulation of contingency plans for responding to pollution damage and threats. Cooperation among nations also must extend to scientific research on marine environmental concerns and formulation of scientific criteria upon which to base environmental standards. States must provide scientific and technical assistance and preferences for special services to developing countries, so that they can better protect the marine environment.

Nations also are obliged to measure and evaluate the risks or effects of pollution. In particular, they must monitor activities they permit or engage in. The results of such studies must be publicized. If an activity is likely to harm the marine environment, then the state with jurisdiction over the activity is required to assess the potential effects and publicize the findings. States are obliged to take national measures to control pollution from land-based sources, seabed activities within and beyond national jurisdiction, dumping, vessels, and atmospheric sources.

Further, nations must follow through on their commitment to environmental protection by actively enforcing national and applicable international standards with regard to all sources of pollution under their jurisdiction. States must ensure that recourse is available in their court systems for claims arising from damage to the marine environment caused by persons under their jurisdiction. States also agree to further develop agreements on liability and compensation. Although warships, auxiliary vessels, and other public vessels are exempted from the environmental protection provisions of the UNCLOS III treaty, states are obligated to ensure that such vessels operate in a manner consistent with the treaty so far as is reasonable and practicable.

In the EEZs, states are obligated to ensure the maintenance of living resources through proper conservation and management measures based on the best scientific evidence available. Conservation principles are echoed in relation to highly migratory species, anadramous stocks, catadromous species, and straddling stocks. States are obligated to take measures necessary for the conservation of living resources on the high seas. Toward this end, they must cooperate with other nations, conduct scientific research, and implement conservation measures to regulate the activities of their nationals. Conservation measures are also required to protect marine mammals on the high seas. There is a duty to protect the habitat of depleted, threatened, or endangered species and other forms of marine life. Also recognized is the concept that certain clearly defined fragile or exceedingly valuable areas should be provided special protection through implementation of unusually stringent environmental protection laws.

STRENGTHENING THE FORCE OF IMO RULES

Perhaps the most important effect of UNCLOS III, once it enters into force, will be to expand and strengthen the power and effect of International Maritime Organization (IMO) rules, codes, guidelines and other "generally accepted international standards." Nations are directed to act cooperatively through the competent international organization to establish and promote the adoption of international rules and standards to prevent, reduce, and control pollution from vessels. Nations also are committed to adopting laws and regulations to control pollution from vessels flying their flag or of their registry. These laws must have the same effect as generally accepted international rules and standards established through the competent international organization or general diplomatic conference.

These provisions suggest that parties to UNCLOS III will be bound to comply with all widely recognized standards, regardless of whether they are also parties to the specific conventions under which the standards are developed. Nations that have ratified UNCLOS III may be expected, therefore, to comply with MARPOL Annex V, whether or not they are parties to Annex V. The IMO is universally regarded as "the competent international organization" in connection with the establishment of standards for the operation of vessels and for vessel-pollution control. Thus, its rules and regulations will continue to be established norms applicable to all UNCLOS III members.

D
Time Line for U.S. Implementation of Annex V

December 29, 1987   President Reagan signs the Marine Plastic Pollution Re-
                     search and Control Act of 1987 (MPPRCA) (P.L. 100-
                     220) into law. The Act extends Annex V mandates to the
                     navigable waters of the United States and the 200-nauti-
                     cal-mile-wide U.S. Exclusive Economic Zone (EEZ).

December 30, 1987   The United States delivers the instrument of ratification
                     to the International Maritime Organization (IMO). The
                     U.S. ratification also triggers the start of the one-year
                     clock that determines when Annex V will take effect
                     internationally. The United States pledges to make "ev-
                     ery reasonable effort" to make the Gulf of Mexico a
                     special area under Annex V.

May 1988   The report of the Interagency Task Force on Persistent
                     Marine Debris is released to the White House Office of
                     Domestic Policy.

September 1988   IMO Marine Environment Protection Committee (MEPC
                     26) approves the *Guidelines for the Implementation of
                     Annex V, Regulations for the Prevention of Pollution by
                     Garbage from Ships.* The United States played an instru-
                     mental role in drafting and finalizing the guidelines.

December 31, 1988   Annex V enters into force in the United States (and
                     worldwide).

August 28, 1989   This is the deadline (set as part of the U.S. Coast Guard
                     Certificate of Adequacy program) for all U.S. ports to
                     arrange to accept foreign garbage that must be quaran-
                     tined under Animal and Plant Health Inspection Service
                     (APHIS) rules.

March 12, 1990   MEPC 29 addresses the proposal to make the Gulf of Mexico a special area under Annex V. Several issues remain to be resolved before the designation can proceed.

May 2, 1990   U.S. Coast Guard (Department of Transportation) issues an interim final rule for record keeping and informational requirements of the MPPRCA. (These are not Annex V requirements.)

September 4, 1990   U.S. Coast Guard issues a final rule implementing Annex V for foreign vessels operating in U.S. waters and for U.S. ships operating in any waters.

November 12, 1990   MEPC 30 adopts amendments proposed by the United States to designate the Antarctic Ocean as a special area under Annex V. MEPC 30 also approves a proposal for designating the Wider Caribbean (as defined by the regional Cartagena Convention) as a special area under Annex V. This action initiates a series of regional preparations that are still in progress.

January 1991   APHIS begins boarding vessels arriving at U.S. ports on behalf of the Coast Guard. APHIS inspectors use four questions to detect Annex V violations, which are referred to the Coast Guard.

February 1991   The United States submits a proposed standard specification for shipboard incinerators to the IMO Marine Safety Committee's Subcommittee on Ship Design and Equipment.

March 1, 1991   U.S. Coast Guard issues a final rule that makes permanent requirements for waste management plans and placards on vessels 26 feet or more in length.

October 1991   The U.S. Gulf of Mexico Program issues a Marine Debris Action Plan for that region. (An Addendum was issued in December 1992.)

October 30, 1992   The MEPC adopts the standard specification for shipboard incinerators.

November 1992   The U.S. Coast Guard issues internal MARPOL 73/78 Annex V Guidance and Procedures in the *Marine Safety Manual*, Chapter 33, section F.

November 1993   The U.S. Congress extends deadlines for U.S. Navy compliance to 1998 for the plastics ban and the year 2000 for special area requirements.

May 19, 1994   U.S. Coast Guard regulations take effect requiring garbage logs on ocean-going, U.S.-flag vessels over 12.2 meters (about 40 feet) long in commercial service.

August 1994   U.S. General Accounting Office issues the first part of a congressionally mandated report on U.S. Navy compliance efforts. (A follow-up report was issued in November 1994.)

Characteristics of Annex V Special Areas

excerpts[1] from:

An Analysis of Proposed Shipborne Waste Handling
Practices Aboard United States Navy Vessels

R. L. Swanson, R. R. Young, and S. S. Ross

Marine Sciences Research Center
State University of New York at Stony Brook
Stony Brook, New York

CHARACTERISTICS OF SPECIAL AREAS

Antarctic Ocean

The East Wind Drift (attributed to the prevailing easterly winds) is a westward-flowing coastal current around most of the continent. Further north, the Southern Ocean is dominated by the Antarctic Circumpolar Current. This strong current flows in an eastward direction between about latitude 40°S and latitude 60°S. Surface flow is driven primarily by the frictional stress of the westerly winds in the region. This stress, together with the Coriolis force, contributes a northward component to the surface current, resulting in the formation of fronts. Below the surface layer, the density structure is in geostrophic balance with the circulation (Pickard and Emery, 1990).

There are three major basins in the Antarctic Ocean: the Atlantic-Indian-Antarctic Basin, the Eastern Indian-Antarctic Basin (also referred to as the Australian-Antarctic Basin or Knox Basin), and the Pacific Antarctic Basin (or Bellingshausen Basin). There is also a single deep-sea trench, the South Sand-

---

[1]These excerpts have been edited for grammar and style; factual accuracy is the sole responsibility of the authors. Copies of the complete paper may be obtained from the Marine Board, National Research Council, 2101 Constitution Avenue, N.W., Washington, D.C. 20418.

wich Trench, in the Antarctic Ocean, on the east of the South Sandwich Island and adjoining the Scotia Ridge. The trench extends 600 miles and reaches a maximum depth of 8,260 meters (m), located between latitude 55°S longitude 32°W and latitude 61°S longitude 27°W (Fairbridge, 1966).

Two to three thousand tourists each year visit the Antarctic. Palmer Station, a U.S. research base on the peninsula, has become such a popular destination that a quota has been introduced (Elder and Pernetta, 1991).

It is estimated that total annual production of plant matter in surface waters south of the Antarctic Convergence is 610 million tonnes.

The Baltic Sea

The Baltic Sea, including the Gulf of Bothnia and the Gulf of Finland, is the largest area of brackish water in the ocean system (Pickard and Emery, 1990). It is brackish because precipitation and runoff greatly exceed evaporation (Sverdrup et al., 1942). Its bottom topography is irregular, with a mean depth of 57 m and a number of basins, the deepest of which is 459 m deep (Pickard and Emery, 1990). The Baltic is connected to the Atlantic Ocean at its southwest end through intricate passages. Its sill depth in the narrows between Gedser and the Darss is about 18 m, leading to the Kattegat and the North Sea (Pickard and Emery, 1990).

Evaporation and precipitation each are estimated at about 47 centimeters (cm) per year, thus canceling one another. Annual river runoff is equivalent to 130 cm of water over the entire sea; however, there are significant year-to-year variations (Pickard and Emery, 1990). Overall general circulation is weak (approximately 1 cm per second) (Pickard and Emery, 1990), as there are no tidal currents to disturb the stratification (Dietrich, 1963).

Because of shallow sill depth, a rejuvenation of the deep water occurs only when large-scale meteorological conditions can override the estuarine circulation. These conditions are not rare: Significant vertical mixing can occur because the Baltic basin is so shallow and broad (Gross, 1967). Thus, it is possible for the residence time of the Baltic to be less than one year, although this is variable.

The area's humid climate aids the development of a density discontinuity layer, thus greatly preventing a thermohaline convection. The Baltic is a two-layer system with a well-mixed upper layer 30–50 m deep in the south, increasing to 60–70 m in the central Baltic (Pickard and Emery, 1990).

Dissolved oxygen may reach saturation levels in the surface layers but is relatively low in deep water. Changes may occur on a decadal time scale and are related to variations of inflowing water to the south. There has been a general trend toward decreasing dissolved oxygen values since the beginning of the twentieth century (Pickard and Emery, 1990). In many of the deep basins which have a residence time of several years, anoxic conditions occur (Pickard and Emery, 1990).

Black Sea

The Black Sea is the archetypical anoxic basin (Pickard and Emery, 1990). The surface circulation is defined by an anticlockwise gyre in each of the east and west basins (Pickard and Emery, 1990). It receives a volume of fresh water via river runoff and precipitation that far exceeds the amount of evaporation; consequently, its salinity is depressed (Pinet, 1992). A sharp halocline stratifies the water column; additionally, summer heating creates a thermocline that further intensifies vertical stratification. There is a pronounced density discontinuity at about 100 m (Pinet, 1992).

Renewal of the deep water of the Black Sea is very slow, because it occurs via water which flows in along the bottom of the Bosporus. This inflow is so small (193 cubic kilometers (km) per year (Dietrich, 1963) in proportion to the total volume of water that renewal below a depth of 30 m is estimated to take about 2,500 years (Sverdrup et al., 1942). Thus, salinity in the deep sea remains low, representing the equilibrium between influx and vertical convection (Dietrich, 1963). Below a depth of about 200 m, the Black Sea contains large amounts of hydrogen sulfide rather than oxygen. With the exception of anaerobic bacteria, water below this depth is inhospitable to life (Pinet, 1992).

Although the residence time of water below the halocline is long, mixing occurs at a faster rate—about once every 100 years, via storm movement of deeper water and surface cooling over the winter, which lessens density stratification (Pinet, 1992).

Caribbean Sea

The Wider Caribbean special area includes the Caribbean Sea and the Gulf of Mexico. The area consists of a number of deep basins separated by major sills (Clark, 1986). The Caribbean Sea is tropical and experiences little seasonal change. Over much of the area there is a permanent thermocline at about 100 m. Upwelling is not a dominant feature, although there are localized areas where bottom water comes to the surface. Because of the permanent thermocline and lack of upwelling, the Caribbean tends to be nutrient-poor, confining fisheries to the shallow waters (Clark, 1986).

There are approximately 60 species of corals in the Caribbean Sea. The second largest coral reef in the world is the 250 km-long barrier reef in the waters off Belize (Pinet, 1992).

The greater Caribbean area attracts about 100 million tourists each year (Clark, 1986), with 3 million of those coming on cruise ships (Elder and Pernetta, 1991).

Gulf of Mexico

The Gulf of Mexico forms the northeastern component of the Wider Caribbean. It is surrounded by the Yucatan Peninsula, Cuba, and the Florida coast and exhibits a wide continental shelf. Its northern shoreline consists mostly of sedimentary material derived from Mississippi River Basin.

The western and southern coasts are characterized by large lagoons separated from the sea by barrier beaches. Residence times for water within lagoons varies widely. In the Great Barrier Reef, times of 0.5–4 days have been estimated for lagoons of 2–10 km in diameter; for Bikini Atoll, 40–80 days, and for the very shallow Fanning Atoll (18 km long but only a few meters deep), periods of up to 11 months have been estimated (Pickard and Emery, 1990).

The salinity of lagoons varies with tidal action, evaporation, and freshwater input from rain and runoff from land (Elder and Pernetta, 1991). The lagoons serve as an important habitat.

The Gulf of Mexico and the Caribbean Sea are connected by the Yucatan Channel (sill depth about 1,600 m). The topography is rugged, with great contrasts between ridges and troughs. This is an area of particular interest for tectonophysical and geophysical studies, due to the presence of pronounced gravity anomalies, volcanism, and strong seismic activity (Neumann and Pierson, 1966).

The Gulf of Mexico may be divided into two halves, based on the character of its circulation. The eastern part is dominated by the Loop Current, whose water originates in the northwestern Caribbean Sea as the Yucatan Current and flows into the central eastern gulf. The Yucatan Current flows over a sill between the Yucatan and western Cuba and deepens to 1,800 m in the Gulf (Pinet, 1992). From there, it veers eastward and exits to the south of Florida. (Water in the eastern Gulf that is deeper than 600 m remains in the basin, trapped by a shallow sill south of Florida [Pinet, 1992].) This current rotates clockwise and has surface speeds of 50–200 cm per second (Pinet, 1992). In contrast, circulation is weak and variable in the western half of the Gulf of Mexico where the clockwise surface flow averages less than 50 cm per second (Pinet, 1992).

Primary productivity in the Gulf of Mexico is generally low, averaging about 25 grams (g) carbon per m^3 per year; however, some areas are much more productive due to upwelling and an inflow of nutrients from the Mississippi River. In the northern Gulf, primary productivity ranges from 250–350 g carbon per m^3 per year.

Almost two-thirds of the United States contributes to freshwater runoff into the Gulf of Mexico, greatly stressing the environment. Most pollutants discharged by U.S. rivers are dispersed in the western Gulf, where levels may build up due to the weakly circulating water of the area (Pinet, 1992).

The Mediterranean Sea

Mediterranean water forms in the northwestern part of the Mediterranean Sea in winter. Cooler winter temperatures and higher-than-normal evaporation, associated with the cold, dry Mistral winds, increases the surface water density such that vertical mixing occurs all the way to the sea floor (2,000 m). Evaporation (about 100 cm per year) exceeds precipitation plus river runoff, so there is a net loss of volume which is made up by inflow of salt water from the Atlantic Ocean (Pickard and Emery, 1990).

The homogeneous Mediterranean water mass has a salinity of more than 38.4 practical salinity units (psu) and a temperature of about 12.8°C (Davis, 1977). Mediterranean water leaves the Straits of Gibraltar at approximately 300–500 m depth, below incoming Atlantic water. Intense mixing occurs at the interface of the Mediterranean and Atlantic waters. The least-mixed Mediterranean water has a salinity of 36.5 psu and a temperature of 11°C (Davis, 1977). Due to its high density, it sinks to about 1,000 m, where it becomes neutrally buoyant and spreads out. This distinctive tongue of mediterranean water can be recognized throughout much of the Atlantic Ocean by its high temperature and salinity profiles. The Mediterranean Sea's relatively long residence time (estimated at 70–100 years) makes it particularly vulnerable to pollution.

North Sea

The North Sea is broad and shallow; thus, it is subject to storm surges (Gross, 1982). At its southern end the North Sea is constricted at the Straits of Dover; however, there is no geographical northern boundary. The south and southeastern parts are less than 50 m deep, and the northern part is 120–145 m deep. The North Sea is not an homogeneous body of water. The residence time is approximately 0.9 year (Otto, 1983).

Overall, there is an excess of precipitation over evaporation. In winter, however, the lee effect of the British Isles produces a net loss by evaporation in the western and southwestern parts of the North Sea. During summer, all parts receive an excess of water due to precipitation. As surface waters become less saline, stratification occurs between the warm, less dense water over the deeper water. During calm weather in the eastern North Sea and German Bight, a thermocline may develop, resulting in reduced oxygen concentrations in the bottom water (Clark, 1986).

Persian Gulf

The average depth of the Persian Gulf is 30 m, with a maximum depth of 90 m. It is so shallow that there is no significant exchange of water between it and the adjacent Gulf of Oman (Sverdrup et al., 1942), although some water does flow into the Arabian Sea (Pickard and Emery, 1990). Because the Persian Gulf is so shallow, it experiences uniformly high levels of salinity (40–70 psu) and wide seasonal changes in sea temperatures (15–38°C). Thorough wind-driven mixing occurs throughout most of the year (International Maritime Organization, 1994).

The Persian Gulf experiences high evaporation and low rainfall rates—a contributing factor to the high salinity of the water. These factors work to restrict biological diversity, and many species live at or near their limits of environmental tolerance (International Maritime Organization, 1994). Under these conditions, any added stress, such as an oil spill or other pollution event, can disproportionately influence the area.

Red Sea

The Red Sea is a rift valley, resulting from the separation of Africa and the Arabian peninsula (Pickard and Emery, 1990). With the exception of the Suez Canal, it is closed to the north. It opens to the Gulf of Aden, Arabian Sea, and the Indian Ocean to the south, through the narrow strait of the Bab al Mandab. There is a sill of about 110 m at the Bab al Mandab (Pickard and Emery, 1990).

There are no rivers flowing into the Sea. Evaporation is high (about 200 cm per year), while precipitation averages about 7 cm per year, making this the most saline large body of ocean water in the world (Pickard and Emery, 1990).

The surface layer is saturated with dissolved oxygen; however, absolute values are low (less than 4 milliliters per liter) due to high temperatures and salinities. Red Sea circulation varies seasonally with the winds. In summer (southwest monsoon) the winds are to the south. Surface flow is southward, with outflow through the Bab al Mandab; additionally, there is a subsurface inflow to the north through that strait. In winter (northeast monsoon) the winds over the southern half of the sea change to the north, and there is a northward surface flow over the entire Red Sea, with a subsurface southward flow through the Bab al Mandab. The outflow is from an intermediate layer to about 100 m. This water can be traced through the Arabian Sea and down the west side of the Indian Ocean (Pickard and Emery, 1990).

Residence time for the surface layer has been estimated at six years; for the deep water, about 200 years (Pickard and Emery, 1990). A notable feature of the Red Sea are the hot brine pools found in some of the deepest parts. Pickard and Emery offer the explanation with fewest arguments. They assert that "this is interstitial water from sediments, or solutions in water of crystallization from solid materials in the sea bottom, released from heating from below and forced out through cracks into the deep basins of the Red Sea."

TABLE E-1 General Physical Characteristics of MARPOL Special Areas

	depth (m)	mean area (10^6km^2)	volume (10^3km^3)	surface temp. range (°C)	surface salinity range (psu)	relative surface water viscosity[a] (%)	residence time (years)
Antarctic Ocean	4,000			−2/+4	34.6	94-110	100
Baltic Sea	86	0.39	33	−2/+15	6-8	65-107	short
Black Sea	1,166	0.46	537	9-25[‡]	18-21[‡]	53-79	2500[*]-3000[§]
Caribbean Sea	2,491	2.8	6,860	25-28	36	51-55	
Gulf of Mexico	1,512	1.5	2,332	20-29	36	50-61	100[*]
Mediterranean Sea	1,494	2.5	3,758	13-26	37-39	54-73	70-100
North Sea	91	0.60	55	5-16	34-35	67-92	0.9[||]
Persian Gulf	25[†]	0.24	10	10-25	38	55-78	long
Red Sea	558	0.45	251	18-32	40-41	47-65	6.0[§]

[a]Assumes 100% at 0 psu and 0°C.

Sources: van der Leeden et al., 1990; [*]Geyer, 1981; [†]Sverdrup et al., 1942; [‡]Pinet, 1992; [§]Pickard and Emery, 1990; [||]Otto, 1983

REFERENCES

Clark, R.B. 1986. Marine Pollution. Oxford: Oxford University Press.

Davis, R.A., Jr. 1977. Principles of Oceanography. Reading, Mass.: Addison-Wesley Publishing Co., Inc.

Dietrich, G. 1963. General Oceanography (translated from German). New York: John Wiley & Sons.

Elder, D. and J. Pernetta (eds.). 1991. Random House Atlas of the Oceans. New York: Random House with World Conservation Union.

Fairbridge, R.W. (ed.). 1966. The Encyclopedia of Oceanography. New York: Reinhold Publishing Corp.

Geyer, R.A. (ed.). 1981. Marine Environmental Pollution, 2. Dumping and Mining. College Station: Texas A&M University.

Gross, M.G. 1967. Oceanography: A View of the Earth. Columbus, Ohio: Charles E. Merrill Books.

Gross, M.G. 1982. Oceanography: A View of the Earth. Englewood Cliffs, N.J.: Prentice-Hall, Inc.

International Maritime Organization (IMO). 1994. Three years on: The Persian Gulf oil spill. IMO News 1:21-24.

Neumann, G. and W.J. Pierson, Jr. 1966. Principles of Physical Oceanography. Englewood Cliffs, N.J.: Prentice-Hall.

Otto, L. 1983. Currents and water balance in the North Sea. Pp. 26-43 in North Sea Dynamics, J. Sundermann and W. Lenz, eds. Berlin: Springer-Verlag.

Pickard, G.L. and W.J. Emery. 1990. Descriptive Physical Oceanography: An Introduction. Oxford: Pergamon Press.

Pinet, P.R. 1992. Oceanography: An Introduction to the Planet Oceanus. St. Paul, Minn.: West Publ.

Sverdrup, H.U., M.W. Johnson, and R.H. Fleming. 1942. The Oceans: Their Physics, Chemistry, and General Biology. Englewood Cliffs, N.J.: Prentice-Hall.

van der Leeden, F., F.L. Troise, and D.K. Todd. 1990. The Water Encyclopedia, 2nd Ed. Chelsea, Mich.: Lewis Publishers.

Ecological Effects of Marine Debris*

GENERAL EFFECTS ON ECOSYSTEMS

Judgments concerning the broad impact of plastic debris on marine ecosystems are speculative at present. The bioaccumulation of plastics through food chains, for example, may be a problem, based on observations of secondary and tertiary ingestion of plastics by certain species: bald eagles preying on parakeet auklets with plastics in their stomachs (Day et al., 1985); Antarctic skuas preying on broad-billed prions in the South Atlantic (Bourne and Imber, 1982); and short-eared owls in the Galapagos Islands preying on blue-footed boobies that had ingested fish containing plastic pellets (Anonymous, 1981).

There is little scientific information available on how debris may affect marine invertebrate species, plant life, or marine habitats in general, aside from observations that debris damages coral reefs, is ingested by squid (Araya, 1983; Machida, 1983), and may present a new habitat niche for encrusting marine species (Winston, 1982).

ENTANGLEMENT OF MARINE SPECIES

A major reason for the heightened concern over marine debris was the increasing number of reports that plastic was causing widespread mortality of marine species. Among the first species to be highlighted was the Northern fur seal. Studies indicated that each year as many as 50,000 of these seals were

---

*Summary prepared by the Committee on Shipborne Wastes as a supplement to Chapter 2.

becoming entangled and dying in plastic debris, primarily fishing nets and strapping bands. Subsequent findings have shown that the increased use and subsequent disposal of plastics in the marine environment is causing widespread mortality among marine mammals, turtles, birds, and fish, either through entanglement or ingestion. However, most of this information is drawn from a few studies, and there has been no attempt to compile the data at one source, nor has there been any extensive effort to monitor trends.

Entanglement in plastic debris poses a potentially serious threat to a number of marine species, at both the individual and population levels. (To date, the threat to populations has been documented only in the Northern fur seal.[1]) Marine mammals, sea turtles, seabirds, and fish have been found entangled in the loops and openings of fishing nets, strapping bands, and other plastic items. Once ensnared, an individual may be unable to swim or feed or may incur open wounds that become infected.

There have been attempts to identify the circumstances that can lead to entanglement. In some cases, random encounters with debris are to blame. For example, an animal may not be able to see or otherwise detect plastic debris, especially fishing gear designed to be nearly transparent in water (Balazs, 1985). However, a number of biological factors appear to increase the risk of entanglement for certain species. Like natural ocean rubble such as sargassum weed and logs, floating plastics attract fish, crustaceans, and other species seeking shelter and concentrated food sources. Marine mammals, turtles, and birds also are attracted to floating debris, and they may become ensnared when attempting to feed. Predators, such as seals and sea birds, are at increased risk of becoming entangled in discarded fishing gear, which may have fish entrapped in netting. Finally, pinnipeds haul themselves out of the water to rest on natural debris, such as floating kelp mats and logs, while young seals are attracted to floating objects as playthings. If such debris includes plastics, entanglement can result.

Due to their behavioral characteristics, seals and sea lions may be the most prone to entanglement. Individuals from at least 15 of the world's 32 species of seals have been observed ensnared in plastic debris; these include several species found in the United States, such as the northern fur seal (Fowler, 1985, 1988; Scordino, 1985), northern sea lion (Calkins, 1985), California sea lion, northern elephant seal, harbor seal (Stewart and Yochem, 1985, 1987), and the Hawaiian monk seal (Henderson, 1984, 1985), which is on the U.S. government's list of endangered species.

---

[1]The Pribilof Islands of Alaska are home to a population of approximately 827,000 Northern fur seals, 71 percent of the estimated total world population of this species. Studies show that the Pribilof population is less than half that observed 40 years ago and is declining at an annual rate of 4 to 8 percent (Fowler and Merrell, 1986). Entanglement in plastic debris is thought to be contributing to the decline.

Most information available on pinniped interactions with debris has been compiled by the National Marine Fisheries Service (NMFS). The effects of entanglement on an individual animal may vary. Most entangled Northern fur seals have been observed with debris around their necks and shoulders. This kind of entanglement, if constricting, may directly impair swimming or feeding. Entangling debris also increases drag during swimming. Consequently, an entangled seal must use more energy to swim than it normally would and therefore must consume more food to compensate; unfortunately, drag inhibits the high-speed swimming required for pursuit of prey and therefore may lead to starvation of the animal. In other cases, abrasion from entangling debris may cause wounds that are susceptible to infection.

Entanglement of breeding animals also can adversely affect their young. In field studies on St. Paul Island, Alaska, nine of 17 female northern fur seals observed entangled in debris never returned to their pups after foraging at sea (Fowler, 1988). The other entangled seals took twice as long to return as did unencumbered female seals (Fowler, 1988).

Sea turtles also appear to be prone to entanglement in plastic debris. In the first comprehensive assessment of this problem, carried out by the NMFS, Balazs (1985) complied a list of 60 cases of sea turtle entanglements worldwide involving green, loggerhead, hawksbill, olive ridley, and leatherback turtles. The debris involved most often was monofilament fishing line. Other cases involved (in order of decreasing frequency) rope, trawl nets, gill nets, and plastic sheets or bags.

As is the case for pinnipeds, entangled sea turtles are unable to carry out basic biological functions such as feeding, swimming, or surfacing to breath; constricting debris also may cause lesions or even necrosis of flippers. According to Bourne (1990), however, there appears to be no evidence that the entanglement of turtles in debris is affecting their numbers, in contrast to the significant effects of other threats such as drowning in shrimp trawls, overfishing, direct harvesting for meat and eggs, disturbance of breeding habitat, and ingestion of debris.

While pinniped and sea turtle entanglement in plastic debris has been documented, accounts of the impact of plastics on birds are entirely anecdotal. There has been no attempt by any agency to collect extensive data on bird mortality due to entanglement in debris. In the past, the entanglement problem has been overshadowed by the magnitude of seabird mortality related to active fishing operations, principally in the high seas drift-net fisheries in the Pacific. For example, the Japanese salmon gillnet fishery, in which more than 2,575 kilometers (km) (1,600 miles) of drift gill net were set each night, reportedly drowned over 250,000 seabirds in U.S. waters each year during a two-month fishing season (King, 1984). An international moratorium was enacted recently on high seas drift-gillnet fishing. Seabirds also are attracted to lost or discarded nets and have been found entangled in large pieces of lost gill nets that continue to ghost fish at sea (Jones and Ferrero, 1985).

Based on anecdotal reports, monofilament fishing line may be the item most often known to entangle birds. During the 1991 and 1992 International Coastal Cleanups coordinated by the Center for Marine Conservation (CMC), 56 of the 120 reported cases of bird entanglements involved fishing lines (Younger and Hodge, 1992; Hodge et al., 1993). It should be noted that these animals were reported on just a fraction of the U.S. coastline within a few hours. Hence, it would seem worthwhile to investigate just how serious a problem monofilament line poses to birds.

In some cases, birds may become entangled in fishing lines when they attempt to eat bait from fishing hooks. An entangled bird trailing line either may be immobilized immediately or may become snagged on a tree or power line, unable to break free. Other items, such as plastic six-pack rings, get stuck around the necks of marine birds and waterfowl when they attempt to dive or feed through the rings. Ospreys, cormorants, and other birds actively collect pieces of nets and fishing line for nest material; this activity can lead to strangulation of both adults and their young (O'Hara and Iudicello, 1987; Podolsky and Kress, 1990).

Little is known about the extent of entanglement among species of cetaceans. The lack of knowledge may be due to the fact that these animals are only found on occasions when they wash ashore, added to the expense of and sometimes lack of expertise for conducting necropsies.

Information on the entanglement of fish in marine debris is largely anecdotal at present. During the 1992 International Coastal Cleanup, volunteers reported approximately 20 cases of entanglement of fish and crustaceans in debris in just three hours. The range of items found to ensnare fish is remarkable. The CMC maintains a photograph library of wildlife interactions with debris that includes pictures of a gar and bluefish in plastic six-pack rings, sharks in plastic straps and cables, a red drum in a plastic vegetable sack, and a billfish with a plastic baby bottle cap on its bill.

Finally, there have been sporadic accounts of debris on coastlines entangling terrestrial species. For example, foxes and rabbits have been observed entangled in nets and other plastic items (Fowler and Merrell, 1986; O'Hara and Younger, 1990). In one case, the skeletal remains of 15 reindeer were found in a Japanese gill net on a beach in Alaska (Beach et al., 1976). But again, this information is not compiled by any agency.

INGESTION OF PLASTICS BY MARINE SPECIES

Along with increasing reports of wildlife entanglement involving plastic debris, there has been growing documentation of a less obvious problem: the ingestion of plastics by marine species. There appears to be some understanding of the factors that may increase the likelihood of plastic ingestion by certain species. For example, winds and currents that tend to concentrate food sources such as fish and plankton also concentrate debris. For some species, floating items actually may resemble authentic food items. Seabirds, for example, are thought to mistake small pieces and fragments of plastic for planktonic organisms, fish eggs, or even the eyes of squid or fish (Day et al., 1985). Plastics covered with fish eggs or encrusting organisms such as barnacles, algae, and bryozoa may even "smell" or "taste" like authentic food. It has been suggested that hungry animals are less likely than are satiated animals to discriminate between natural foods and look-alike debris and are more likely to eat the plastic items (Balazs, 1985).

Perhaps the most highly publicized example of plastic ingestion has been the consumption of plastic bags or sheeting by sea turtles that are thought to mistake these items for jellyfish, squid, and other prey. In the only comprehensive review of this subject, Balazs (1985) reported five species of sea turtles known to ingest plastics: green, loggerhead, leatherback, hawksbill, and Kemp's ridley. Of the items ingested, plastic bags and sheets were most common (32 percent of 79 cases) followed by tar balls (20.8 percent) and plastic particles (18.9 percent). On San Jose Island, Texas, Amos (1993) reported that 20 to 30 percent of plastic containers that wash ashore exhibit bite marks from turtles.

Plotkin and Amos (1990) reported ingestion of plastic and other man-made debris by 46 percent of 76 sea turtles stranded on Texas beaches in an 18-month period. Items found in turtle guts included plastic bags, foamed plastic "peanuts," balloons, strapping band fragments, polypropylene rope, as well as miscellaneous plastic pieces. In some cases, the stomach and gut were completely impacted with plastic. The angular shapes and the rigidity of some of the plastic pieces are not dissimilar to fragments of natural prey that must be excreted, such as crustacean carapaces and sea-pen stalks.

Recent studies suggest that young turtles that congregate to feed in the open ocean at areas of convergence are particularly prone to ingesting plastics. The downwelling in these areas concentrates not only turtle food but also plastic debris. For all turtles species, with the exception of the leatherback (which is rarely seen in immature stages), reports of immature animals that have ingested debris are more common than are reports of adults (Balazs, 1985). Carr (1987) noted that plastic pellets found in the stomachs of dead juvenile sea turtles are similar in size and shape to sargassum weed, which concentrates in areas of convergence and provides both shelter and sources of food for turtles.

The effect of plastics ingestion on sea turtle longevity and reproductive potential is unknown. It is thought that ingested plastics may cause mechanical blockage of the digestive tract, starvation, reduced absorption of nutrients, and ulceration. Buoyancy caused by plastics also could inhibit diving activities needed for pursuit of prey and escape from predators (Balazs, 1985). For several reasons—the prevalence of plastic ingestion among sea turtles, the significant lesions and mortality caused by ingested items, and the fact that all species of sea turtles are threatened with extinction—the effects of ingestion of debris on sea turtles is considered a research priority (Sileo, 1990).

The ingestion of plastic debris by seabirds also has received attention in recent years. The first documented report of plastic ingestion by a seabird, a Layson albatross, was in the 1960s (Kenyon and Kridler, 1969). Today, at least 80 of the world's 280 seabird species are known to ingest plastic debris (Harrison, 1983). This tendency appears to be closely related to bird feeding habits, with diving birds having the highest incidence of plastic ingestion. Most bird species also exhibit preferences for certain types of plastic based on debris color, shape, or size. For example, the parakeet auklet, which feeds primarily on planktonic crustaceans, was found to ingest large amounts of light-brown plastic particles that are similar in size and shape to its crustacean prey. Some birds also feed plastics to their young. In one study, all of the 300 Layson albatross chicks examined on Midway Islands of Hawaii (located more than 1,600 km [1,000 miles] northwest of the nearest populated Hawaiian islands) had ingested plastic debris, including plastic fragments, toys, bottle caps, balloons, condoms, and cigarette lighters (Sileo et al., 1990).

Although many birds naturally digest and regurgitate hard, nonfood items such as fish bones and bottom substrate, some researchers believe that large quantities of ingested plastics may cause intestinal blockage or a false feeling of satiation or may reduce absorption of nutrients, thus robbing the animal of needed nutrition (Day et al., 1985). Suffocation, ulceration, or intestinal injury could be caused by jagged edges on plastics or grinding of these items against intestinal walls. Long-term effects of plastics ingestion may include physical deterioration due to malnutrition, decreased reproductive performance, and the inability to maintain energy requirements (Day et al., 1985).

Limited data are available concerning ingestion of plastic debris by marine mammals, although information from marine parks and zoos suggests that debris ingestion has the potential to be a direct cause of mortality (Walker and Coe, 1990). Several species of wild cetaceans have been found to ingest plastics, primarily in the form of bags and sheeting (Martin and Clarke, 1986; Barros et al., 1990; Walker and Coe, 1990). Because most of this information was obtained through studies of dead animals that had stranded, the actual cause of death is uncertain. In Texas, however, a stranded pygmy sperm whale, which was taken into captivity, died later from the effects of plastic garbage bags, a bread wrapper, and a corn chip bag ingested while in the wild (O'Hara et al., 1987).

Analyses of the stomach contents of sperm whales at an Icelandic whaling station from 1977 to 1981 revealed plastic drinking cups and children's toys as well as large pieces of fishing nets. Because sperm whales readily ingest and subsequently regurgitate the hard parts of prey, principally fish bones and cephalopod beaks, small pieces of plastic are thought to pose no significant problem. But in one case, an ingested fishing net weighing 139 pounds was considered to be large enough to cause eventual starvation of the sperm whale.

Other marine mammals that have died as a result of ingestion of debris include a northern elephant seal and a Steller sea lion (Mate, 1985). Walker and

Coe (1990) also point out that, at least in the case of some species of cetaceans, debris that sinks continues to pose a threat to wildlife; the sperm whale, Baird's beaked whale, and the grey whale, all species of odontocete cetaceans that spend some time feeding on the bottom, are known to ingest non-buoyant debris.

The value of using existing procedures to compile and maintain a database on wildlife interactions with debris is demonstrated by a recent report on plastic ingestion by the West Indian manatee, an endangered species. In the southeastern United States, dead manatees routinely are salvaged to determine cause of death and collect biological information. In Florida, personnel from the U.S. Fish and Wildlife Service and the University of Miami have performed systematic necropsies on dead manatees. Using this information, Beck and Barros (1991) found that of 439 manatees necropsied between 1978 and 1986, 63 (14.4 percent) had ingested debris. Pieces of monofilament fishing line were the most common debris items ingested (49 manatees). Other items included string, twine, rope, fish hooks or wire, paper, cellophane, synthetic sponges, rubber bands, plastic bags, and stockings.

Finally, Hoss and Settle (1990) compiled a list based on existing literature and their own work of at least 20 fish species reported to ingest plastics. This list included reports of larva, juvenile, and adults from benthic to pelagic habitats. Adults had ingested a wide variety of items, including rope, plastic pellets, packaging, sheeting, cups, cigar holders, a bottle, and colored fragments. Barr (1990) reported plastics in 12 percent of the yellow fin tuna and 3 percent of the blue fin tuna caught off the coast of Virginia. Higher percentages of plastics found in these and other pelagic species have been attributed to more frequent association of these fish with areas where debris concentrates, such as in drift lines.

GHOST FISHING

A major problem that ultimately could affect marine ecosystems, as well as create a major economic concern, is ghost fishing—the capability of lost or discarded fishing gear to continue to catch finfish and shellfish species indefinitely. Unfortunately, this is a difficult problem to study and there are few quantitative data on the subject. Because individuals fishing in the United States are not required to report lost fishing gear, there is no way to determine and monitor the total amount of lost fishing gear and its potential impacts on U.S. fishery resources.

However, the potential for impact on fishery resources and economics can be demonstrated for one segment of the fishing industry—the inshore lobster fishery of Maine. For this fishery, it has been estimated that 25 percent of all traps are lost each year, and that each lost trap can continue to catch up to 1.2 kilograms (2.5 pounds [lbs.]) of lobster (Smolowitz, 1978). While this may not seem significant, the cumulative effect could be; of the 1,787,795 lobster pots used in Maine's inshore fishery in 1987, nearly 450,000 traps were lost. Accordingly, those lost traps had the potential to catch more than 499 metric tons (MT) (1.1. million lbs.) of lobster valued at approximately $2.7 million dollars (1987 landings for this fishery were 10.160 MT [22.4 million lbs.] valued at $54.5 million [Natural Resources Consultants, 1990]).

Gillnets have also been found to have a significant potential to ghost fish. According to one estimate, lost gillnets can fish at a 15 percent effectiveness rate for up to eight years (Natural Resources Consultants, 1990). These lost nets not only pose a threat to marine wildlife in general but also can deplete species, including striped bass populations in the north Atlantic and south Atlantic, red drum in the Gulf of Mexico, and salmon and lake trout in the Great Lakes.

At present, the effects of ghost fishing related to U.S. commercial fisheries cannot be addressed due to the inadequacy of available information. There are no data on the number of gear units deployed in various fisheries, the number lost, or the capability of various types of gear to ghost fish (Natural Resources Consultants, 1990). The cumulative effects of lost gear also need to be considered. In addition, the effects of the increasing use of plastic or plastic-coated wire traps need to be examined, as these trends could prolong the capability of traps to ghost fish.

DATA ON ECOLOGICAL EFFECTS

Limited data are available on the ecological impacts of marine debris, and the information that has been collected is uneven and incomplete. The broad impact of plastics on ecosystems is unknown. Pinniped and sea turtle entanglement in plastic debris have been documented, but accounts of the impact of plastics on birds, fish, marine mammals, and terrestrial species are largely anecdotal. Furthermore, most available data on wildlife entanglements with debris is drawn from a few studies, and there has been no attempt to compile the data at one source, nor has there been any extensive effort to monitor trends.

Some data has been compiled by the NMFS, which collects information on entanglements involving certain species. In fact, the only way in which information on wildlife interactions with debris is formally exchanged among researchers and agencies, and in some manner compiled, is through the workshops and resultant proceedings coordinated by the NMFS. It is clear that research on the ecological effects of marine debris would be facilitated by a centralized system for keeping track of relevant data on all species.

The value of systematically compiling and maintaining a database on debris interactions with wildlife is demonstrated by the West Indian manatee program.

REFERENCES

Amos, A.F. 1993. Solid waste pollution of Texas beaches: a Post-MARPOL Annex V study, Vol 1: Narrative. OCS Study MMS 93-0013. Available from the public information unit of the U.S.

Department of the Interior, Minerals Management Service, Gulf of Mexico OCS Region, New Orleans, La. July.

Anonymous. 1981. Galapagos tainted by plastic pollution. Geo 3:137.

Araya, H. 1983. Fishery biology and stock assessment of Ommastrephes bartrami in the North Pacific Ocean. Mem. of the National Museum in Victoria (Australia) 44:269–283.

Balazs, G.H. 1985. Impacts of ocean debris on marine turtles: entanglement and ingestion. Pp. 387-429 in Proceedings of the Workshop on the Fate and Impact of Marine Debris, 27–29 November 1984, Honolulu, Hawaii (Vol. I), R.S. Shomura and H.O. Yoshida, eds. NOAA-TM-NMFS-SWFC-54. Available from the Marine Entanglement Research Program of the National Marine Fisheries Service (National Oceanic and Atmospheric Administration), Seattle, Wash.

Barr, C. 1990. Preliminary evaluation of feeding of blue fin tuna (Thunnus thynnus) and yellowfin tuna (Thunnus alabacares) off the coast of Virginia. Gloucester Point, Va.: Virginia Institute of Marine Science.

Barros, N.B., D.K. Odell, and G.W. Patton. 1990. Ingestion of plastic debris by stranded marine mammals in Florida. P. 746 in Proceedings of the Second International Conference on Marine Debris, 2–7 April 1989, Honolulu, Hawaii (Vol. I), R.S. Shomura and M.L. Godfrey, eds. NMFS NOAA-TM-NMFS-SWFSC-154. Available from the Marine Entanglement Research Program of the National Marine Fisheries Service (National Oceanic and Atmospheric Administration), Seattle, Wash.

Beach, R.J., T.C. Newby, R.O. Larsen, M. Penderson and J. Juris. 1976. Entanglement of an Aleutian reindeer in a Japanese fish net. Murrelet 57(3):66.

Beck, C.A. and N. B. Barros. 1991. The impact of debris on the Florida manatee. Marine Pollution Bulletin 22(10):508-510. October.

Bourne, W.R.P. (chair). 1990. Report of the working group on entanglement of marine life. Pp. 1207-1215 in Proceedings of the Second International Conference on Marine Debris, 2–7 April 1989, Honolulu, Hawaii (Vol. II), R.S. Shomura and M.L. Godfrey, eds. NOAA-TM-NMFS-SWFSC-154. Available from the Marine Entanglement Research Program of the National Marine Fisheries Service (National Oceanic and Atmospheric Administration), Seattle, Wash.

Bourne, W.R.P. and M.J. Imber. 1982. Plastic pellets collected by a prion on Gough Island, central South Atlantic Ocean. Marine Pollution Bulletin 13(1):20-21.

Calkins, D.G. 1985. Steller sea lion entanglement in marine debris. Pp. 308-314 in Proceedings of the Workshop on the Fate and Impact of Marine Debris, 27–29 November 1984, Honolulu, Hawaii, R.S. Shomura and H.O. Yoshida, eds. NOAA-TM-NMFS-SWFC-54. Available from the Marine Entanglement Research Program of the National Marine Fisheries Service (National Oceanic and Atmospheric Administration), Seattle, Wash.

Carr, A. 1987. Impact of nondegradable marine debris on the ecology and survival outlook of sea turtles. Marine Pollution Bulletin 18(6B):352-356.

Day, R.H., D.S. Wehle, and F.C. Coleman. 1985. Ingestion of plastic pollutants by marine birds. Pp. 344-386 in Proceedings of the Workshop on the Fate and Impact of Marine Debris, 27–29 November 1984, Honolulu, Hawaii, R.S. Shomura and H.O. Yoshida, eds. NMFS NOAA-TM-NMFS-SWFC-54. Available from the Marine Entanglement Research Program of the National Marine Fisheries Service (National Oceanic and Atmospheric Administration), Seattle, Wash.

Fowler, C.W. 1985. An evaluation of the role of entanglement in the population dynamics of Northern fur seals. Pp. 291-307 in Proceedings of the Workshop on the Fate and Impact of Marine Debris, 27–29 November 1984, Honolulu, Hawaii, R.S. Shomura and H.O. Yoshida, eds. NOAA-TM-NMFS-SWFC-54. Available from the Marine Entanglement Research Program of the National Marine Fisheries Service (National Oceanic and Atmospheric Administration), Seattle, Wash. December.

Fowler, C.W. 1988. A review of seal and sea lion entanglement in marine fishing debris. Pp. 16-63 in Proceedings of the North Pacific Rim Fishermen's Conference on Marine Debris, October 13–

16, 1987, Kailua-Kona, Hawaii, D. Alverson and J.A. June, eds. Seattle, Wash.: Natural Resources Consultants. October.

Fowler, C.W. and T.R. Merrell. 1986. Victims of plastic technology. Alaska Fish and Game 18(2):34-37.

Harrison, P. 1983. Seabirds: An Identification Guide. Boston, Mass.: Houghton Mifflin Co.

Henderson, J.R. 1984. Encounters of Hawaiian monk seals with fishing gear at Lisianski Island, 1982. Marine Fisheries Review 46(3):59-61.

Henderson, J.R. 1985. A review of Hawaiian monk seal entanglements in marine debris. Pp. 326-335 in Proceedings of the Workshop on the Fate and Impact of Marine Debris, 27–29 November 1984, Honolulu, Hawaii, R.S. Shomura and H.O. Yoshida, eds. NOAA-TM-NMFS-SWFC-54. Available from the Marine Entanglement Research Program of the National Marine Fisheries Service (National Oceanic and Atmospheric Administration), Seattle, Wash.

Hodge, K., J. Glen and D. Lewis. 1993. 1992 International Coastal Cleanup Results. Washington, D.C.: Center for Marine Conservation.

Hoss, D.E. and L.R. Settle. 1990. Ingestion of plastics by fishes. Pp. 693-709 in Proc. of the Second International Conference on Marine Debris, 2–7 April 1989, Honolulu, Hawaii (Vol. I), R.S. Shomura and M.L. Godfrey, eds. NOAA-TM-NMFS-SWFSC-154. Available from the Marine Entanglement Research Program of the National Marine Fisheries Service (National Oceanic and Atmospheric Administration), Seattle, Wash. December.

Jones, L.L. and R.C. Ferrero. 1985. Observations of net debris and associated entanglements in the North Pacific Ocean and Bering Sea, 1978-84. Pp. 183-196 in Proceedings of the Workshop on the Fate and Impact of Marine Debris, 27–29 November 1984, Honolulu, Hawaii, R.S. Shomura and H.O. Yoshida, eds. NOAA-TM-NMFS-SWFC-54. Available from the Marine Entanglement Research Program of the National Marine Fisheries Service (National Oceanic and Atmospheric Administration), Seattle, Wash.

Kenyon, K.W. and E. Kridler. 1969. Laysan albatross swallow ingestible matter. Auk 86:339-343.

King, W.B. 1984. Incidental Mortality of Seabirds in Gillnets in the North Pacific. Pp. 709-715 in Status and Conservation of the World's Seabirds, J.P. Croxall, P.G.H. Evans, and R.W. Schreiber, eds. ICBP Tech. Pub. No. 2. London: International Council for Bird Preservation.

Machida, S. 1983. A brief review of the squid fishery by Hoyo Maru No. 67 in southeast Australian waters in 1979/80. Mem. of the National Museum in Victoria (Australia). 44:291-295. Cited in Walker, W.A. and J.M. Coe. 1990. Survey of marine debris ingestion by odontocete cetaceans. Pp. 747-774 in Proc. of the Second International Conference on Marine Debris, 2–7 April 1989, Honolulu, Hawaii (Vol. I), R.S. Shomura and M.L. Godfrey, eds. NOAA-TM-NMFS-SWFSC-154. Available from the Marine Entanglement Research Program of the National Marine Fisheries Service (National Oceanic and Atmospheric Administration), Seattle, Wash. December.

Martin, A.R. and M.R. Clarke. 1986. The diet of sperm whales (Physeter macrocephalus) captured between Iceland and Greenland. Journal of the Marine Biological Association of the United Kingdom 66:779-790.

Mate, B.R. 1985. Incidents of marine mammals encounters with debris and active fishing gear. Pp. 453-457 in Proceedings of the Workshop on the Fate and Impact of Marine Debris, 27–29 November 1984, Honolulu, Hawaii, R.S. Shomura and H.O. Yoshida, eds. NOAA-TM-NMFS-SWFC-54. Available from the Marine Entanglement Research Program of the National Marine Fisheries Service (National Oceanic and Atmospheric Administration), Seattle, Wash.

Natural Resources Consultants. 1990. Survey and Evaluation of Fishing Gear Loss in Marine and Great Lakes Fisheries of the United States. Final report prepared for the Marine Entanglement Research Program of the National Marine Fisheries Service, Seattle, Wash.

O'Hara, K.J. and S. Iudicello. 1987. Plastics in the Ocean: More than a Litter Problem. Report prepared for the U.S. Environmental Protection Agency by the Center for Marine Conservation, Washington, D.C.

O'Hara, K.J., L. Maraniss, J. Deichman, J. Perry, and R. Bierce. 1987. 1986 Texas Coastal Cleanup Report. Washington, D.C.: Center for Environmental Education (now the Center for Marine Conservation).

O'Hara, K.J. and L. Younger. 1990. Cleaning North America's Beaches: 1989 Beach Cleanup Results. Washington, D.C.: Center for Marine Conservation.

Plotkin, P.E. and A.F. Amos. 1990. Entanglement in and ingestion of marine debris by sea turtles stranded along the south Texas Coast. Pp. 79-82 in Proceedings of the Eighth Annual Workshop on Sea Turtle Biology and Conservation. NOAA-TM-NMFS-SEFC-214. Available from the Southeast Fisheries Center, National Marine Fisheries Service, Miami, Fla.

Podolsky, R.H. and S.W. Kress. 1990. Plastic debris incorporated into double-crested cormorant nests in the Gulf of Maine. P. 692 in Proc. of the Second International Conference on Marine Debris, 2–7 April 1989, Honolulu, Hawaii (Vol. I), R.S. Shomura and M.L. Godfrey, eds. NOAA-TM-NMFS-SWFSC-154. Available from the Marine Entanglement Research Program of the National Marine Fisheries Service (National Oceanic and Atmospheric Administration), Seattle, Wash.

Scordino, J. 1985. Studies of fur seal entanglement, 1981–1984, St. Paul Island, Alaska. Pp. 278-290 in Proceedings of the Workshop on the Fate and Impact of Marine Debris, 27–29 November 1984, Honolulu, Hawaii, R.S. Shomura and H.O. Yoshida, eds. NOAA-TM-NMFS-SWFC-54. Available from the Marine Entanglement Research Program of the National Marine Fisheries Service (National Oceanic and Atmospheric Administration), Seattle, Wash.

Sileo, L. (chair). 1990. Report of the working group on ingestion. Pp. 1226-1231 in Proceedings of the Second International Conference on Marine Debris, 2–7 April 1989, Honolulu, Hawaii (Vol. II), R.S. Shomura and M.L. Godfrey, eds. NOAA-TM-NMFS-SWFSC-154. Available from the Marine Entanglement Research Program of the National Marine Fisheries Service (National Oceanic and Atmospheric Administration), Seattle, Wash.

Sileo, L., P.R. Sievert, M.D. Samuel, and S.I. Fefer. 1990. Prevalence and characteristics of plastic ingested by Hawaiian seabirds. Pp. 665-681 in Proceedings of the Second International Conference on Marine Debris, 2–7 April 1989, Honolulu, Hawaii (Vol. I), R.S. Shomura and M.L. Godfrey, eds. NOAA-TM-NMFS-SWFSC-154. Available from the Marine Entanglement Research Program of the National Marine Fisheries Service (National Oceanic and Atmospheric Administration), Seattle, Wash.

Smolowitz, R.J. 1978. Trap design and ghost fishing: Discussion. Marine Fisheries Review 40(5-6):59-67.

Stewart, B.S. and P.K. Yochem. 1985. Entanglement of pinnipeds in net and line fragments and other debris in the Southern California Bight. Pp. 315-325 in Proceedings of the Workshop on the Fate and Impact of Marine Debris, 27–29 November 1984, Honolulu, Hawaii, R.S. Shomura and H.O. Yoshida, eds. NOAA-TM-NMFS-SWFC-54. Available from the Marine Entanglement Research Program of the National Marine Fisheries Service (National Oceanic and Atmospheric Administration), Seattle, Wash.

Stewart, B.S. and P.K. Yochem. 1987. Entanglement of pinnipeds in synthetic debris and fishing net fragments at San Nicolas and San Miguel islands, California, 1978–1986. Marine Pollution Bulletin 18:336-339.

Walker, W.A. and J.M. Coe. 1990. Survey of marine debris ingestion by odontocete cetaceans. Pp. 747-774 in Proceedings of the Second International Conference on Marine Debris, 2–7 April 1989, Honolulu, Hawaii (Vol. I), R.S. Shomura and M.L. Godfrey, eds. NOAA-TM-NMFS-SWFSC-154. Available from the Marine Entanglement Research Program of the National Marine Fisheries Service (National Oceanic and Atmospheric Administration), Seattle, Wash.

Winston, J.E. 1982. Drift plastic—an expanding niche for a marine invertebrate? Marine Pollution Bulletin 13(10):348-351.

Younger, L. and K. Hodge. 1992. 1991 International Coastal Cleanup Results. Washington, D.C.: Center for Marine Conservation.

Index

Z